top-down calculus

calculus

A CONCISE COURSE

COMPUTERS AND MATH SERIES
ISSN 0888-2193

Series Editor
Marvin Marcus, *University of California at Santa Barbara*

1. *Discrete Mathematics: A Computational Approach Using BASIC**
   Marvin Marcus
2. *Problem Solving in Apple Pascal**
   Lowell A. Carmony, Robert J. McGlinn, Ann Miller Millman, and Jerry P. Becker
3. *Apple Pascal: A Self-Study Guide for the Apple II Plus, IIe, and IIc**
   Lowell A. Carmony, Robert J. McGlinn, Ann Miller Millman, and Jerry P. Becker
4. *Combinatorics for Computer Science**
   Gill Williamson
5. *Introduction to Computer Mathematics**
   Russell Merris
6. *Macintosh Pascal**
   Lowell A. Carmony and Robert L. Holliday
7. *Computing Without Mathematics: BASIC, Pascal Applications**
   Jeffrey Marcus and Marvin Marcus
8. *An Introduction to Pascal and Precalculus**
   Marvin Marcus
9. *Learning Apple FORTRAN**
   Donald J. Geenen
10. *MacAlgebra: BASIC Algebra on the Macintosh**
    Marvin Marcus
11. *Top-Down Calculus*
    S. Gill Williamson

*These previously-published books are in the *Computers and Math Series* but they are not numbered within the volume itself. All future volumes in the *Computers and Math Series* will be numbered.

top-down calculus

A CONCISE COURSE

S. GILL WILLIAMSON

University of California
at San Diego

COMPUTER SCIENCE PRESS

Computer Science Press, Inc.
1803 Research Boulevard
Rockville, Maryland 20850

   1   2   3   4   5   6   91 90 89 88 87

Library of Congress Cataloging-in-Publication Data

Williamson, S. Gill (Stanley Gill)
   Top-down calculus.

   Bibliography: p.
   Includes index.
   1. Calculus. I. Title.
QA303.W497   1987   515   87-302
ISBN 0-88175-072-7

CONTENTS

PREFACE

Top-Down Calculus is a textbook for a first course in differential and integral calculus. It is directed toward students in engineering, the sciences, mathematics, and computer science. Of particular concern to the author are students pursuing studies in the mathematical aspects of computer science. The material covered is suitable for anyone who has a serious professional need to know calculus.

The principal goal of this book is to bring students fairly quickly to a level of technical competence and intuitive understanding of calculus that is adequate for applying the subject to "real world" problems. To accomplish this, there are some simple rules regarding exercises that should be carefully followed. The exercises are, with a few exceptions, organized along a pattern of PROTOTYPAL EXERCISE SET → SOLUTIONS → VARIATIONS OF INCREASING DIFFICULTY. Students should, alone or with the help of the instructor, study the PROTOTYPAL EXERCISE SET and the SOLUTIONS. When each solution is learned, they should immediately make some modification, *however minor*, of the original problem and then work the modified problem. Students should then begin the task of working through the VARIATIONS. They can communicate their solutions of these variations to the instructor as quizzes, homework, or through "workshop" sessions where they solve problems in class under the supervision of the instructor.

The title of this book, *Top-Down Calculus*, comes from the style of programming called "Top-Down Programming." In this approach to software development, the first step is to gain a conceptual overview of the entire task and then, by progressively becoming more and more detailed, develop working code. That is the general philosophy of this book. In the beginning, the fundamental concepts of calculus are developed in a graphical or geometric setting. The geometric properties of derivatives are used to develop a corresponding intuitive understanding of the all-important chain rule in the first few pages of Chapter 1. Having at our disposal powerful tools understood geometrically, we do not hesitate to turn them quickly into analytical and technical tools to develop the calculus of algebraic and transcendental functions.

The time required to cover the material in this book depends largely on the extent to which students work the variations on the exercises. For the last several years, the first four chapters have served as the basis for a course taught in two six-week summer sessions. For computer science students, the entire book, including Chapter 5, can be the basis of a two- or three-quarter course followed by a solid one-quarter course in combinatorial analysis and the analysis of algorithms. In Chapter 5, Infinite Series, the rigorousness of the presentation is increased slightly. This chapter, although still largely intuitive, is intended to provide a transition to a course in the analysis of algorithms or to a more advanced course in mathematical analysis.

Throughout *Top-Down Calculus* we encourage educated guessing and the gathering of empirical evidence. The key ideas of calculus are looked at critically in the context of simple computational experiments. The students are encouraged to emulate this approach. Some of the problems in the harder variations are by no means trivial and require careful computational technique and organization. This is an important part of what should be learned from a beginning course in calculus.

As mentioned above, the lower division needs of computer science and mathematics-computer science students have motivated a number of features in our presentation of calculus. The lower division requirements of such students are heavy, including several programming languages and, ideally, a solid course in discrete mathematics that must compete for time with the calculus sequence. These time restraints create the need for a concise but technically solid calculus course. Such a course should be organized in such a way as to develop the technical skills required for such topics as the analysis of algorithms, combinatorial generating functions, and asymptotic analysis. For computer science students, there is a need for a greater emphasis on the calculus of logarithmic and exponential functions than in the standard calculus course, and an earlier and more comprehensive treatment of infinite series. All of these issues have been addressed in this book.

S. Gill Williamson

ACKNOWLEDGMENTS

I would like to express my appreciation to the teaching assistants who have so patiently helped hundreds of students learn to work the problems in this book. In particular, I would like to thank Zhi-Cheng Gao, Ho Hong, Pat Morandi, and Lisa Taylor. Jill Warn and the U.C.S.D. Summer Session staff have gone out of their way to be helpful in distributing the various versions of this book to summer session students. Neola Crimmins' technical help and advice in the preparation of difficult parts of the original manuscript were indispensable. I am especially grateful to my colleague and friend Ron Evans for his numerous thoughtful comments and observations.

Chapter 1

LINEAR FUNCTIONS AND DERIVATIVES

Linear Functions Are The Foundation Of Calculus

A "linear function" is a function whose graph is a straight line. Almost every calculus book has a section or two on linear functions. We shall describe such functions in this chapter and leave it to the reader to supplement our discussion by reading in his or her precalculus textbook. Every linear function has an equation of the form $f(x) = ax + b$. In such an equation, x is called the "independent variable" or simply the "variable." For example, $f(x) = 2x + 1$ is a linear function. The graph of this function and the graphs of two other linear functions are shown in FIGURE 1.1. The graph of the function $f(x) = 2x + 1$ is, by definition, the set of all points in the plane of the form $(x, 2x + 1)$. Thus (1, 3) and (2, 5) are points on the graph of $2x + 1$ but (1, 2) and (2, 4) are not on the graph of $2x + 1$. In the linear function $f(x) = ax + b$, the number a is called the "slope of the function f" and the number b is called the "intercept of the function f" or the "vertical intercept of the function f." A discussion of these terms and the alternative ways of describing linear functions will be found in your algebra or precalculus textbook.

The Notation Of Calculus Demands Critical Attention

One of the major problems in learning calculus is the notation, which cynics say "ranges from bad to horrible." We should begin now to think about

FIGURE 1.1 The Graphs of Three Linear Functions: f, g, and h

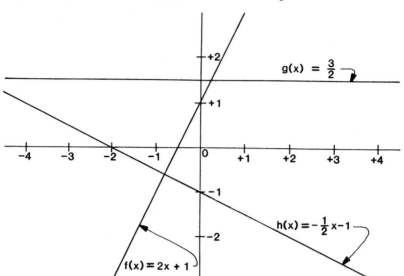

notation in a critical way. Take a look at FIGURE 1.1 again. The graph of the function f(x) = 2x + 1 is the "set of all points in the plane of the form (x, 2x + 1)." If you look in the previous sentence at the statement in quotes, there is no mention of the symbol f. Thus, if you had been asked to graph p(x) = 2x + 1 or β(x) = 2x + 1 you would have produced exactly the same graph! The only difference would be in how you would *refer* to the graph that you produced. If you have graphed f(x) = 2x + 1 you might say "Bill, put your finger on the graph of f." If you graphed p(x) = 2x + 1 or β(x) = 2x + 1 instead, then you would use p or β in making such a statement in place of f. One standard way of specifying a linear function is to use "y" rather than "f" as we have done. Of course, it makes no difference which letter you use except in referring to the linear function (as we have just noted). Thus, instead of f(x) = 2x + 1, one might write y(x) = 2x + 1. The notation "f(x)" tells us that f is the name of the function and x is the name of the variable. One could also write f = 2x + 1 or y = 2x + 1 and the function and variable names would be equally well specified. Oddly enough, one rarely sees f = 2x + 1, but y = 2x + 1 is very common. This sort of arbitrary notational tradition is quite common in calculus and is a source of confusion to the beginner.

Playing The Envelope Game

There is a little game that the beginning calculus student can play that will help keep the question of notation in perspective. This game is called THE ENVELOPE GAME. We shall play it from time to time. For linear functions the game goes like this: Imagine you have a linear function in an envelope. Before you open up the envelope try and describe what it is that you will see on the inside. The object of the game is to list the various ways that a linear function might be described. One possibility is that the envelope contains a piece of graph paper with the graph of the linear function (a straight line, of course) on it. Another possibility is that one sees an equation of the form y = ax + b. Can you think of other possible contents of the envelope? What you should begin to understand from THE ENVELOPE GAME is the distinction between a mathematical object or concept and the manner used to specify or describe a particular instance of this object or concept. In the case at hand, the concept is that of a linear function which can be described in many different ways. A computer scientist might refer to these various ways of describing the same basic object as "different data structures."

Locally, Linear Functions Approximate Nonlinear Functions

THE PROPERTIES OF LINEAR FUNCTIONS ARE THE FOUNDATION OF ALL OF CALCULUS. There is a simple intuitive reason for this, which is shown in FIGURE 1.2. In FIGURE 1.2 we see the graph of a nonlinear function f. The function is nonlinear because its graph is not a straight line. Imagine that we choose three points on the graph of f and look at the curve under a microscope at these points. The three points we have selected are called A′, B′, and C′ and are shown in FIGURE 1.2(a). The circles centered at these points represent the field of view of the microscope. In FIGURE 1.2(b) we see the view under the microscope centered at point B′. What we see is (essentially!) a straight line segment. This straight line segment defines a straight line as indicated by the dotted line of FIGURE 1.2(b). Like any straight line, it is the graph of a linear function of the form y = ax + b. By inspection, it appears that a = − 1/3 in this case. What do you think b is for this straight line (a rough guess will do)? Actually, from the point of view of calculus, the value of b is not too important. It is the value of a that gets all of the attention. The value of a (in this case − 1/3) is called *the derivative of f at B′*. Thus, to compute the derivative of a function f at a point B′ on the graph of f one focuses a microscope at the point B′,

FIGURE 1.2   The Derivative of a Function at a Point

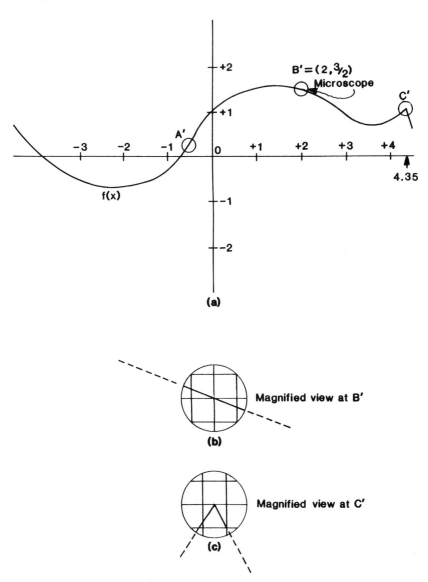

ups the magnification until the portion of the graph under the microscope looks like a straight line segment, and computes the slope of this line. The resulting number is *the derivative of f at B'*.

Let's play THE ENVELOPE GAME. Imagine that you have an envelope and inside it is the derivative of a function f at a point B' on the graph of the function. Before opening up the envelope, describe what it is that you are going to see (Answer: a number). These ideas are summarized in SLOPPY DEFINITION 1.3.

The Intuitive Definition Of The Derivative

1.3 SLOPPY DEFINITION Let f be a function and let B′ be a point on the graph of f. Focus a microscope on the point B′ and increase the magnification until the portion of the graph in the field of view of the microscope looks like a straight line segment. The slope of this straight line segment is *the derivative of f at B′*.

A Function May Not Have A Derivative At Certain Points

One problem with SLOPPY DEFINITION 1.3 is shown at the point C′ in FIGURE 1.2(a). At this point, the graph of f has a sharp "spike" or "cusp." No matter how much we increase the magnification of the microscope, we never see a straight line segment! Thus, SLOPPY DEFINITION 1.3 doesn't work for such a point. We say, in this case, that "f does not have a derivative at C′." The functions that are studied in calculus have very few bad points such as C′ where the derivative does not exist. Some functions that naturally occur do have such points (for example, $f(x) = |x|$ has such a point at $x = 0$ and $f(x) = |x| + |x - 1|$ has two such points). Notice that the function shown in FIGURE 1.2 has infinitely many "good points" where the derivative exists but only one point C′ where the derivative does not exist (there may be more points not shown in FIGURE 1.2, but you get the idea!).

Compute The Derivative At Lots Of Points To Get The Derivative Function

We have been discussing the derivative of a function f *at a point* on the graph of f. Look now at FIGURE 1.4. There we see a function f(x). At each point on the graph of f(x) we have attempted to compute the derivative of f using SLOPPY DEFINITION 1.3. At the point $(-4, +3.6)$, which is on the graph of f, we thought the derivative was about -0.4. At the point $(-1, +1)$ on the graph of f we thought that the derivative was about -1.0. At $(-7, +4.5)$ the derivative was 0, etc. Of course, there are infinitely many such points and we can't compute derivatives for all of them. We computed derivatives for a number of such points and then drew a smooth curve through them to obtain the graph of f′(x) shown in FIGURE 1.4. This new function is called *the derivative function of f*.

FIGURE 1.4   The Derivative Function f′ of a Function f

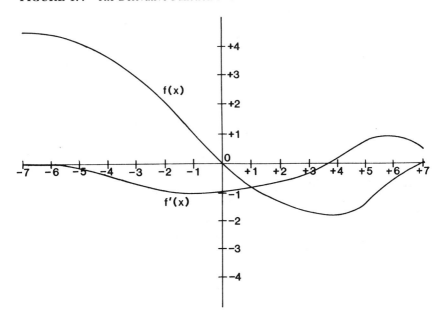

1.5   DEFINITION   Let f be a function. For each point (x, f(x)) on the graph of f that has a derivative, compute that derivative and call it f′(x). The function f′(x) is called *the derivative function of f*.

Computing Derivative Functions Graphically

1.6   EXERCISES

(1)   Draw the graph of the derivative functions f′, g′, and h′ for each of the three linear functions of FIGURE 1.1.

(2)   Draw the graph of the derivative function f′ for the function f of FIGURE 1.2. The point C′ is a problem as there is no derivative at that point. Draw carefully what the graph of f′ looks like for points of the graph of f to the left of C′ (values of x < 4.35) and to the right of C′ (values of x > 4.35).

1.7   IMPORTANT PROPERTIES OF LINEAR FUNCTIONS   We now take a look at some simple but extremely important properties of linear functions as they relate to calculus. To emphasize when we are talking about a *linear* function we shall use a "tilde" over the symbol representing that function. Thus f̃ will be a linear function.

Suppose that $\tilde{f}(x)$ = ax + b and $\tilde{g}(x)$ = cx + d are two linear functions. Define $\tilde{s}(x)$ = $\tilde{f}(x)$ + $\tilde{g}(x)$ to be the *sum* of $\tilde{f}$ and $\tilde{g}$. Thus $\tilde{s}(x)$ = (a + c)x + (b + d). The important thing to notice here is that THE SLOPE OF THE SUM OF TWO LINEAR FUNCTIONS IS THE SUM OF THEIR SLOPES. It is evident from the calculation that we just did that the sum of two linear functions is again a linear function. This is not the case for the product of linear functions!

If a linear function is multiplied by any real number r, then we again obtain a linear function. For example, $\tilde{rf}(x)$ = rax + rb. Note that the slope of the new linear function is r times the slope of the function $\tilde{f}$. Thus we have that THE SLOPE OF A REAL NUMBER TIMES A LINEAR FUNCTION IS THAT REAL NUMBER TIMES ITS SLOPE.

For Linear Functions, The Slope Of A Composition Is The Product Of The Slopes

The above two properties of linear functions are really pretty simple. We now consider a much deeper property of linear functions. Understanding this property *well* will be the key to a good grade in calculus!

Adding linear functions produces *another linear function* whose slope is the *sum* of the slopes of the two original linear functions. What sort of operation on linear functions produces *another linear function* whose slope is the *product* of the slopes of the two original linear functions? Taking products of linear functions won't do the job. For example, (ax + b) (cx + d) = acx^2 + (ad + bc)x + bd, which is not even linear (it's quadratic). It turns out that the operation that we need is *composition of functions*. The operation of composition of functions is not restricted to linear functions but we shall start with that case as it is particularly easy to follow. Let $\tilde{f}$ and $\tilde{g}$ be as in the previous paragraph. The *composition* of $\tilde{f}$ and $\tilde{g}$ is the function $\tilde{f}(\tilde{g}(x))$ obtained by replacing each occurrence of the variable x in $\tilde{f}(x)$ by the entire expression $\tilde{g}(x)$. Thus, if $\tilde{f}(x)$ = ax + b and $\tilde{g}(x)$ = cx + d then $\tilde{f}(\tilde{g}(x))$ = a(cx + d) + b = acx + (ad + b). From this calculation we see that the composition of two linear functions is again a linear function and THE SLOPE OF THE COMPOSITION OF TWO LINEAR FUNCTIONS IS THE PRODUCT OF THEIR SLOPES. Let us summarize these important properties of linear functions:

(1) IF A REAL NUMBER IS MULTIPLIED TIMES A LINEAR FUNC-TION, THEN THE NEW FUNCTION IS LINEAR AND ITS SLOPE IS THE PRODUCT OF THE REAL NUMBER AND THE SLOPE OF THE ORIGINAL FUNCTION.

(2) THE SUM OF TWO LINEAR FUNCTIONS IS A LINEAR FUNC-TION. THE SLOPE OF THE NEW FUNCTION IS THE SUM OF THE SLOPES OF THE ORIGINAL FUNCTIONS.

(3) THE COMPOSITION OF TWO LINEAR FUNCTIONS IS A LIN-EAR FUNCTION. THE SLOPE OF THE NEW FUNCTION IS THE PROD-UCT OF THE SLOPES OF THE ORIGINAL FUNCTIONS.

We now must take a look at the meaning of composition of functions f and g where f and g may not be linear. Consider DEFINITION 1.8.

But We Can Compose Nonlinear Functions Too

1.8   DEFINITION   Let f and g be functions (real valued). Define a function h whose value h(x) at any real number x is gotten by first evaluating g at x to obtain the real number g(x) and then evaluating f at g(x) to obtain f(g(x)). The function h is called the *composition of f and g*. We write $h(x) = f(g(x))$.

In calculus we deal primarily with functions whose "domain" and "range" are real numbers. These functions are called *real valued* and can be graphed using the standard horizontal and vertical real number lines (as in FIGURES 1.1 and 1.2, for example). Thus, when we form the composition of two real valued functions we obtain another real valued function.

It is now time to play THE ENVELOPE GAME with DEFINITION 1.8. Suppose we have an envelope and inside it are two functions f and g and their composition $h(x) = f(g(x))$. What are we going to see when we open the envelope? One possibility is that f and g and h are described by "formulas" or "closed expressions." Another possibility is that f and g and hence h are given as graphs. As an example of the formula type description, we might have $f(x) = 2x^3 + 5x^2 - 3x + 2$ and $g(x) = \sin(x) + \sqrt{x}$. Then we find that

$$f(g(x)) = 2(\sin(x) + \sqrt{x})^3 + 5(\sin(x) + \sqrt{x})^2 - 3(\sin(x) + \sqrt{x}) + 2.$$

As the above expression shows, the composition of two functions f and g given as formulas or closed expressions can be done just as with linear functions by replacing each occurrence of x in the expression for f(x) by the whole expression for g(x). There is a problem that might occur here. Suppose $f(x) = \sqrt{x}$ and $g(x) = x^3$. Let h be the composition of f and g. If we try to evaluate $h(-1)$ then we first form $g(-1) = -1$ and then try to evaluate $f(-1)$, which is the square root of -1. Thus, $h(-1)$ is not defined as a real number. In general, this sort of thing happens frequently. There will be real numbers x for which g(x) is defined but h(x) is not defined. This would be the case with the composition g(f(x)) using the g and f of the previous paragraph (used there to compute f(g(x))). You should easily be able to find values of x for which this function g(f(x)) is not defined. To find *all* such values is a little more work! Some exercises are given at the end of the chapter for the reader to practice these ideas. To consider the case where the two functions f and g being composed are given graphically, look at the *very important* example shown in FIGURE 1.9.

Composing Functions Graphically

In FIGURE 1.9 we see three functions f, g, and h given graphically. The function h is the composition of f and g: h(x) = f(g(x)). Computing the composition of functions given in graphical form can be a rather tedious process. In principle, one must compute f(g(x)) at each point x. Thus, for x = 3, we first compute g(3) by looking at the graph of g. It seems that g(3) = 2. We now compute f(g(3)) = f(2), which is about -1. Thus, by definition, h(3) = -1. If we do that for enough points then we can draw the graph of h as has been done in FIGURE 1.9(c). In FIGURE 1.9(a) a microscope has been placed at the point A$'$ = (3, g(3)) = (3, 2) on the graph of g. The straight line segment (or nearly so) that appears in the field of view (indicated by the circle) has been extended to obtain a straight line $\tilde{g}(x) = (1/3)x + 1$. This means, by SLOPPY DEFINITION 1.3, that slope($\tilde{g}$) = 1/3 is g$'$(3). Similarly, in FIGURE 1.9(c) a microscope has been put at the point B$'$ = (g(3), f(g(3))) = (2, -1). The line segment in the field of view at B$'$ has been extended to a straight line $\tilde{f}(x) = (-3/2)x + 2$.

Composing Linear Approximations To Functions

Suppose that two students, Larry (for Linear) and Nancy (for Nonlinear) are given the task of composing functions from FIGURE 1.9. Larry is going to compose $\tilde{f}$ and $\tilde{g}$ to obtain the function $\tilde{h}$ of FIGURE 1.9(c). Nancy is going to compose f and g to obtain the function h of FIGURE 1.9(c). Of course, their work is going to be quite different for most values of x. Note, however, that the graphs of $\tilde{g}$ and g coincide (essentially) in the circle about A$'$ and the graphs of $\tilde{f}$ and f coincide (essentially) in the circle about B$'$. Thus when Larry and Nancy are dealing with values in these circled regions they will essentially get the same answer for the composition as is shown in the circled region about C$'$ in FIGURE 1.9(c).

Larry, being a good student, has learned IMPORTANT PROPERTIES 1.7 (in particular number 3). Thus, he knows that the slope of $\tilde{f}$ times the slope

FIGURE 1.9   Graphical Composition and the Chain Rule

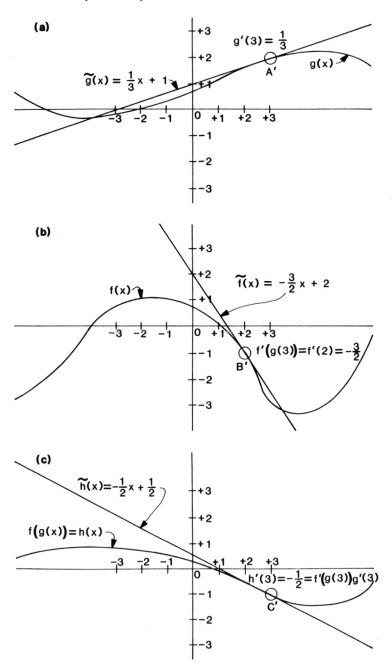

of $\tilde{g}$ is the slope of $\tilde{h}$ (in this case $-3/2$ times $1/3$ equals $-1/2$). Nancy, also a good student, has learned SLOPPY DEFINITION 1.3, which she applies to the points A′, B′, and C′. She concludes correctly that the slope of $\tilde{g}$ is the derivative g′(3) of g at the point A′, the slope of $\tilde{f}$ is the derivative f′(g(3)) = f′(2) of f at the point B′, and the slope of $\tilde{h}$ is the derivative h′(3) of h at the point C′. Putting their observations together, Larry and Nancy conclude that h′(3) = f′(g(3))g′(3).

The same process would have worked for any x, not just x = 3 (the functions would have to be defined at x and g(x), of course, and have derivatives). Thus we would discover that h′(x) = f′(g(x))g′(x). This observation is called *the chain rule* or *the composite function rule* and is without doubt the single most important fact that must be learned and thoroughly understood by the beginning calculus student. We state this rule in IMPORTANT THEOREM 1.10.

The Most Important Concept To Master . . .

1.10 IMPORTANT THEOREM (THE CHAIN RULE) Let f(x) and g(x) be functions and let h(x) = f(g(x)) be the composition of f and g. Then, the derivative h′(x) is equal to f′(g(x))g′(x).

Many beginning calculus students would be much happier if IMPORTANT THEOREM 1.10 stated h′(x) = f′(x)g′(x) but, alas, this is not true in general. It is not true for x = 3 in FIGURE 1.9, as the reader should verify.

Linearity Of The Derivative: (rf(x) + sg(x))′ = rf′(x) + sg′(x)

As we have just seen, IMPORTANT PROPERTY 1.7(3) of linear functions gives rise to IMPORTANT THEOREM 1.10 on derivatives. What rules of derivatives follow from IMPORTANT PROPERTIES 1.7(1) and (2)? Fortunately, these rules are so simple that most beginning calculus students apply them automatically. There are times, however, when these rules need to be articulated clearly in solving a problem so we shall now discuss them briefly.

Take a look at FIGURE 1.11. There we see the graphs of functions f(x), g(x), and h(x). These functions have been constructed so that h(x) = f(x) + g(x). As with FIGURE 1.9, microscopes have been placed on these graphs at points A′, B′, and C′. These points all correspond to x = +1. Thus A′ = (1, f(1)) = (1, 3), B′ = (1, g(1)) = (1, −2), and C′ = (1, h(1)) = (1, 1). In the microscope, each curve looks like a straight line segment. These straight line segments are extended to give the lines $\tilde{f}$, $\tilde{g}$, and $\tilde{h}$ as shown in

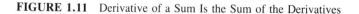

FIGURE 1.11   Derivative of a Sum Is the Sum of the Derivatives

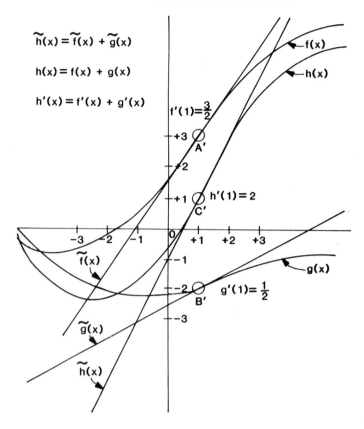

FIGURE 1.11. Clearly, $\tilde{h} = \tilde{f} + \tilde{g}$. By IMPORTANT PROPERTY 1.7(2), slope($\tilde{h}$) = slope($\tilde{f}$) + slope($\tilde{g}$) and hence by SLOPPY DEFINITION 1.3, $h'(1) = f'(1) + g'(1)$, as can be seen in the example of FIGURE 1.1. Instead of using $x = 1$ we could have used any x in this argument (provided that the relevant derivatives were defined) and we would obtain $h'(x) = f'(x) + g'(x)$. This rule states that *the derivative of a sum* (in this case $h'(x)$) *is the sum of the derivatives* (in this case $f'(x) + g'(x)$). The rule corresponding to IMPORTANT PROPERTY 1.7(1) is even easier: if $h(x) = rf(x)$ where h and f are functions and r is a real number, then $h'(x) = rf'(x)$. The reader should explain this rule graphically as was done for the other two rules in FIGURE 1.9 and FIGURE 1.11 respectively. We state these three rules in RULES FOR DERIVATIVES 1.12.

1.12   RULES FOR DERIVATIVES

(1) CONSTANT MULTIPLE RULE: If f(x) is a function and r is a constant (i.e., a real number) and h(x) = rf(x) then h'(x) = rf'(x).

(2) SUM RULE: If f(x) and g(x) are functions and h(x) = f(x) + g(x) then h'(x) = f'(x) + g'(x).

(3) CHAIN RULE: If f(x) and g(x) are functions and h(x) = f(g(x)) then h'(x) = f'(g(x))g'(x).

The Envelope Game Again. . .

Understanding the RULES FOR DERIVATIVES 1.12 has been the basic object of this chapter. The next chapter will be devoted to the problem of "computing derivatives." To begin our thinking about this problem, let's play THE ENVELOPE GAME once again. Suppose we are given a function f(x) graphically, as in FIGURE 1.4. We are given an envelope and inside the envelope is the derivative function f'(x). What is it that we are going to see when we open the envelope? Most likely it will be another graph, just as in FIGURE 1.4. If this were all that was involved in the study of derivatives, calculus would be a trivial subject. Given any function f(x), we would draw its graph and compute f'(x) as in FIGURE 1.4. This would work for any function. For example, it would work for the function f(x) = x². You may have already had enough experience with calculus to know, however, that if your calculus instructor asked you to find the derivative of f(x) = x² and your answer was a graph, he or she would probably "flip out." The expected answer would be "f'(x) = 2x." In other words, given a function f(x) as a formula or "closed expression" and an envelope containing f'(x), one would expect to open the envelope and find another formula or "closed expression." So given f(x) = x² the envelope should contain f'(x) = 2x.

If f(x) = x² Then f'(x) = 2x. Here's Why. . .

Every calculus book proves that if f(x) = x² then f'(x) = 2x. To understand how this interesting fact relates to SLOPPY DEFINITION 1.3, take a look at FIGURE 1.13. In FIGURE 1.13(a) we see a portion of the graph of f(x) − x². In the spirit of SLOPPY DEFINITION 1.3, we have put our microscope at the point A' = (1, 1) on the graph of f. The slope of the straight line segment in the field of view of the microscope seems to be about 2. If f'(x) = 2x is the correct formula then this is as it should be as f'(1) = 2.

FIGURE 1.13 Derivative of f(x) = x²

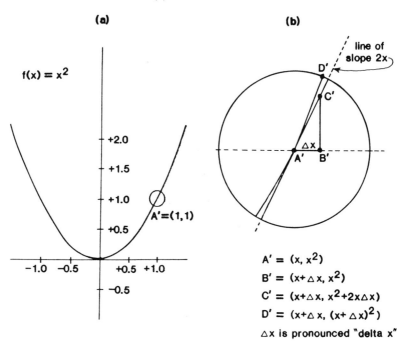

(a)

(b)

$A' = (x, x^2)$
$B' = (x+\triangle x, x^2)$
$C' = (x+\triangle x, x^2+2x\triangle x)$
$D' = (x+\triangle x, (x+\triangle x)^2)$

$\triangle x$ is pronounced "delta x"

The Square Of A Small Number Is Even Smaller

In FIGURE 1.13(b) we are taking a more careful look at the field of view of the microscope at the point A'. To be a little more general, we are looking at A' = (x, x²) (instead of just x = 1 as shown in FIGURE 1.13(a)) where x is some value near 1. In FIGURE 1.13(b) the line joining A' to C' is a straight line segment of slope 2x. The distance from A' to B' is Δx. The symbol "Δx" is used in calculus to stand for a "small number which is to be added to x to change it slightly." The line joining A' to D' is a part of the graph of f(x) = x² and is not, of course, exactly a straight line segment.

To see how far off from being straight this latter line is, let's compute the distance the point D' is above the point C'. The second coordinate of C' is x² + 2xΔx and the second coordinate of D' is (x + Δx)² = x² + 2xΔx + (Δx)². The difference between these two coordinates is (Δx)², which is the height of D' above C'. We have in mind that Δx is small, say .001. In this case, (Δx)² = .000001, is much smaller still. The ratio of the second of these two quantities to the first is Δx = .001. Referring to FIGURE 1.13(b), we have shown that the ratio of the distance between D′ and C′ to the distance between A′ and B′ is Δx. Thus, this ratio goes to zero as Δx goes to zero. This is the analytic statement of the fact that the smaller the field of view (which is roughly Δx) the more the part of the graph in the field of view appears to be a straight line.

The Difference Quotient: Δf = f(x + Δx) − f(x) Over Δx

Look again at FIGURE 1.13(b). If the line from A′ to D′ was a straight line segment then its slope would be $\dfrac{f(x + \Delta x) - f(x)}{\Delta x} =$ $\dfrac{(x + \Delta x)^2 - x^2}{\Delta x} = 2x + \Delta x$. The quantity $\dfrac{f(x + \Delta x) - f(x)}{\Delta x}$ is called the "difference quotient of f with respect to Δx." As we have just seen, the difference quotient is almost $f'(x)$ and the smaller Δx is the closer this difference quotient is to $f'(x)$. Mathematicians would express this fact by saying that "the limit of $\dfrac{f(x + \Delta x) - f(x)}{\Delta x}$ as Δx tends to zero is $f'(x)$." This statement assumes that f has a derivative at x. This important idea is summarized in ERUDITE OBSERVATION 1.14.

1.14   ERUDITE OBSERVATION   If f is any function that has a derivative at x then the difference quotient $\dfrac{f(x + \Delta x) - f(x)}{\Delta x}$ is very close to $f'(x)$ for small enough values of Δx. In other words, the limit of $\dfrac{f(x + \Delta x) - f(x)}{\Delta x}$ as Δx tends to zero is $f'(x)$. In symbols we may write

$$\lim_{\Delta x \to 0} \frac{f(x + \Delta x) - f(x)}{\Delta x} = f'(x).$$

Another common notation for ERUDITE OBSERVATION 1.14 is to define $\Delta f = f(x + \Delta x) - f(x)$. We then may write $\lim\limits_{\Delta x \to 0} \dfrac{\Delta f}{\Delta x} = f'(x)$. In the case where $f(x) = x^2$ we found that $\dfrac{\Delta f}{\Delta x} = 2x \ | \ \Delta x$ and thus, $f'(x) = 2x$.

In the next chapter we shall be concerned with techniques for computing derivatives. Most of what the beginning calculus student has to learn is precisely "techniques of computation." The major difficulty in this task is the profusion of notation brought about by the many different subjects to

which calculus has been applied. We close this chapter by reviewing the ideas
we have discussed thus far through the study of a number of specific examples.

1.15   EXAMPLES OF RULES FOR DERIVATIVES 1.12

$\dfrac{d}{dx}$ f Is $\dfrac{df}{dx}$ Is f'(x) In Differential Notation

(1) We now know how to compute the derivative function of $f(x) = x^2$.
The answer is $f'(x) = 2x$. Another standard notation for the derivative function
$f'(x)$ is $\dfrac{df}{dx}$. Thus if $f(x) = x^2$ then $\dfrac{df}{dx} = 2x$. The notation $f'(x)$ tells us that
f' is a function of x. In the notation $\dfrac{df}{dx}$ it is the x in dx that tells us that the
variable is x. Sometimes one sees $\dfrac{df}{dx}(x)$ or $\dfrac{df(x)}{dx}$ if one wants to emphasize
that the variable is x. Another way of saying that the derivative of x^2 is 2x
is to write $(x^2)' = 2x$ or $\dfrac{d}{dx}x^2 = 2x$. Unfortunately, one must get used to
all of these notations!

Constant Functions Have Zero Derivative Functions

(2) A function $f(x) = r$ where r is a number is called a "constant function."
Thus $f(x) = 2$ is a constant function. The graph of $f(x) = 2$ is a line parallel
to and two units above the horizontal axis. The function $g(x)$ of FIGURE 1.1
is a constant function with value 3/2.

Beginning students sometimes confuse constant functions with constants
(i.e., numbers). There is an important difference. The function $f(x) = 2$ is
a rule that assigns to every real number x the number 2 in the same sense
that $f(x) = x^2$ assigns to every x the value x^2. The graph of $f(x) = 2$ is the
set of all pairs (x, 2). The graph of $g(x) = 3/2$ of FIGURE 1.1 is the set of
all pairs (x, 3/2) as shown. The numbers 2 or 3/2 are not the same as the
functions $f(x) = 2$ or $g(x) = 3/2$. Any constant function, such as $f(x) = 2$,
has slope zero at every point of its graph. Thus $f'(x) = 0$ for such a function.

In practice, the possibility of being seriously confused by the difference
between constants and constant functions is very small. If one sees a statement
such as $\dfrac{d}{dx}(2) = 0$ then one knows that the constant function with value 2 is
being differentiated (i.e., having its derivative taken) and its derivative func-

tion is the constant function with value 0. Another way of saying the same thing is $(2)' = 0$. No matter how large the value of the constant function, its derivative function is still the zero function. Thus, $(1,000,000)' = 0$.

The Derivative Of A Linear Function Is Its Slope

(3) In EXERCISE 1.6, the reader was asked to compute the derivative functions of the three linear functions of FIGURE 1.1. The answers are $f'(x) = 2$, $g'(x) = 0$, and $h'(x) = -1/2$. In general, if $f(x) = ax + b$ is a linear function then $f'(x) = a$ is the constant function with value equal to the slope of $f(x)$. Thus, $\frac{d}{dx}(23x + 45) = 23$, $(-12x - 124)' = -12$, and $\frac{d}{dx}x = 1$. This latter result can also be written $(x)' = 1$ and, probably because it is so simple, is sometimes a source of confusion to the beginner.

The Derivative Of A Sum Is. . .

(4) We now take a look at RULES FOR DERIVATIVES 1.12. Applying 1.12(1), we compute $(2x^2)' = 2(x^2)' = 2(2x) = 4x$. In our alternative notation, $\frac{d}{dx}(2x^2) = 2\frac{d}{dx}(x^2) = 2(2x) = 4x$. In a similar fashion, $(45x^2)' = 90x$, $(-4x^2)' = -8x$. Applying 1.12(2) we can compute $(2x^2 + 4x)' = (2x^2)' + (4x)'$ which, by 1.12(1), is $2(x^2)' + 4(x)' = 4x + 4$. In our alternative notation we would have $\frac{d}{dx}(2x^2 + 4x) = \frac{d}{dx}(2x^2) + \frac{d}{dx}(4x) = 4x + 4$. As stated, RULE 1.12 applies to the sum of two functions, but it is obviously valid for 3, 4, or any finite sum of functions. Thus $(e(x) + f(x) + g(x))' = e'(x) + (f(x) + g(x))' = e'(x) + f'(x) + g'(x)$. As an example, $(5x^2 + 6x + 9)' = (5x^2)' + (6x)' + (9)' = 10x + 6 + 0 = 10x + 6$.

Using The Chain Rule Requires Some Guesswork

(5) Rules 1.12(1) and 1.12(2) illustrated in example (4) above are easily mastered by the beginning student and usually applied correctly with little thought involved. The CHAIN RULE, 1.12(3), is a different matter! As we have already stated, it is the most important rule of calculus, at least in the beginning. The CHAIN RULE concerns functions of the form $f(g(x))$ which are compositions of two functions $f(x)$ and $g(x)$. The CHAIN RULE is then used to compute the derivative $(f(g(x))'$ of this function. The CHAIN RULE requires that you compute first $f'(x)$ and then compose this with $g(x)$ to get

$f'(g(x))$. This is then multiplied times $g'(x)$ to get $f'(g(x))\,g'(x)$, which is the correct answer.

That sounds easy enough. For example, let $f(x) = x^2$ and let $g(x) = 9x^2 - 6x + 4$. Then $f(g(x)) = (9x^2 - 6x + 4)^2$. We compute that $f'(x) = 2x$ and $g'(x) = 18x - 6$ and hence $f'(g(x))g'(x) = 2(9x^2 - 6x + 4)(18x - 6)$. In practice, there is an additional complication: $f(g(x))$ may be given but not $f(x)$ and $g(x)$. Suppose we are asked to differentiate $(3x^2 + 3x + 3)^{2/3}$. We think we would like to use the CHAIN RULE, but what are $f(x)$ and $g(x)$? The answer is that we must guess what they are! In this case we could take $f(x) = x^2$ and $g(x) = (3x^2 + 3x + 3)^{1/3}$ or we could take $f(x) = x^{2/3}$ and $g(x) = 3x^2 + 3x + 3$. Both work to give the expression $(3x^2 + 3x + 3)^{2/3}$ for $f(g(x))$. In terms of applying the CHAIN RULE, the latter is a much better choice. To apply the CHAIN RULE in this case we need to know that $(x^r)' = rx^{r-1}$ for any real number r (not just $r = 2$ as we have already shown). This formula will be derived in the next chapter and can just be accepted for now. Using this formula, we see that $f'(x) = (2/3)x^{-1/3}$. Obviously by now, $g'(x) = 6x + 3$ and so

$$f'(g(x))g'(x) = (2/3)(3x^2 + 3x + 3)^{-1/3}(6x + 3),$$

which is the correct derivative. We shall continue worrying about the CHAIN RULE in the next example.

The Chain Rule In Differential Notation

(6) Suppose we are going to apply the CHAIN RULE to $f(g(x))$ where $f(x) = x^3$. We would first find $f'(x) = 3x^2$ and then substitute $g(x)$ for x to obtain $f'(g(x)) = 3(g(x))^2$, which we may write more simply as $f'(g) = 3g^2$. If instead of writing $f(x) = x^3$ we had written $f(g) = g^3$ and differentiated with respect to g to get $\dfrac{df}{dg} = 3g^2$ we would have obtained the same result directly. For this reason, $f'(g(x))g'(x)$ may be written $\dfrac{df}{dg}\dfrac{dg}{dx}$. The CHAIN RULE may now be stated $\dfrac{df}{dx} = \dfrac{df}{dg}\dfrac{dg}{dx}$.

But We Still Must Guess A Lot. . .

(7) In applying the CHAIN RULE in the form $\dfrac{df}{dx} = \dfrac{df}{dg}\dfrac{dg}{dx}$ we are faced with the same difficulties as in example (5) above. Suppose we are asked to find the derivative of $(2x^2 + 1)^3 + 8(2x^2 + 1)^2 + 6(2x^2 + 1) + 3$.

We would like to think of this expression as $f(g(x))$ for some f and g, but what are f and g? Again we must guess. In this case it looks like a natural choice for g is the function $2x^2 + 1$ which occurs throughout this expression. With this choice for g, $f = g^3 + 8g^2 + 6g + 3$. We compute $\dfrac{df}{dg} = 3g^2 + 16g + 6$ and $\dfrac{dg}{dx} = 4x$. Thus, $\dfrac{df}{dx} = (3g^2 + 16g + 6)4x = (3(2x^2 + 1)^2 + 16(2x^2 + 1) + 6)4x$. Of course, we could use any two distinct symbols for f and g. If we had used v and w instead the formula would be $\dfrac{dv}{dx} = \dfrac{dv}{dw}\dfrac{dw}{dx}$.

$$\frac{d}{dx}\,g^r = rg^{r-1}\frac{dg}{dx}$$

(8) As we remarked in example (5) above, $(x^r)' = rx^{r-1}$ for any real number r. We shall derive this result in the next chapter. This means that for any function $g(x)$, $((g(x))^r)' = r(g(x))^{r-1}g'(x)$. This follows from the CHAIN RULE with $f(x) = x^r$. In our alternative notation

$$\frac{d}{dx}(g(x))^r = r(g(x))^{r-1}\frac{d}{dx}g(x).$$

It is fun to apply this rule over and over again to see what complicated expressions one can derive. For example, take $g(x) = x^3 + 1$ and $r = .20$. Then $\dfrac{d}{dx}(x^3 + 1)^{.2} = .2(x^3 + 1)^{-.8}3x^2$. Let $g(x) = ((x^3 + 1)^{.2} + 1)$ and $r = .5$. Then $\dfrac{d}{dx}((x^3 + 1)^{.2} + 1)^{.5}) = .5((x^3 + 1)^{.2} + 1)^{-.5}.2(x^3+1)^{-.8}3x^2$.

Now we can repeat this process a few more times and you will amaze your friends with the complex functions you can differentiate (i.e., find the derivative function of). It is at this point that the beginning student, amazed by the complexity of the calculations, forgets completely what the subject is about! This is the time to go back and look at FIGURES 1.2 and 1.4. Also, rethink the CHAIN RULE and what it means in terms of FIGURE 1.9. No matter how complicated the expressions, these are the ideas that underlie the formulas we have been deriving in these examples.

Differentiating Different Expressions For The Same Function

(9) As our final example, we shall differentiate the two expressions $(x^2 + x^3)^2$ and $x^4 + 2x^5 + x^6$. Actually, these two expressions represent the same function, as the second is what we get by computing the square indicated

by the first. Since these two expressions represent the same function, the expressions obtained by differentiating them must also represent the same function. Using the CHAIN RULE, we compute $\dfrac{d}{dx}(x^2 + x^3)^2 =$ $2(x^2 + x^3)(2x + 3x^2)$ and, using the rule for differentiating sums, we obtain $\dfrac{d}{dx}(x^4 + 2x^5 + x^6) = 4x^3 + 10x^4 + 6x^5$. These two expressions look different at first glance but we know they are the same as functions (i.e., give the same value for each value of x) because they are just different descriptions of the derivative of the same function. In fact, multiplying the terms of the first expression gives the second.

This idea can be used to prove a very useful general identity for derivatives. To do this, replace x^2 by $f(x)$ and x^3 by $g(x)$ where $f(x)$ and $g(x)$ are any two functions that can be differentiated ("differentiable functions"). We shall try and compute $\dfrac{d}{dx}(f(x) + g(x))^2$ in two different ways as above. First, using the CHAIN RULE we obtain

$$\cdot\frac{d}{dx}(f(x) + g(x))^2 = 2(f(x) + g(x))(f'(x) + g'(x))$$

$$= 2f(x)f'(x) + 2(f'(x)g(x) + f(x)g'(x)) + 2g(x)g'(x).$$

Second, by first computing the square, we obtain

$$\frac{d}{dx}((f(x))^2 + 2f(x)g(x) + (g(x))^2)$$

$$= 2f(x)f'(x) + 2(f(x)g(x))' + 2g(x)g'(x).$$

The Product Rule

By setting these two expressions equal to each other and canceling common terms we obtain the important "product rule"

$$(f(x)g(x))' = f'(x)g(x) + f(x)g'(x)$$

In our derivation of this result, we mixed our two notations for the derivative. This is fine as long as the meaning is clear. In our alternative notation, our new formula for the derivative of a product of two functions becomes

$$\frac{d}{dx}(f(x)g(x)) = \left(\frac{d}{dx}f(x)\right)g(x) + f(x)\left(\frac{d}{dx}g(x)\right).$$

We shall have much more to say about this important formula in the next chapter.

So Now Memorize It: $(fg)' = f'g + fg'$

We now give some exercises. Following the exercises we give the solutions. Try to work each exercise first without looking at the solution. If you do look at the solution to an exercise, *immediately* make up on your own a variation of that exercise and work your variation. Finally, we give a complete set of variations of these exercises for you to practice with. In the Appendix, Supplementary Reading and Exercises, you will find related reading assignments and exercises from various standard calculus books. If you have access to one of these books, you should test yourself by looking at this material.

Read The Previous Paragraph Again And Resolve To Do It!

1.16 EXERCISES

Some Routine Work With Compositions

(1) Find the compositions indicated below. In the cases indicated, specify the values for which the composite function is defined.

(a) Find $h(x) = f(g(x))$ where $f(x) = 2x^3 + 3$ and $g(x) = (-x - 1)^3$.

(b) Find $h(x) = f(g(x))$ where $f(t) = 2t^3 + 3$ and $g(x) = (-x - 1)^3$.

(c) Find $u(v(x))$ where $u = 3v^4 + 2v^3 + 3v + 6$ and $v(x) = -x^2 + 1$.

(d) Find $p(q(x))$ where $p = \dfrac{1}{2y^3 + 3}$ and $q = x^{1/2}$.

(e) Find $h(z) = f(g(z))$ where $f(g) = \dfrac{g + 1}{g - 1}$ and $g(z) = \sqrt{z}$. For what values of z is $h(z)$ defined?

(f) Find $h(k(y))$ and $k(h(y))$ where $h(y) = y^2$ and $k(y) = y^{1/2}$. For what values of y are these functions defined?

(g) Find $h(k(y))$ and $k(h(y))$ where $h(y) = y^3$ and $k(y) = y^{1/3}$. For what values of y are these functions defined?

Now Some Guesswork. . .

(2) Find functions $f(g)$ and $g(x)$ such that $h(x) = f(g(x))$ in each of the following cases. There are many possible correct answers in each case but generally only one "natural" choice.

    (a) $h(x) = (x^2 + 1)^{1/3} + (x^2 + 1)^{-1/3}$

    (b) $h(x) = (x + 1)^3 + x^2 + 2x + 1$

    (c) $h(x) = \dfrac{x^2 + 2x}{(x + 1)^5}$

    (d) $h(x) = \dfrac{x^2 + 2x + 5}{x^2 + 2x + 6}$. For this case, try $g(x) = x^2 + 2x + 6$, $g(x) = x^2 + 2x + 5$, and $g(x) = x + 1$.

There's More Than One Way To Do It. . .

(3) In each of the following cases a function $h(x)$ is given. For each of the specified functions $g(x)$, find $f(g)$ such that $f(g(x)) = h(x)$.

    (a) For $h(x) = \dfrac{1}{(x^{8/5} - x^2)^{1/2}}$ find $f(g)$ when $g(x) = x^{8/5} - x^2$ and $g(x) = x^{1/5}$.

    (b) For $h(x) = \dfrac{1}{\sqrt{9 - 4x^2}}$ find $f(g)$ when $g(x) = 9 - 4x^2$ and $g(x) = (9 - 4x^2)^{1/2}$, and $g(x) = 2x/3$.

    (c) For $h(x) = \dfrac{1}{\sqrt{1 - 9x^2}}$ find $f(g)$ when $g(x) = 1 - 9x^2$ and $g(x) = 3x$.

    (d) For $h(x) = (16 - 2x^2)^{-1/2}$ find $f(g)$ when $g(x) = 16 - 2x^2$ and $g(x) = x/\sqrt{8}$.

    (e) For $h(x) = \dfrac{1}{(x^{4/3} + x^{2/3})^2}$ find $f(g)$ when $g(x) = x^{4/3} + x^{2/3}$, $g(x) = x^{2/3}$, and $g(x) = x^{2/3} + 1/2$.

    (f) For $h(x) = \dfrac{1}{(x - x^{4/7})^5}$ find $f(g)$ when $g(x) = x - x^{4/7}$ and $g(x) = x^{1/7}$.

    (g) For $h(x) = \dfrac{4x + x^{1/2} + 1}{4x + x^{1/2} + 5}$ find $f(g)$ when $g(x) = 4x + x^{1/2} + 5$ and $g(x) = 2x^{1/2} + 1/4$.

Now For Some Graphs. . .

(4) For each of the following choices of f(g) and g(x), sketch the graph of h(x) = f(g(x)). Be reasonably accurate but don't nitpick!

(a) $f(g) = g^{1/2}$ and $g(x) = \sin(x)$

(b) $f(g) = g^{1/3}$ and $g(x) = \sin(x)$

(c) $f(g) = |g|$ and $g(x) = \sin(x)$

At Last, Some Derivatives To Try

(5) Compute the following derivatives:

(a) $\dfrac{d}{ds}s =$

(b) $\dfrac{d}{dx}\left(x^{1/2} + \dfrac{1}{(2x + 1)^{1/3}}\right) =$

(c) $\dfrac{d}{dx}(x^3 + 5x + 9)^{10} =$

(d) $\dfrac{d}{dx}\left(\dfrac{6}{x^3 + 2x + 1}\right)^{10} =$

(e) If $f(x) = \dfrac{1}{\sqrt{x^3 + 9x^5}}$ then $f'(x) =$

(f) $\dfrac{d}{dt}(t^{3/2} + t^{-3/2}) =$

(g) Find $F'(2)$ if $F(x) = 8x^3 + 3x^{-1}$.

(h) If $D(x) = (A(x) + B(x) + C(x))^2$ find $D'(2)$ if $A(2) = B(2) = 1$, $C(2) = 2$, $A'(2) = B'(2) = 1/2$, and $C'(2) = 3$.

(i) Find the equation of the line tangent to $y = x^3 + 3x^2 + 3$ at x = 1.

(j) Find the equation of the line normal to the curve $y = x^{1/3}$ at x = 8.

(k) For $f(x) = \dfrac{1}{2x - 1}$ find $f'(x)$ and the second derivative $f''(x)$.

(l) For f(x) as in (i) above, find the "third derivative" $\dfrac{d^3}{dx^3}f(x)$.

(m) Find $\left[\dfrac{d}{dt}w(t^3 + t^2 + t + 1)\right]_{t=1}$ if $w'(4) = 6$.

1.17   SOLUTIONS TO EXERCISE 1.16

(1)   **(a)**   For each occurrence of x in $f(x) = 2x^3 + 3$ we substitute the expression $(-x - 1)^3$. Thus we write $h(x) = 2((-x - 1)^3)^3 + 3 = 2(-x - 1)^9 + 3$.

(b)   For each occurrence of t in $f(t) = 2t^3 + 3$ we substitute the expression $(-x - 1)^3$. Obviously, we get the same answer as in (a) above. The choice of symbol for the variable in the f function does not affect the resulting function $f(g(x))$.

(c)   We get $u(v(x)) = 3(-x^2 + 1)^4 + 2(-x^2 + 1)^3 + 3(-x^2 + 1) + 6$. This function is a polynomial of degree 8 in x and could, if we do some tedious algebra, be written as a sum of powers of x with integer coefficients. There will be situations where this sort of algebra will be important, but not here! The answer is best left in this form in the absence of any direct motivation for doing otherwise.

(d)   We have $p(q(x)) = \dfrac{1}{2(x^{1/2})^3 + 3} = \dfrac{1}{2x^{3/2} + 3}$.

(e)   We substitute $\sqrt{z}$ for each occurrence of g in $\dfrac{g + 1}{g - 1}$ to obtain $f(g(z)) = \dfrac{\sqrt{z} + 1}{\sqrt{z} - 1}$. The function $\sqrt{z}$ is defined (as a function from real numbers to real numbers) only for z a nonnegative real number. If $z = 1$ then the denominator of $f(g(z))$ is zero. Thus $f(g(z))$ is defined for all nonnegative real numbers z except $z = 1$.

(f)   First, $h(k(y)) = (y^{1/2})^2 = y$. In reverse order we find $k(h(y)) = (y^2)^{1/2} = y$. The function $i(y) = y$ is the "identity" function (it does nothing to y). The function $i(y)$ is the linear function with slope 1 passing through the origin. In general, functions $h(y)$ and $k(y)$ such that $h(k(y)) = k(h(y)) = y$ for all y are called "compositional inverses of each other." We must be a little bit careful about the statement "for all y" in this definition. In our example, the function $h(y) = y^2$ is defined for all y but the function $k(y) = y^{1/2}$ is defined only for nonnegative real numbers (we are ignoring complex numbers at this point). Thus, the composition $k(h(y))$ is defined for all real numbers but the composition $h(k(y))$ is defined only for nonnegative real numbers. Technically, the statement $h(k(y)) = k(h(y))$ can be made only for non-negative real numbers and hence h and k are compositional inverses for all *nonnegative* real numbers. In general, when we say that two functions h and k are compositional inverses we mean that $h(k(y)) = k(h(y))$ = y for all y in some specified common domain of definition of h and k. In our example this "common domain of definition" is the set of all nonnegative real numbers.

(g)   As in (f) above, $h(k(z)) = k(h(z)) = z$. In this case, however, this relation is defined for all real numbers z because $z^{1/3}$ is defined for all real numbers and, of course, so is z^3. In general, if p is an odd integer then z^p and $z^{1/p}$ are compositional inverses for all z. If p is even (and nonzero) they are compositional inverses for all nonnegative z.

(2)   **(a)** $f(g) = g^{1/3} + g^{-1/3}$ and $g(x) = x^2 + 1$ is the most natural choice. Also $f(g) = g^{2/3} + g^{-2/3}$ and $g(x) = (x^2 + 1)^{1/2}$, $f(g) = g + g^{-1}$ and $g(x) = (x^2 + 1)^{1/3}$, etc. will work. There are infinitely many possibilities. . .

(b) $h(x) = (x + 1)^3 + (x + 1)^2$ so we may take $f(g) = g^3 + g^2$ and $g(x) = x + 1$ as the most natural choice.

Remember This Trick! Completing The Square

(c)   The numerator in h(x) is the expression $x^2 + 2x$. This expression is the sum of the first two terms of $(x + 1)^2 = x^2 + 2x + 1$ and thus can be written as $x^2 + 2x = (x + 1)^2 - 1$. This little trick is called "completing the square." It can be applied to any expression $ax^2 + bx$ by observing that $(\sqrt{a}x + b/2\sqrt{a})^2 = ax^2 + bx + b^2/4a$ and hence that

$$ax^2 + bx = (\sqrt{a}x + b/2\sqrt{a})^2 - b^2/4a.$$

You should learn this trick thoroughly and try a number of examples until you feel comfortable with it! Applying it to our immediate problem gives $f(g) = \dfrac{g^2 - 1}{g^5} = g^{-3} - g^{-5}$ and $g(x) = x + 1$.

(d)   Using the "complete the square" trick described in (c) above, we replace $x^2 + 2x$ by $(x + 1)^2 - 1$ in both the numerator and denominator of h(x). Doing this we see that, in the case where $g(x) = x + 1$, we have $f(g) = \dfrac{g^2 + 4}{g^2 + 5}$. If $g(x) = x^2 + 2x + 6$ then $f(g) = (g - 1)/g$ or $1 - g^{-1}$. If $g(x) = x^2 + 2x + 5$ then $f(g) = g/(g + 1)$.

Integration By Substitution Is What We'll Call It Later. . .

(3) As a general remark about this problem, the CHAIN RULE is very important in both differential calculus (that's what we are studying now) and integral calculus (we'll study that a little later). The tricks for writing a given function h(x) as $f(g(x))$ are a little different in these two subjects. That's the motivation for the different choices for $g(x)$ in these problems.

(a) For $g(x) = x^{8/5} - x^2$, $f(g) = g^{-1/2}$. For $g(x) = x^{1/5}$, $f(g) = g^{-4} (1 - g^2)^{-1/2}$.

(b) The notation $\dfrac{1}{\sqrt{9 - 4x^2}}$ is a bad one and should be replaced by $\dfrac{1}{(9 - 4x^2)^{1/2}}$ or simply $(9 - 4x^2)^{-1/2}$. When $g(x) = 9 - 4x^2$ then $f(g) = g^{-1/2}$. When $g(x) = (9 - 4x^2)^{1/2}$ then $f(g) = g^{-1}$. When $g(x) = 2x/3$ then $x = 3g/2$ and thus $(9 - 4x^2)^{-1/2} = (9 - 4(3g/2)^2)^{-1/2} = (9 - 9g^2)^{-1/2} = \dfrac{1}{3}(1 - g^2)^{-1/2} = f(g)$. This is an important trick in "integral calculus."

(c) When $g(x) = 1 - 9x^2$ then $f(g) = g^{-1/2}$. If $g(x) = 3x$ then $x = g/3$ so $(1 - 9x^2)^{-1/2} = (1 - 9(g/3)^2)^{-1/2} = (1 - g^2)^{-1/2}$. Compare this latter "change of variable" with that of (b) above.

(d) When $g(x) = 16 - 2x^2$ then $f(g) = g^{-1/2}$. When $g(x) = x/\sqrt{8}$ then $x = \sqrt{8}g$ and $(16 - 2x^2)^{-1/2} = (16 - 2(\sqrt{8}g)^2)^{-1/2} = \dfrac{1}{4}(1 - g^2)^{-1/2}$.

(e) When $g(x) = x^{4/3} + x^{2/3}$ then $f(g) = g^{-2}$. When $g(x) = x^{2/3}$ then $x = g^{3/2}$ and $f(g) = (g^2 + g)^{-2}$. Completing the square we find that $f(g) = ((g + 1/2)^2 - 1/4)^2$. If $g = x^{2/3} + 1/2$ then $f(g) = (g^2 - 1/4)^{-2}$.

(f) When $g(x) = x - x^{4/7}$ then $f(g) = g^{-5}$. When $g(x) = x^{1/7}$ then $f(g) = (g^7 - g^4)^{-5} = g^{-20}(g^3 - 1)^{-5}$.

(g) If $g(x) = 4x + x^{1/2} + 5$ then $f(g) = \dfrac{g - 4}{g} = 1 - 4g^{-1}$. For the second part, note that $4x + x^{1/2} = (2x^{1/2} + 1/4)^2 - 1/16$. Here we are again completing the square using $x^{1/2}$ as the variable. Thus for $g(x) = 2x^{1/2} + 1/4$ we get $f(g) = \dfrac{g^2 + 15/16}{g^2 + 79/16}$.

When Graphing, Do A Rough Approximation First

(4)

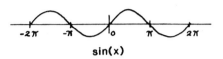

sin(x)

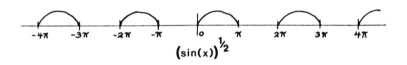

$(\sin(x))^{1/2}$

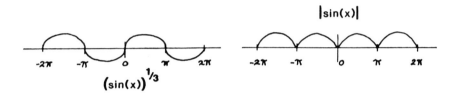

$(\sin(x))^{1/3}$

$|\sin(x)|$

(5) Be sure to try the VARIATIONS for additional practice in taking derivatives.

(a) $\dfrac{d}{ds}s = \dfrac{d}{dt}t = \dfrac{d}{dy}y = \dfrac{d}{dw}w = \dfrac{d}{dx}x = \ldots = 1.$

(b) $\dfrac{d}{dx}\left(x^{1/2} + \dfrac{1}{(2x+1)^{1/3}}\right) = \dfrac{d}{dx}x^{1/2} + \dfrac{d}{dx}(2x+1)^{-1/3} =$
$(1/2)x^{-1/2} + (-1/3)(2x+1)^{-4/3}(2) = (1/2)x^{-1/2} - (2/3)(2x+1)^{-4/3}.$
The second term was computed using the CHAIN RULE applied to
$f(g(x))$ with $f(g) = g^{-1/3}$ and $g(x) = 2x + 1$ (and hence $g'(x) = 2$).

(c) Apply the CHAIN RULE to $f(g(x))$ where $f(g) = g^{10}$ and $g(x)$
$= x^3 + 5x + 9$. The answer is $10(x^3 + 5x + 9)^9(3x^2 + 5)$.

(d) We can write this as $(6^{10})\dfrac{d}{dx}(x^3 + 2x + 1)^{-10}$ and use the CHAIN

RULE for $f(g(x))$ with $f(g) = g^{-10}$ and $g(x) = x^3 + 2x + 1$ to obtain
$6^{10}(-10)(x^3 + 2x + 1)^{-11}(3x^2 + 2))$. Another, and probably more
common, way a calculus student would attempt to work this problem

is to apply the CHAIN RULE with $g(x) = \dfrac{6}{x^3 + 2x + 1}$ and $f(g) = g^{10}$ to get

$$10\left(\frac{6}{x^3 + 2x + 1}\right)^9 \frac{d}{dx}\left(\frac{6}{x^3 + 2x + 1}\right).$$

The latter derivative $\dfrac{d}{dx}\left(\dfrac{6}{x^3 + 2x + 1}\right)$ would then be computed with a second application of the CHAIN RULE. The reader should carry this out and verify that the answer is the same as the one just given.

(e) We should first get rid of the awkward square root notation and write this function as $(x^3 + 9x^5)^{-1/2}$. Use the CHAIN RULE with $g(x) = x^3 + 9x^5$ and $f(g) = g^{-1/2}$ to get $(-1/2)(x^3 + 9x^5)^{-3/2}(3x^2 + 45x^4)$.

(f) The answer is $(3/2)t^{1/2} + (-3/2)t^{-5/2}$.

(g) $F'(x) = 24x^2 + 3(-1)x^{-2} = 24x^2 - \dfrac{3}{x^2}$ and thus $F'(2) = 96 - 3/4 = 95.25$.

(h) We again use the CHAIN RULE with $f(g) = g^2$ and $g(x) = A(x) + B(x) + C(x)$ so that $D(x) = f(g(x))$. Thus, $D'(x) = 2(A(x) + B(x) + C(x))(A'(x) + B'(x) + C'(x))$ and $D'(2) = 2(1 + 1 + 2)(1/2 + 1/2 + 3) = 32$.

(i) If you look at FIGURE 1.9(a), the line $\tilde{g}$ is tangent to the function g at the point $(3, 2)$ on the graph of g. In general, if $g(x)$ is a function that has a derivative at the point $P = (x, g(x))$ then the *tangent to g at P* is the unique straight line passing through P and having slope $g'(x)$. In this problem, we are asked to find the tangent to $y(x) = x^3 + 3x^2 + 3$ at the point $x = 1$. This is a typical way of stating this sort of problem in which the statement "at $x = 1$" really means "at the point $(1, y(1)) = (1, 7)$." We must find the equation of the straight line passing through the point $(1, 7)$ and having slope $y'(1) = 9$. This is the line $y = 9x - 2$.

(j) The normal line to a curve at a point P is the line passing through P and perpendicular to the tangent line to the curve at P. In this problem, $P = (8, 2)$. The tangent line to $y = x^{1/3}$ at P has slope $y'(8) = (1/3)8^{-2/3} = 1/12$ and its equation is given by $y = (1/12)x + 4/3$. You should know that the slope of the line normal to a line of slope m is $-1/m$. Thus the normal line we are looking for has slope -12 and passes through the point $(8, 2)$. Its equation is $y = -12x + 98$, which is the answer to this problem.

(k) Write $f(x) = (2x - 1)^{-1}$. We find that $f'(x) = (-1)(2x - 1)^{-2}(2x - 1)' = -2(2x - 1)^{-2}$. The function $f'(x)$ can again be differentiated. The first derivative of $f'(x)$ is called the *second derivative* of $f(x)$ and is denoted by $f''(x)$ or by $\dfrac{d^2}{dx^2}$. In this problem, we compute $f''(x) = (-2(2x - 1)^{-2})' = (-2)(-2)(2x - 1)^{-3}(2x - 1)' = 8(2x - 1)^{-3}$.

(l) The third derivative of $f(x)$ is the derivative of $f''(x)$. The third derivative is denoted by $f'''(x)$ or by $\dfrac{d^3}{dx^3}$. In this problem we compute $(8(2x - 1)^{-3})' = (8)(-3)(2x - 1)^{-4}(2x - 1)' = -48(2x - 1)^{-4}$. In general, the nth derivative of $f(x)$ is obtained by differentiating $f(x)$ n times. The nth derivative is denoted by $f^{(n)}(x)$ or by $\dfrac{d^n}{dx^n}f(x)$. Can you give a formula for $\dfrac{d^n}{dx^n}(2x + 1)^{-1}$? In general, it will happen that a function with a first derivative at a point may not have an nth derivative at that point for some n. For example, $f(x) = x^{4/3}$ has first derivative function $f'(x) = (4/3)x^{1/3}$ and second derivative $f''(x) = (4/9)x^{-2/3}$. At $x = 0$ we have $f'(0) = 0$ but $f''(0)$ is not defined.

(m) This problem illustrates the "bracket notation" for the two-step process of *first* computing a derivative function and *second* evaluating that function at a certain value. The notation $\left[\dfrac{d}{dx}f(x)\right]_{x=a}$ means first compute $f'(x)$ and then evaluate $f'(a)$. Some students like to do it the other way around by first finding $f(a)$ and then computing the derivative with respect to x of the constant $f(a)$. The answer is of course always zero! This is not the way to go. In our particular problem, we compute

$$\frac{d}{dt}w(t^3 + t^2 + t + 1) = w'(t^3 + t^2 + t + 1)(t^3 + t^2 + t + 1)'$$

$$= w'(t^3 + t^2 + t + 1)(3t^2 + 2t + 1).$$

Substituting $t = 1$ this becomes $w'(4)(6) = 6w'(4)$. We are using the CHAIN RULE without knowing explicitly what w is (sometimes this is a very useful trick). We could go no further except for the fact that we have (conveniently!) been given that $w'(4) = 6$ so the final answer is 36.

Do You See How The Variations Parallel The Original Problems?

1.18   VARIATIONS ON EXERCISE 1.16

(1) Find the compositions indicated below. In the cases indicated, specify the values for which the composite function is defined.

(a) Find $h(x) = f(g(x))$ where $f(x) = 2x^5 - 3$ and $g(x) = (-x + 1)^{1/3}$.

(b) Find $h(x) = f(g(x))$ where $f(t) = 2t^5 - 3$ and $g(x) = (-x + 1)^3$.

(c) Find $u(v(x))$ where $u = 2v^5 + 2v^3 + 3v + 9$ and $v(x) = (-x^2 - 1)^{-1}$.

(d) Find $p(q(x))$ where $p = \dfrac{1}{2y^{-3} - 3}$ and $q = x^{-1/2}$.

(e) Find $h(z) = f(g(z))$ where $f(g) = \dfrac{g^3 + 1}{g^3 - 1}$ and $g(z) = \sqrt{z}$. For what values of z is $h(z)$ defined?

(f) Find $h(k(y))$ and $k(h(y))$ where $h(y) = y^6$ and $k(y) = y^{1/6}$. For what values of y are these functions defined?

(g) Find $h(k(y))$ and $k(h(y))$ where $h(y) = y^5$ and $k(y) = y^{1/5}$. For what values of y are these functions defined?

(2) Find functions $f(g)$ and $g(x)$ such that $h(x) = f(g(x))$ in each of the following cases. There are many possible correct answers in each case but generally only one "natural" choice.

(a) $h(x) = (x^2 + 1)^{1/3} + (x^4 + 2x^2 + 1)^{-1/3}$

(b) $h(x) = (x^{1/2} + 1)^3 + x + 2x^{1/2} + 9$

(c) $h(x) = \dfrac{4x^2 + 4x + 9}{(2x + 1)^5}$

(d) $h(x) = \dfrac{x^3 + 3x^2 + 3x + 5}{x^2 + 2x + 6}$. For this case, try $g(x) = x + 1$.

(3) In each of the following cases a function $h(x)$ is given. For each of the specified functions $g(x)$, find $f(g)$ such that $f(g(x)) = h(x)$.

(a) For $h(x) = \dfrac{1}{(x^{3/2} - x^{2/3})^2}$ find $f(g)$ when $g(x) = x^{2/3}$ and $g(x) = x^{3/2}$.

(b)   For $h(x) = \dfrac{1}{\sqrt{4 - 9x^2}}$ when $g(x) = (4 - 9x^2)^{1/2}$, $g(x) = 3x$,

and $g(x) = 3x/2$.

(c)   For $h(x) = \dfrac{1}{\sqrt{1 - 8x^3}}$ find $f(g)$ where $g(x) = 1 - 8x^3$ and

$g(x) = 2x$.

(d)   For $h(x) = (10 - 2x^2)^{-1/2}$ find $f(g)$ when $g(x) = 10 - 2x^2$
and $g(x) = x/\sqrt{5}$.

(e)   For $h(x) = \dfrac{1}{(x^{4/3} + x^{2/3} + 5/4)^2}$ find $f(g)$ when $g(x) =$

$x^{4/3} + x^{2/3}$, $g(x) = x^{2/3}$, and $g(x) = x^{2/3} + 1/2$.

(f)   For $h(x) = \dfrac{1}{(x - x^{4/3})^3}$ find $f(g)$ where $g(x) = x - x^{4/3}$ and

$g(x) = x^{1/3}$.

(g)   For $h(x) = \dfrac{4x + 2x^{1/2} + 1}{4x + 2x^{1/2} + 5}$ find $f(g)$ where $g(x) =$

$4x + 2x^{1/2} + 5$ and $g(x) = 2x^{1/2} + 1/2$.

(4)   For each of the following choices of $f(g)$ and $g(x)$, sketch the graph of
$h(x) = f(g(x))$.

(a) $f(g) = \sin(g)$ and $g(x) = 1/x$

(b) $f(g) = \sin(g)/g$ and $g(x) = 1/x$

(c) $f(g) = \tan(g)$ and $g(x) = 1/x$

(5)   Take each of (a) through (m) of EXERCISE 1.16 (5) and vary it slightly.
Change constants, exponents, and variable names. Write down the problems
thus created and work them. Try some of your classmate's problems also.

1.19   VARIATIONS ON EXERCISE 1.16

(1)   Find the compositions indicated below. In the cases indicated, specify
the values for which the composition is defined.

(a)   Find $h(x) = f(g(x))$ where $f(x) = 2x^2 + 4$ and $g(x) = (-x + 1)^5$.

(b)   Find $h(x) = f(g(x))$ where $f(t) = 3t^2 - 2$ and $g(x) = (x - 1)^2$.

(c)   Find $u(v(x))$ where $u(v) = 3v^5 - 2v^2 + v - 1$ and $v(x) = 2x^3 - 14$.

(d) Find $p(q(x))$ where $p = \dfrac{1}{q^4 - q}$ and $q = x^{1/3}$.

(e) Find $h(z) = f(g(z))$ where $f(g) = \dfrac{g - 2}{g + 2}$ and $g(z) = \sqrt{z} - 3$.
For what values of z is $h(z)$ defined?

(f) Find $h(k(y))$ and $k(h(y))$ where $h(y) = y^4$ and $k(y) = y^{1/4}$. For what values of y are these functions defined?

(g) Find $h(k(y))$ and $k(h(y))$ where $h(y) = y^{1/5}$ and $k(y) = y^5$. For what values of y are these functions defined?

(2) Find functions $f(g)$ and $g(x)$ such that $h(x) = f(g(x))$ in each of the following cases. There are many possible answers but usually one "natural" answer.

(a) $h(x) = (x - 1)^3 - 3(x - 1)^2 + \sqrt{x - 1} + 4$

(b) $h(x) = (x + 2)^2 + x^2 + 4x + 5$

(c) $h(x) = \dfrac{x^2 + 9x + 4}{(x + 3)^2}$

(d) $h(x) = \dfrac{x^2 - 1}{x^2 + 1}$. Try $g(x) = x^2$, $g(x) = x^2 - 1$, $g(x) = x^2 + 1$.

(3) In each of the following cases a function $h(x)$ is given. For each of the specified functions $g(x)$, find $f(g)$ such that $f(g(x)) = h(x)$.

(a) For $h(x) = \sqrt{x^2 - x^3}$ find $f(g)$ when $g(x) = x^2 - x^3$ and $g(x) = x^{1/4}$.

(b) For $h(x) = (25 - 4x^2)^3 - 4$ find $f(g)$ when $g(x) = 25 - 4x^2$, and $g(x) = 2x/5$.

(c) For $h(x) = (1 - 8x^3)^{-1/2}$ find $f(g)$ when $g(x) = 1 - 8x^3$, and $g(x) = x/2$.

(d) For $h(x) = 1/(x^{1/5} - x^{3/5})^4$ find $f(g)$ when $g(x) = x^{1/5} - x^{3/5}$, $g(x) = x^{1/5}$, and $g(x) = x^{1/5} + 1$.

(e) For $h(x) = (3 - 2x^2)^{5/2}$ find $f(g)$ when $g(x) = 3 - 2x^2$, and $g(x) = \sqrt{2}x/\sqrt{3}$.

(f) For $h(x) = (x^{2/9} - x + 1)^{-2}$ find $f(g)$ when $g(x) = x^{2/9} - x$, and $g(x) = x^{1/9}$.

(g) For $h(x) = \dfrac{x + 3\sqrt{x} - 1}{x + 3\sqrt{x} + 4}$, find $f(g)$ when $g(x) = x + 3\sqrt{x}$, and $g(x) = \sqrt{x} + 3/2$.

(4) For each of the following choices of f(g) and g(x), sketch the graph of h(x) = f(g(x)). Be reasonably accurate, but don't worry about being perfect.

    (a) $f(g) = g^2$, and $g(x) = \cos(x)$

    (b) $f(g) = 2g + 1$ and $g(x) = \cos(x)$

    (c) $f(g) = \sqrt{g}$ and $g(x) = \cos(x) + 1$

(5) Take each of (a) through (m) of EXERCISE 1.16 (5) and vary it slightly. Change constants, exponents, and variable names. Write down the problems thus created and work them. Try some of your classmate's problems also.

1.20 VARIATIONS ON EXERCISE 1.16

(1) Find the compositions indicated below. In the cases indicated, specify the values for which the composition is defined.

    (a)   Find h(x) = f(g(x)) where $f(x) = 3x^4 + 4x^2 + 1$ and g(x) = x + 2.

    (b)   Find h(x) = f(g(x)) where $f(x) = \dfrac{1}{x - 2}$ and g(x) = x - 2.

    (c)   Find r(s(x)) where $r(s) = \sqrt{s} - 4s + 1$ and s(x) = 2x.

    (d)   Find m(n(x)) where $m(n) = (n + 4)^3 - 1$ and n(x) = 14x - 37.

    (e)   Find h(x) = f(g(x)) where $f(x) = \dfrac{(x + 1)^2 - 1}{(x + 1)^2 + 1}$ and g(x) = $2x^2 + 3$.

    (f)   Find h(u) = f(g(u)) where $f(g) = 2/(g - 3)$ and $g(u) = \sqrt{u + 2}$. For what values of u is h(u) defined?

    (g)   Find f(g(x)) and g(f(x)) where $f(x) = (x + 2)^{-2}$ and $g(x) = x^3 - 1$. For what values of x are f(g((x)) and g(f(x)) defined?

    (h)   Find f(g(x)) and g(f(x)) where $f(x) = \sqrt{x}$ and $g(x) = x^2 + x$. For what values of x are these functions defined?

(2) Find functions f(g) and g(x) such that h(x) = f(g(x)) in each of the following cases. There are many possible answers but generally only one "natural" choice.

    (a) $h(x) = \sqrt{x^2 + 4x}$

    (b) $h(x) = (x + 1)^3 - 3(x + 1)^2 + 4$

    (c) $h(x) = (x - 2)^2 + 2(x^2 - 4x + 5)$

(d) $h(x) = \dfrac{x^2 - 2x + 3}{x^2 - 2x + 5}$. Try $g(x) = x^2 - 2x + 3$, and $g(x) = (x - 1)$.

(3)   In each of the following cases a function is given. For each of the specified functions $g(x)$, find $f(g)$ such that $f(g(x)) = h(x)$.

(a)   For $h(x) = (1/x^{1/3} - 1/x^{2/3})^3$ find $f(g)$ when $g(x) = 1/x^{1/3} - 1/x^{2/3}$, $g(x) = x^{1/3}$.

(b)   For $h(x) = 1/\sqrt{x^2 + 2}$ find $f(g)$ when $g(x) = \sqrt{x^2 + 2}$, $g(x) = \sqrt{x}$.

(c)   For $h(x) = (1 - 4x^2)^{1/2}$ find $f(g)$ when $g(x) = 1 - 4x^2$, $g(x) = 2x$.

(d)   For $h(x) = x^2 + 2x + 5$ find $f(g)$ when $g(x) = \sqrt{x}$, $g(x) = x + 1$.

(e)   For $h(x) = x^{2/3} - x^{4/3} + 1$ find $f(g)$ when $g(x) = x^{2/3} - x^{4/3}$, $g(x) = x^{1/3}$.

(f)   For $h(x) = \sqrt{x^{1/2} + x^2}$ find $f(g)$ when $g(x) = x^{1/2}$, $g(x) = x^2$.

(g)   For $h(x) = \dfrac{x^2 + x + 1}{x^2 + x + 2}$ find $f(g)$ when $g(x) = x^2 + x$, $g(x) = 2x$.

(4)   For each of the following choices of $f(g)$ and $g(x)$, sketch a fairly accurate graph of $f(g(x))$:

(a) $f(g) = -g + 2$ and $g(x) = \sin(x) + 1$

(b) $f(g) = \sqrt{g}$ and $g(x) = 3\sin(x)$

(c) $f(g) = |g|$ and $g(x) = \cos(x)$

(5)   Compute the following derivatives:

(a) $\dfrac{d}{dt}(t) =$

(b) $\dfrac{d}{dx}(x + 2/(x - 1)^{2/3}) =$

(c) $\dfrac{d}{dx}(x^2 + x + 1)^5 =$

(d) $\dfrac{d}{dx}\left(\left(\dfrac{2}{x^2 + 1}\right)^8\right) =$

(e)   If $g(t) = \sqrt{3t^2 - 4}$ then $g'(t) =$

(f)   Find $\dfrac{df}{dx}$ if $f(x) = \dfrac{1}{(3x^2 - 5x^2)^2}$.

(g) $\dfrac{d}{dy}(y^4 - (y + 1)^3)$

(h)   If $C(x) = (A(x) - 3B(x))^4$ find $C'(-2)$ if $A(-2) = B(-2) = 1$, $A'(-2) = 0$, and $B'(-2) = -2$.

(i)   Find the equation of the line tangent to the curve $y = \sqrt{x^2 - 1}$ at $x = 2$.

(j)   Find the equation of the line normal to the curve $y = x^4 + 2$ at $x = -1$.

(k)   For $f(x) = 1/\sqrt{x + 1}$ find $f'(x)$ and $f''(x)$.

(l)   For $g(t) = t^3 + 2t^2 + t + 4$ find the "third derivative" $\dfrac{d^3g(t)}{dt^3}$.

(m)   Find $\left[\dfrac{d}{dx} f(x^2 + x + 1)\right]_{x=0}$ if $f'(1) = 2$.

1.21   VARIATIONS ON EXERCISE 1.16

(1)   Find the compositions indicated below. In the cases indicated, specify the values for which the composition is defined.

(a)   Find $h(x) = f(g(x))$ where $f(x) = (x - 1)^2 + 4x + 1$ and $g(x) = x^3 + 2x$.

(b)   Find $h(x) = f(g(x))$ where $f(s) = 3s^3 + 5$ and $g(x) = (-x + 4)^3$.

(c)   Find $h(x) = s(t(x))$ where $s(x) = x^4 - x^3 + 2x + 1$ and $t(x) = x^2 + 2x - 1$.

(d)   Find $P(q(y))$ where $P(q) = (q^2 - 1)/(q^2 + 1)$ and $q(y) = \sqrt{y}$.

(e)   Find $h(x) = f(g(x))$ where $f(g) = \dfrac{g}{g - 1}$ and $g(x) = x^{1/3}$. For what values of x is $h(x)$ defined?

(f)   Find $h(k(t))$ and $k(h(t))$ where $h(t) = t^2 - 1$ and $k(t) = \sqrt{t + 1}$. For what values of t are these functions defined?

(g)   Find $f(g(x))$ and $g(f(x))$ if $f(x) = x^2 + x$ and $g(x) = x - 1$.

(2)   Find functions $f(g)$ and $g(x)$ such that $h(x) = f(g(x))$ in each of the following cases:

(a) $h(x) = (x^2 + x - 1)^3 + \sqrt{x^2 + x - 1} + 4$

(b) $h(x) = (x^3 - 1)^2 + x^3 - 2$

(c) $h(x) = ((x - 2)^2 - 1)/((x - 2)^2 + 1)$

(d) $h(x) = (x^2 + 8x + 3)/((x - 4)^2 + 1)$

(3)   In each of the following cases a function $h(x)$ is given. For each of the specified functions $g(x)$, find $f(g)$ so that $f(g(x)) = h(x)$.

(a)   For $h(x) = (x + 3)^2 - x + 2$, find $f(g)$ when $g(x) = x + 3$.

(b)   For $h(x) = x^2 + x^3$, find $f(g)$ when $g(x) = x^2$ and $g(x) = x^{1/2}$.

(c)   For $h(x) = \sqrt{2x^2 - 3}$, find $f(g)$ when $g(x) = 2x^2 - 3$ and $g(x) = 2x^2$.

(d)   For $h(x) = x/(x^{1/3} + x^2)$, find $f(g)$ when $g(x) = x^{1/3}$ and $g(x) = x^{1/3} + 1$.

(e)   For $h(x) = (3 + x + 2\sqrt{x})^{5/2}$, find $f(g)$ when $g(x) = \sqrt{x}$, and $g(x) = 1 + x + 2\sqrt{x}$.

(f)   For $h(x) = (x^{1/2} - x^{-1/2})/(x^{1/2} + x^{-1/2})$, find $f(g)$ when $g(x) = x^{1/2}$ and $g(x) = x^{-1/2}$

(4)   For each of the following choices of $f(g)$ and $g(x)$, sketch the graph of $h(x) = f(g(x))$. Be fairly accurate.

(a) $f(g) = g^2 + 1$ and $g(x) = |x|$

(b) $f(g) = |g|$ and $g(x) = \cos(x)$

(c) $f(g) = \sqrt{g}$ and $g(x) = |x|$

(5)   Compute the following derivatives:

(a) $\dfrac{d}{dx}(3) =$

(b) $\dfrac{d}{dx}(x^4 + 3x^3 + 2x^2 + x - 1) =$

(c) $\dfrac{d}{dt}(t^4 + 1)^8 =$

(d) $\dfrac{d}{ds}\left(\dfrac{2}{s^2 + 1}\right)^3 =$

(e)   Find $g'(x)$ if $g(x) = \sqrt{\sqrt{x} + 1}$.

(f) $\dfrac{d^2}{dx^2}(x - x^{-1}) =$

(g)   Find $g'(x)$ and $g''(x)$ if $g(x) = 1/x^2$.

(h)   If $f(x) = (g(x))^2 + 1$ find $f'(2)$ if $g(2) = 1$, $g'(2) = -1$.

(i)   Find the equation of the line tangent to $y = \sqrt{x}$ at $x = 4$.

(j)   Find the line normal to the curve $y = 2x^2 - 3x + 1$ at $x = -1$.

(k)   Find $g^{(3)}(0)$ if $g(x) = x^4 - x^3 + x^2 - x + 1$.

1.22   VARIATIONS ON EXERCISE 1.16

(1)   Find the compositions indicated below. In the cases indicated, specify the values for which the composition is defined.

(a)   Find $h(x) = f(g(x))$ where $f(x) = 2x/\sqrt{1 - x^2}$ and $g(x) = 1 + x^2$.

(b)   Find $g(f(x))$ where $g(t) = 3 + \cos^2 t$ and $f(x) = x^2 - 1$.

(c)   Find $r(s(t))$ where $r(y) = 1/y + 2/y^2 - 3/y^4$ and $s(t) = 2 + 1/t$.

(d)   Find $a(b(x))$ where $a(x) = \sin^2 x + 1$ and $b(x) = 1 + \cos x$.

(e)   Find $h(t) = f(g(t))$ where $f(t) = \sqrt{t}$ and $g(t) = -t$. For what values of t is $h(t)$ defined?

(f)   Find $f(g(x))$ and $g(f(x))$ where $f(x) = \sqrt{x}$ and $g(x) = \sin x$. For what values of x between $-\pi$ and π are these two functions defined?

(g)   Find $f(g(x))$ and $g(f(x))$ where $f(x) = -x^2$ and $g(x) = |x|$.

(2)   Find functions $f(g)$ and $g(x)$ such that $h(x) = f(g(x))$ in each of the following cases:

(a) $h(x) = \sin^2(x^2 + 4x + 1)$

(b) $h(x) = \sqrt{2x^2 + 1} - 4$

(c) $h(x) = (2x - 3)^2 + 5x - 3$

(d) $h(x) = 2x^2 - 18x + 4 - (x - 3)^{2/3}$

(3)   In each of the following cases a function $h(x)$ is given. For each of the specified functions $g(x)$, find $f(g)$ so that $h(x) = f(g(x))$.

(a)   For $h(x) = \sin^3(x^2 + 1)$ find $f(g)$ when $g(x) = x^2 + 1$ and $g(x) = \sin(x^2 + 1)$.

(b)   For $h(x) = x^2 - x^{-2}$ find $f(g)$ when $g(x) = x^2$ and $g(x) = x^{-1/3}$.

(c)   For $h(x) = \sqrt{(x - 1)^2 + 3}$ find $f(g)$ when $g(x) = (x - 1)$ and $g(x) = (x - 1)^2$.

(d) For $h(x) = \cos(\sqrt{x + 1})$ find $f(g)$ when $g(x) = x + 1$ and $g(x) = \sqrt{x + 1}$.

(e) For $h(x) = \sqrt{(x^2 - 4)^3}$ find $f(g)$ when $g(x) = x^2$ and $g(x) = x^2 - 4$.

(f) For $h(x) = \dfrac{1}{3x^2 + 4x + 1}$ find $f(g)$ when $g(x) = 3x^2 + 4x + 1$ and $g(x) = 3x^2 + 4x - 1$.

(4) For each of the following choices of $f(g)$ and $g(x)$, sketch a reasonably accurate graph of $f(g(x))$:

    (a) $f(g) = \sin(g)$ and $g(x) = 2x + \pi$

    (b) $f(g) = g^2 + 1$ and $g(x) = \cos x$

    (c) $f(g) = 1 - 3g$ and $g(x) = |x|$

(5) Compute the following derivatives:

    (a) $\dfrac{d}{dx}((1 + x^2)^{-1}) =$

    (b) $\dfrac{d}{dz}(2z^2 - 4/z)^2 =$

    (c) If $g(x) = \sqrt{2x^3 + 1}$ then $g'(1) =$

    (d) $\dfrac{d}{dd}(2) =$

    (e) $\dfrac{d}{dx}\left(\dfrac{2}{1 + |x|}\right) =$

    (f) Find $f'(x)$, $f''(x)$, and $f(x) = (x - 1)^{10}$.

    (g) $\dfrac{d^2}{dt^2}(1 + t^{-1})^{-1} =$

    (h) Find the equation of the line tangent to $y = 1/x$ at $x = -1$.

    (i) Find the equation of the line normal to $y = \sqrt{x} - 1$ at $x = 2$.

    (j) If $f(x) = \sqrt{(g(x))^2 + 1}$ find $f'(0)$ if $g(0) = 1$, $g'(0) = 2$.

    (k) Find $\dfrac{d^3 f(x)}{dx^3}$ if $f(x) = x^4 - 2x^2$.

Chapter 2

COMPUTING DERIVATIVES

We Must Enlarge The Class Of Functions We Can Differentiate

In Chapter 1 we developed the basic conceptual ideas of calculus. Our approach is mostly intuitive, but EXERCISE 1.16 should give a hint of the power of these ideas. In this chapter we shall concentrate on techniques of differentiation. Our first task is to enlarge the class of basic functions that we know how to differentiate. We shall also learn a few more rules for differentiation.

To review, the keystone of our approach to derivatives is an understanding of RULES FOR DERIVATIVES 1.12, particularly the all-important CHAIN RULE. In EXAMPLE 1.15(5) we stated without proof the rule $(x^r)' = rx^{r-1}$. This important rule should be memorized and is valid for any real number r. In EXAMPLE 1.15(8) we used this rule together with the CHAIN RULE to state the rule $((g(x))^r)' = r(g(x))^{r-1}g'(x)$, valid for any differentiable function g. This rule is a very common special case of the CHAIN RULE. Another very useful rule for computing derivatives was derived in EXAMPLE 1.15(9). This rule is called the PRODUCT RULE and states that $(f(x)g(x))' = f'(x)g(x) + f(x)g'(x)$. If you have not carefully studied EXERCISE 1.16 you should do so now.

As our first task in this chapter, we shall prove the validity of the rule $(x^r)' = rx^{r-1}$. Our proof consists of a series of very short "lemmas" dealing with special cases of this formula. By piecing together these lemmas, we obtain a proof of the general result. A student interested only in the techniques of differentiation could memorize the result $(x^r)' = rx^{r-1}$ and skip the proof of this formula. In this particular case, however, we have chosen our lemmas such that they themselves illustrate important ways of applying our rules of differentiation. For this reason it is probably worthwhile that all students study this series of lemmas.

Now We Prove That $(x^r)' = rx^{r-1}$ For Any r

2.1 LEMMA For any nonnegative integer n, $(x^n)' = nx^{n-1}$.

Proof: For n = 0, we have $(x^0)' = (1)' = 0 = 0x^{0-1} = 0x^{-1}$. Technically, $0x^{-1}$ is not defined for $x = 0$, but we interpret $0x^{-1}$ to be the function which is zero for all real numbers (the zero function). For n = 1 we have $(x^1)' = (x)' = 1 = 1x^{1-1} = 1x^0$. We explained in FIGURE 1.13 why $(x^2)' = 2x$. Thus we know that the formula $(x^n)' = nx^{n-1}$ is valid for n = 0, 1, and 2. The proof for general n is by induction. Assume that $(x^{n-1})' = (n-1)x^{n-2}$ is known to be true. Write $x^n = x^{n-1}x$ and use the product rule with $f(x) = x^{n-1}$ and $g(x) = x$. We obtain $(x^n)' = (f(x)g(x))' = f'(x)g(x) + f(x)g'(x) = (n-1)x^{n-2}x + x^{n-1}(1) = nx^{n-1}$. This proves the formula $(x^n)' = nx^{n-1}$ for all nonnegative integers n.

If $h(x) = f(g(x))$, Knowing h' And f' Gives g'

2.2 LEMMA Let m be a positive integer and let $r = 1/m$. Then, $(x^r)' = rx^{r-1}$.

Proof: Here is a trick worth remembering! We apply the CHAIN RULE to $h(x) = f(g(x))$ where $g(x) = x^{1/m}$ and $f(x) = x^m$. Thus the formula $h(x) = f(g(x))$ becomes $x = (x^{1/m})^m$. The CHAIN RULE $h'(x) = f'(g(x))g'(x)$ becomes $1 = m(x^{1/m})^{m-1}(x^{1/m})'$. Solving for $(x^{1/m})'$ we obtain $(x^{1/m})' = (1/m)x^{(1/m)-1}$, which is the identity to be proved.

The trick we used in LEMMA 2.2 is an important one. We applied the CHAIN RULE to $h(x) = f(g(x))$ knowing $f'(x)$ and $h'(x)$ but not knowing $g'(x)$. The identity $h'(x) = f'(g(x))g'(x)$ is then used to solve for $g'(x)$.

2.3 LEMMA Let m and n be positive integers and let $r = n/m$. Then, $(x^r)' = rx^{r-1}$.

Proof: We apply the CHAIN RULE to $h(x) = f(g(x))$ where $h(x) = x^{n/m}$, $f(x) = x^n$, and $g(x) = x^{1/m}$. The identity $h(x) = f(g(x))$ becomes $x^{n/m} = $

$(x^{1/m})^n$. Applying the CHAIN RULE, $h'(x) = f'(g(x))g'(x)$ becomes $(x^{n/m})' = n(x^{1/m})^{n-1}(x^{1/m})'$. Using the identity $(x^{1/m})' = (1/m)x^{(1/m)-1}$ from LEMMA 2.2, we obtain $(x^{n/m})' = (n/m)x^{(n/m)-1}$, which was to be proved.

Irrational Numbers Are Approximated By Rationals

At this point we have shown that the formula $(x^r)' = rx^{r-1}$ is valid for any "nonnegative rational number r." A rational number is a number that can be expressed as a ratio of two integers n/m. For example, 2/3, 0/1, 257/1011, $-23/45$, and 34/343 are rational numbers. The number $-23/45$ is a negative rational number, the others are nonnegative. Every integer is a rational number. The number π is known to be an irrational number (to prove this takes some work). The number $\sqrt{2}$ is irrational (this is easy to show). An irrational number such as π can be approximated to within any degree of accuracy by rational numbers. For example, π can be approximated by 3.14, 3.14159, 3.1415926, 3.141592653589793238, ... Numbers such as 3.14159 (terminating decimal expansions) are rational numbers because they can be written as ratios of integers (3.14159 = 314159/100000, for example). Thus, by what we have shown, if r = 3.14159 then $(x^r)' = rx^{r-1}$. If you think about the geometric meaning of the derivative as the slope of a curve at a point, you can see that $(x^\pi)' = \pi x^{\pi-1}$ since this formula must hold for all rational approximations to π. The same argument works for any nonnegative irrational number s, not just π, so we must have $(x^s)' = sx^{s-1}$ for all nonnegative real numbers. To treat the case of negative exponents we need the next lemma.

2.4 LEMMA Let r = -1. Then $(x^r)' = rx^{r-1}$.

Proof: Apply the product rule to $h(x) = f(x)g(x)$ with $h(x) = 1$, $f(x) = x^{-1}$, and $g(x) = x$. The identity $h'(x) = f'(x)g(x) + f(x)g'(x)$ becomes $0 = (x^{-1})' x + (x^{-1})(1)$. Solving for $(x^{-1})'$ gives $(x^{-1})' = (-1)x^{-2}$, which was to be shown.

By putting LEMMA 2.1 to 2.4 together we obtain the result we were seeking, THEOREM 2.5.

2.5 THEOREM For any real number r, $(x^r)' = rx^{r-1}$.

Proof: We know the result is true for any *nonnegative* real number s. Suppose r = $-s$ is a negative real number. Then $x^r = (x^s)^{-1}$ and by LEMMA 2.4 and the CHAIN RULE,

$$(x^r)' = (-1)(x^s)^{-2}(x^s)' = (-1)x^{-2s}sx^{s-1} = (-s)x^{-s-1} = rx^{r-1}$$

as was to be shown.

The Quotient Rule

We remind the reader again of the principal application of THEOREM 2.5 as described in EXAMPLE 1.15(8), namely, $((g(x))^r)' = r(g(x))^{r-1}g'(x)$ for any differentiable function $g(x)$. Using this latter formula together with the product rule, we can easily obtain our final general rule of differentiation, the "QUOTIENT RULE." Consider the quotient $\dfrac{f(x)}{g(x)}$. We shall derive a formula for computing the derivative $\dfrac{d}{dx}\dfrac{f(x)}{g(x)}$. First, we write $\dfrac{d}{dx}\dfrac{f(x)}{g(x)} = (f(x)(g(x))^{-1})'$. This is just a change of notation. Now, apply the product rule to this latter expresion to obtain $f'(x)(g(x))^{-1} + f(x)(-1)(g(x))^{-2}g'(x)$. Putting this latter expression over a common denominator $(g(x))^2$, we obtain

$$\left(\frac{f(x)}{g(x)}\right)' = \frac{f'(x)g(x) - f(x)g'(x)}{(g(x))^2}.$$

The above formula is called the "QUOTIENT RULE" for obvious reasons. The QUOTIENT RULE should be memorized. Here are some other ways the QUOTIENT RULE is commonly written:

$$\left(\frac{f}{g}\right)' = \frac{f'g - fg'}{g^2}$$

$$\frac{d}{dx}\left(\frac{f}{g}\right) = \frac{\dfrac{df}{dx}g - f\dfrac{dg}{dx}}{g^2}$$

$$d\left(\frac{f}{g}\right) = \frac{(df)g - f(dg)}{g^2}$$

As already stated, you should memorize one of these forms of the QUOTIENT RULE. The last one above is the most abbreviated form. Each occurrence of "d" needs to be "divided by" dx to get the previous form. This is called "differential notation." At this stage we can think of this simply as a memory aid for remembering the QUOTIENT RULE. Said in words, the differential form of the QUOTIENT RULE sounds like "dee f over g is dee f times g minus f times dee g divided by g squared." This may or may not help you remember the QUOTIENT RULE!

2.6   EXAMPLES OF THE QUOTIENT RULE

(1)   Let's try to differentiate $\dfrac{(x^2 + 1)^{1/2}}{(x^3 + 2)^{2/3}}$. We use the QUOTIENT RULE

$$\left(\frac{f(x)}{g(x)}\right)' = \frac{f'(x)g(x) - f(x)g'(x)}{(g(x))^2}$$

where $f(x) = (x^2 + 1)^{1/2}$ and $g(x) = (x^3 + 2)^{2/3}$. Using the CHAIN RULE, we find

$$f'(x) = (1/2)(x^2 + 1)^{-1/2}(2x) = x(x^2 + 1)^{-1/2}$$

and

$$g'(x) = (2/3)(x^3 + 2)^{-1/3}(3x^2) = 2x^2(x^3 + 2)^{-1/3}.$$

Thus,

$$\left(\frac{f(x)}{g(x)}\right)' = \frac{x(x^2 + 1)^{-1/2}(x^3 + 2)^{2/3} - (x^2 + 1)^{1/2}2x^2(x^3 + 2)^{-1/3}}{(x^3 + 2)^{2/3}}.$$

(2)   You may or may not know that $\dfrac{d}{dx}\sin(x) = \cos(x)$. It's true and we shall see why below. It is also true that $\dfrac{d}{dx}\cos(x) = -\sin(x)$. We can use these facts together with the QUOTIENT RULE to compute $\dfrac{d}{dx}\tan(x) = \dfrac{d}{dx}\dfrac{\sin(x)}{\cos(x)}$. We use the QUOTIENT RULE in the form

$$\frac{d}{dx}\left(\frac{f}{g}\right) = \frac{\dfrac{df}{dx}g - f\dfrac{dg}{dx}}{g^2}$$

where $f(x) = \sin(x)$, $g(x) = \cos(x)$, $\dfrac{df}{dx} = \cos(x)$, and $\dfrac{dg}{dx} = -\sin(x)$. Substituting these expressions into the QUOTIENT RULE above, we obtain

$$\frac{d}{dx}\tan(x) = \frac{\cos(x)\cos(x) - \sin(x)(-\sin(x))}{(\cos)(x))^2}.$$

Recalling that $(\cos(x))^2 + (\sin(x))^2 = 1$ for all x, we have that

$$\frac{d}{dx}\tan(x) = (\cos(x))^{-2}.$$

Memorize The Basic Rules

At this point we have learned all of the general rules for differentiation that are required of the beginning calculus student. We now summarize these rules. You should memorize them all!

2.7 DIFFERENTIATION RULES TO MEMORIZE

(1) CHAIN RULE

$$(f(g(x))' = f'(g(x))g'(x)$$

$$\frac{d}{dx}f(g) = \frac{df}{dg}\frac{dg}{dx}$$

(2) PRODUCT RULE

$$(f(x)g(x))' = f'(x)g(x) + f(x)g'(x)$$

$$\frac{d}{dx}(fg) = \frac{df}{dx}g + f\frac{dg}{dx}$$

(3) QUOTIENT RULE

$$\left(\frac{f(x)}{g(x)}\right)' = \frac{f'(x)g(x) - f(x)g'(x)}{(g(x))^2}$$

$$\frac{d}{dx}\left(\frac{f}{g}\right) = \frac{\dfrac{df}{dx}g - f\dfrac{dg}{dx}}{g^2}$$

We shall work a number of problems below in order to practice RULES 2.7. Before doing so, however, it is important that we enlarge the class of basic functions that we know how to differentiate. We shall now learn how to differentiate two important classes: the "trigonometric" functions and the "exponential" functions.

Derivative Of sin(x) Is cos(x)

Look at FIGURE 2.8. There we see a circle of radius one. The angle POQ is defined by specifying the point Q on the circle. As indicated in FIGURE

FIGURE 2.8   Derivative of sin(x) is cos(x)

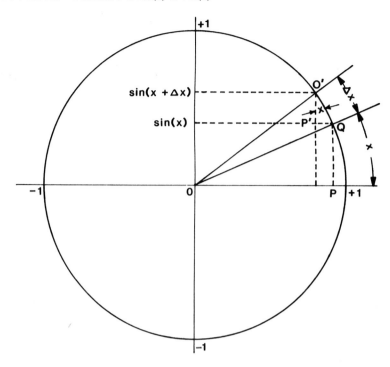

2.8, the angle POQ intercepts an arc of length x on the circle of radius 1. We say, therefore, that the "measure of the angle POQ is x radians." The reader who feels uneasy about measuring angles in radians rather than degrees should consult a precalculus or trigonometry book for some review. Unless otherwise stated, we shall always measure angles in radians (this is common practice in calculus). Recall that *by definition* the coordinates of the point Q are $(\cos(x), \sin(x))$. The point $\sin(x)$ corresponding to the second coordinate of Q is shown on the vertical axis of FIGURE 2.8. We want to compute the derivative $(\sin(x))'$ using ERUDITE OBSERVATION 1.14. Thus we shall first compute the "difference quotient" $\dfrac{\sin(x + \Delta x) - \sin(x)}{\Delta x}$ and try to understand its value for small values of Δx.

Now is where FIGURE 2.8 really comes in handy! Note in FIGURE 2.8 that we have gone a little further along the circle from the point Q to a point O', distance Δx further to be exact. If Δx is small then the arc of the circle from Q to O' looks like a straight line. Pretend it is a straight line in FIGURE 2.8. Note that the "segment" $O'Q$ is the hypotenuse of the little right triangle $O'P'Q$. This little right triangle is congruent to the big right triangle OPQ (one way to see this is to observe that the segment $O'P'$ is perpendicular to OP and $O'Q$ is perpendicular to OQ). Thus, the angle $P'O'Q$ also has measure x radians. Referring to FIGURE 2.8. again, we see that

$$(\sin(x))' \approx \frac{\sin(x + \Delta x) - \sin(x)}{\Delta x} = \frac{\text{length}(O'P')}{\text{length}(O'Q)} \approx \cos(x).$$

Actually, the above formula is not quite correct because the "segment" $O'Q$ is not quite a straight line segment. The smaller Δx is, however, the closer $O'Q$ is to being a straight line segment. Mathematicians express this fact by saying

$$\lim_{\Delta x \to 0} \frac{\sin(x + \Delta x) - \sin(x)}{\Delta x} = (\sin(x))' = \cos(x).$$

We formally state this important (and somewhat amazing!) result in THEOREM 2.9.

2.9 THEOREM Let $f(x) = \sin(x)$ where x is measured in radians. Then, $f'(x) = \cos(x)$.

By inspecting FIGURE 2.8 you will easily see that $\sin(x + \pi/2) = \cos(x)$ for all x. We can think of this fact as $\cos(x) = f(g(x))$ where $f(x) = \sin(x)$ and $g(x) = x + \pi/2$. Applying the CHAIN RULE gives $(\cos(x))' = f'(g(x))g'(x) = \cos(x + \pi/2) (x + \pi/2)' = \cos(x + \pi/2)$. Again, by looking at FIGURE 2.8 you can easily see that $\cos(x + \pi/2) = -\sin(x)$. This proves that

$$\frac{d}{dx}\cos(x) = -\sin(x)$$

By combining the above rules for differentiating $\sin(x)$ and $\cos(x)$ with the rule for differentiating $\tan(x)$ given in EXAMPLE 2.6(2), we obtain the BASIC TRIGONOMETRIC DERIVATIVES 2.10. These formulas should be memorized.

Memorize These Trigonometric Derivatives

2.10 BASIC TRIGONOMETRIC DERIVATIVES

$$\frac{d}{dx}\sin(x) = \cos(x)$$

$$\frac{d}{dx}\cos(x) = -\sin(x)$$

$$\frac{d}{dx}\tan(x) = (\cos(x))^{-2}$$

Exponential And Logarithmic Derivatives

The second class of functions that we must learn how to differentiate are the exponential functions. In particular, we shall look at functions $f_a(x) = a^x$ where a is a real number, $a > 1$. The first thing to notice about exponential functions is that they get large very fast. Consider $f_2(x) = 2^x$ compared with our old friend x^2. At $x = 2$ these two functions have the same value. At $x = 16$, $2^{16} = 65,536$ but $(16)^2$ is only 256. By the time $x = 50$, 2^{50} is greater than 10^{15} while $(50)^2$ is just 2500. For negative values of x we note that a^x tends to zero as x tends to minus infinity. Check the values of 2^{-2}, 2^{-16}, and 2^{-50}. The constant a is called the "base" of the exponential function a^x. We have assumed that $a > 1$. For all such a, a^x gets large ("goes to infinity") as x gets large. The bigger the value of a, the more rapidly a^x gets large. Take a look at FIGURE 2.11, which shows the graphs of the three functions $f_{10}(x) = 10^x$, $f_{2.7}(x) = (2.7)^x$, and $f_{1.2}(x) = (1.2)^x$. A careful understanding of FIGURE 2.11 will be very important for our discussion of exponential functions.

They All Pass Through (0,1)

In FIGURE 2.11, we are looking at the graphs of our three functions in the interval $-0.7 < x < +0.7$. All three of these graphs pass through the point (0,1) as shown. This is true for any exponential function: $f_a(0) = a^0 = 1$. One question that will occur to the reader is "How were the graphs of FIGURE 2.11 computed?" These graphs were computed on a personal computer using BASIC. The command PRINT X ^ Y will print X^Y for numbers X and Y. Of course the numbers X and Y must be rational numbers of a size acceptable to the given computer or hand calculator you are using. How does one program a computer to evaluate strange expressions such as $(5.987241)^{3.98713}$? This is a technical question about numerical methods that is beyond the scope of this book. To get some feeling for this sort of thing, note that a number such as $a^{3.987123}$ can be written as a^3 times $a^{.987123}$. You already know how to compute a^3 for any rational number. Thus, if you want to get serious about such computations you must learn how to compute a^x for

FIGURE 2.11   Three Exponential Functions

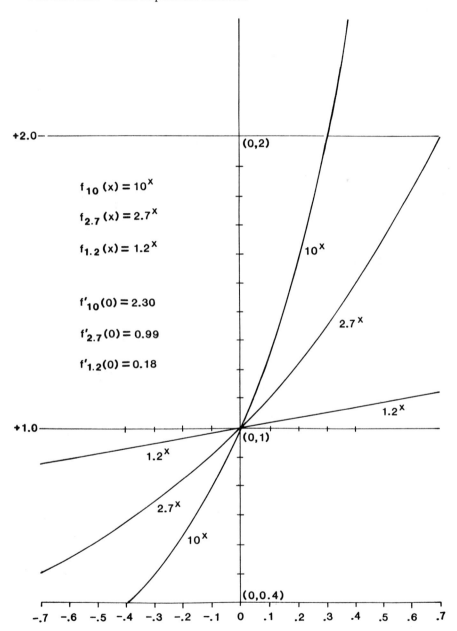

$0 < x < 1$. The interested reader can pursue this topic further by obtaining some beginning texts on numerical mathematics.

At (0,1), y = (2.7)ˣ Has Slope Almost 1

Look again at FIGURE 2.11. Although $f_a(0) = 1$ for all a, the slope $f_a'(0)$ increases as a increases. For example, $f_{1.2}'(0) = 0.18$ while $f_{10}'(0) = 2.30$ (both values are approximate). A little thought reveals that for $1.2 < a < 10$, $f_{1.2}'(0) < f_a'(0) < f_{10}'(0)$. Thus as the base a varies between 1.2 and 10, $f_a'(0)$ varies between 0.18 and 2.3. To the mathematician, this means that there is some particular number a between 1.2 and 10 such that $f_a'(0) = 1$. As a "guess" at this number, we tried a = 2.7 in FIGURE 2.11. This turns out to be a pretty good guess as $f_a'(0) = 0.99$. Here are some attempts to improve on this guess:

Value of a	Value of $f_a'(0)$
2.70000000	.993251773
2.71000000	.996948635
2.71800000	.999896315
2.71820000	.999969897
2.71828000	.999999328
2.71828182	.999999997

So what is this mysterious number a such that $f_a'(0) = 1$? From the above, a = 2.71828182 is *extremely* close to this number. Mathematicians give this mysterious number the name "e" as specified in DEFINITION 2.12.

The Irrational Number e = 2.71828182 . . .

2.12   DEFINITION   Let $f_a(x) = a^x$ where $a > 1$. The unique number a such that $f_a'(0) = 1$ is denoted by e. Thus, $f_e'(0) = 1$. The value of e is about 2.7. For most practical purposes, e = 2.71828182.

In fact, mathematicians have shown that the number e is irrational and hence can never be represented by a terminating or repeating decimal expansion. This means that the exact value of e can never be stored in a computer (only rational numbers can be stored in a computer).

The Compositional Inverse Of aˣ Is logₐ(x)

In CHAPTER 1, in connection with 1.17(f) and (g) (SOLUTIONS TO EXERCISE 1.16), we discussed the idea of "compositional inverses." The compositional inverse of the function $f_a(x) = a^x$ is denoted by $\log_a(x)$ and is called the "logarithm base a" function. The idea is very simple and is illustrated in FIGURE 2.13. First, compare FIGURE 2.13 with FIGURE 2.11.

FIGURE 2.13 Logarithmic Functions

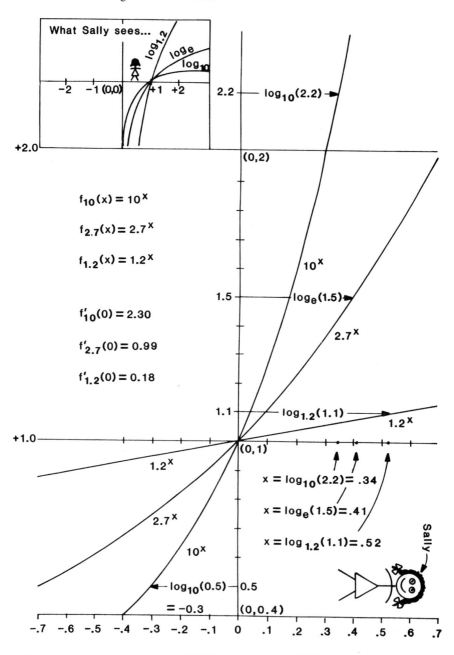

Note that the graphs shown in FIGURE 2.13 and FIGURE 2.11 are the same $(10^x, (2.7)^x, $ and $(1.2)^x)$. Remember, that the function a^x is positive for all values of x. Look at FIGURE 2.13 and pick some point on the vertical axis, say 2.2 as shown. The horizontal distance from that point to the curve a^x is *by definition* $\log_a(2.2)$. In FIGURE 2.13 the horizontal distance from the point 2.2 to the curve 10^x is shown and seems to be about 0.34. Thus, *by definition*, $\log_{10}(2.2) = 0.34$. At the point 1.5 on the vertical axis the horizontal distance to the curve $(2.7)^x$ is shown and is about 0.41. This means that $\log_{2.7}(1.5) = 0.41$. As 2.7 is approximately e, we have written $\log_e(1.5)$ for $\log_{2.7}(1.5)$ in FIGURE 2.13. In general, for any positive real number y, to compute $\log_a(y)$ graphically, find y on the vertical axis and go horizontally to the curve a^x. This horizontal directed distance is *by definition* $\log_a(y)$. By "directed" distance we mean distances to the left are negative. Thus, referring to FIGURE 2.13, $\log_{10}(0.5) = -0.3$. For any nonnegative real number y, the point in the plane with coordinates $(\log_a(y),y)$ is by definition on the curve $f_a(x) = a^x$. We summarize these ideas in DEFINITION 2.14.

Definition Of $\log_a(x)$

2.14 DEFINITION

Let $f_a(x) = a^x$ where $a > 1$. For each $y > 0$ define $\log_a(y)$ to be the unique real number such that $(\log_a(y),y)$ is on the graph of a^x.

Look once again at FIGURE 2.13. Note that $10^{(\log_{10}(2.2))} = 2.2$. Taking $\log_{10}(2.2) = .34$, note also that $\log_{10}(10^{.34}) = .34$. In general, we have the following basic properties of $\log_a(x)$ and a^x.

2.15 COMPOSITIONAL PROPERTIES OF LOGARITHMIC AND EXPONENTIAL FUNCTIONS Let $f_a(x) = a^x$, $a > 1$, and let $g_a(x) = \log_a(x)$ where f_a is defined for all x and g_a is defined for all positive x. Then

    (1)   For all x, $g_a(f_a(x)) = \log_a(a^x) = x$

    (2)   For all positive x, $f_a(g_a(x)) = a^{(\log_a(x))} = x$.

$\log_a(a^x) = x$, $a^{(\log_a(x))} = x$

The reader should, using FIGURE 2.13, check out the validity of COMPOSITIONAL PROPERTIES 2.15 for various values of x. When checking $\log_a(a^x) = x$ start with x on the horizontal axis and when checking $a^{(\log_a(x))} = x$ start with x on the vertical axis. The graphs shown in FIGURE 2.13 were initially drawn for the three functions 10^x, $(2.7)^x$, and $(1.2)^x$ but they can equally well be regarded as the graphs for $\log_{10}(x)$, $\log_{2.7}(x)$, and $\log_{1.2}(x)$ if we understand that the dependent variable x now ranges along the positive vertical axis. Of course, tradition has it that when graphing a function the horizontal axis represents the dependent variable. Look at the stick figure SALLY in FIGURE 2.13. Imagine that she is slightly behind the page and is viewing the three graphs as logarithmic functions of the dependent variable represented on the vertical axis. For her, the dependent variable axis is oriented "properly" with values increasing to her right and decreasing to her left. Her view of the three logarithmic functions is shown in the upper left-hand corner. You should note carefully the shape of these functions and fix their general structure and relationship with each other in your mind.

Derivatives Of a^x And log_a(x)

We now must learn how to compute the derivatives of the two functions a^x and $\log_a(x)$. Using the trick we learned in LEMMA 2.2, it will be very easy to find formulas for these derivatives. Before we give proofs, however, we should take an advance look at the answers to make sure we understand the concepts involved. Recall the number e of DEFINITION 2.12 (e is about 2.7). Here are the derivatives of a^x and $\log_a(x)$:

$$\frac{d}{dx}a^x = \log_e(a)a^x$$

$$\frac{d}{dx}\log_a(x) = \frac{1}{(\log_e(a))x} \ .$$

It is obvious from the above two expressions for the derivatives of a^x and $\log_e(x)$ that we need to be able to compute the numbers of the form $\log_e(a)$ for the various possible bases a. For values of a between 1 and 2 we could use the graph for $(2.7)^x$ of FIGURE 2.13. As shown in FIGURE 2.13, $\log_e(1.5) = 0.41$ (we are approximating e by 2.7 here). This means that

$$\frac{d}{dx}(1.5)^x = 0.41(1.5)^x$$

$$\frac{d}{dx}\log_{1.5}(x) = \frac{1}{(0.41)x} \ .$$

There are, of course, much better ways to compute numbers $\log_e(a)$. If you have access to a computer with the language BASIC then the command PRINT LOG(a) will return $\log_e(a)$. Thus, PRINT LOG(1.5) will return 0.405465108, which rounds off to 0.41, which we obtained from the graph of $(2.7)^x$ in FIGURE 2.13. The function $\log_e(x)$ is called the "natural" or "Naperian" logarithm and is sometimes denoted by ln(x). You may have a calculator with a function key labeled "ln." If so that's it!

$(e^x)' = e^x$, $(ln(x))' = 1/x$

Before proving the above formula for the derivatives of a^x and $\log_a(x)$ we should remember that $\log_e(e) = 1$ and thus take note of the important special cases when $a = e$:

$$\frac{d}{dx}e^x = e^x$$

$$\frac{d}{dx}\log_e(x) = \frac{1}{x}.$$

We now must derive the above formulas. First, consider $f_a(x) = a^x$. The derivative $f_a'(x)$ is, by ERUDITE OBSERVATION 1.14, approximately equal (for very small values of Δx) to the difference quotient

$$\frac{f_a(x + \Delta x) - f_a(x)}{\Delta x} = \frac{a^{x+\Delta x} - a^x}{\Delta x}.$$

The Proof

By the elementary properties of exponents

$$\frac{a^{x+\Delta x} - a^x}{\Delta x} = \frac{a^{\Delta x} - 1}{\Delta x}a^x.$$

But the expression $\dfrac{a^{\Delta x} - 1}{\Delta x}$ is approximately equal to $f_a'(0)$ for small values of Δx. More precisely,

$$\lim_{\Delta x \to 0}\frac{a^{\Delta x} - 1}{\Delta x} = f_a'(0).$$

We summarize this result in THEOREM 2.16.

2.16   THEOREM   Let $f_a(x) = a^x$ where $a > 1$. Then $f_a'(x) = f_a'(0)a^x$.

If you think about THEOREM 2.16 a little bit, you will realize that it is a remarkable result. It says that the derivative function of a^x is again a^x *except for a constant multiple* which is the slope of the graph of a^x at the point $(0,1)$. Of course, to compute $f_a'(0)$ we don't have to graph a^x and measure the slope at $(0,1)$ because it turns out that $f_a'(0) = \log_e(a)$. We now explain why this is so.

The Slope Of a^x At $(0,1)$ Is $\log_e(a)$

We know that $f'_e(0) = 1$ because that is the way we defined the number e. Thus applying THEOREM 2.16 with a = e we get $\dfrac{d}{dx}e^x = e^x$.

By PROPERTIES 2.15, $a = e^{(\log_e(a))}$ and hence $a^x = e^{(\log_e(a))x}$. We write $h(x) = f(g(x))$ where $h(x) = a^x$, $f(g) = e^g$, and $g(x) = (\log_e(a))x$. The CHAIN RULE says that $\dfrac{dh}{dx} = \dfrac{dh}{dg}\dfrac{dg}{dx}$ so we have

$$\frac{d}{dx}a^x = e^g(\log_e(a)) = (\log_e(a))e^{(\log_e(a))x} = (\log_e(a))a^x.$$

By comparing the above formula with THEOREM 2.16, which states that $\dfrac{d}{dx}a^x = f'_a(0)a^x$, we see that $f'_a(0) = \log_e(a)$. These observations are summarized in THEOREM 2.17.

2.17 THEOREM   Let $f_a(x) = a^x$ where $a > 1$. Then $f'_a(x) = (\log_e(a))a^x$.

Compositional Inverses And The Chain Rule Gives $(\log_a(x))'$

We are now in a position to easily obtain the derivative function of $\log_a(x)$. The method we shall use is very similar to the method used in proving LEMMA 2.2. We shall apply the CHAIN RULE to $h(x) = f(g(x))$ where $f(g) = a^g$, $g(x) = \log_a(x)$, and $h(x) = f(g(x)) = a^{(\log_a(x))} = x$ (by PROPERTIES 2.15). For these functions $\dfrac{dh}{dx} = \dfrac{df}{dg}\dfrac{dg}{dx}$ becomes

$$1 = (\log_e(a))a^g\frac{d}{dx}(\log_a(x)).$$

Solving for $\dfrac{d}{dx}(\log_a(x))$ we obtain THEOREM 2.18.

2.18 THEOREM   For $a > 1$, $\dfrac{d}{dx}(\log_a(x)) = \dfrac{1}{(\log_e(a))x}$.

Note that if $a = e$ in THEOREM 2.18 then, using the fact that $\log_e(e) = 1$, we have $\dfrac{d}{dx}(\log_e(x)) = 1/x$. We summarize these results in 2.19.

2.19 EXPONENTIAL AND LOGARITHMIC DERIVATIVES

(1) $\dfrac{d}{dx}a^x = (\log_e(a))a^x$

(2) $\dfrac{d}{dx}e^x = e^x$

(3) $\dfrac{d}{dx}\log_a(x) = \dfrac{1}{(\log_e(a))x}$

(4) $\dfrac{d}{dx}\log_e(x) = \dfrac{1}{x}.$

2.20 EXAMPLES OF COMPUTING DERIVATIVES

A Lesson To Be Learned—Simplify First

(1) As our first example, we shall compute

$$\frac{d}{dx}\left(\frac{x - \sqrt{x}}{x + \sqrt{x}}\right)^{1/2}$$

We don't like the notation $\sqrt{x}$, so we write this as

$$\frac{d}{dx}\left(\frac{x - x^{1/2}}{x + x^{1/2}}\right)^{1/2}.$$

By the CHAIN RULE this becomes

$$(1/2)\left(\frac{x - x^{1/2}}{x + x^{1/2}}\right)^{-1/2}\frac{d}{dx}\left(\frac{x - x^{1/2}}{x + x^{1/2}}\right).$$

Now we use the QUOTIENT RULE to obtain

$$\frac{d}{dx}\left(\frac{x - x^{1/2}}{x + x^{1/2}}\right) = \frac{(x - x^{1/2})'(x + x^{1/2}) - (x - x^{1/2})(x + x^{1/2})'}{(x + x^{1/2})^2}.$$

Using the fact that $(x + x^{1/2})' = 1 + (1/2)x^{-1/2}$ and $(x - x^{1/2})' = 1 - (1/2)x^{-1/2}$ we may substitute these expressions into the above identities to obtain a very messy exact expression for the derivative. In the "real world" one can encounter derivatives that are quite messy but, because they are important for some calculation, are still appreciated. In the case of the derivative just completed, some clever algebra will simplify our answer considerably. In this case, however, we have made a serious blunder from the beginning by not simplifyng the expression we were differentiating. Let's try again, writing (divide numerator and denominator by $\sqrt{x}$):

$$\left(\frac{x - \sqrt{x}}{x + \sqrt{x}}\right)^{1/2} = \left(\frac{x^{1/2} - 1}{x^{1/2} + 1}\right)^{1/2}.$$

Now we have

$$\frac{d}{dx}\left(\frac{x^{1/2} - 1}{x^{1/2} + 1}\right)^{1/2} = (1/2)\left(\frac{x^{1/2} - 1}{x^{1/2} + 1}\right)^{-1/2} \frac{d}{dx}\left(\frac{x^{1/2} - 1}{x^{1/2} + 1}\right).$$

Using the QUOTIENT RULE we now compute

$$\frac{d}{dx}\left(\frac{x^{1/2} - 1}{x^{1/2} + 1}\right) = \frac{(1/2)x^{-1/2}(x^{1/2} + 1) - (x^{1/2} - 1)(1/2)x^{-1/2}}{(x^{1/2} + 1)^2}.$$

Some easy algebra applied to the numerator of this expression gives

$$\frac{d}{dx}\left(\frac{x^{1/2} - 1}{x^{1/2} + 1}\right) = \frac{1}{x^{1/2}(x^{1/2} + 1)^2}.$$

The final answer is that

$$\frac{d}{dx}\left(\frac{x - \sqrt{x}}{x + \sqrt{x}}\right)^{1/2} = (1/2)\left(\frac{x^{1/2} - 1}{x^{1/2} + 1}\right)^{-1/2} \frac{1}{x^{1/2}(x^{1/2} + 1)^2}.$$

$(sec(x))' = tan(x) sec(x)$ Memorize!

(2)   Compute $\dfrac{d}{dx}sec(x)$. The secant function $sec(x)$ is $(cos(x))^{-1}$. The cosecant function $csc(x)$ is $(sin(x))^{-1}$. One would have thought that $sec(x)$ should have been $(sin(x))^{-1}$ and $csc(x)$ should have been $(cos(x))^{-1}$ as that would have been easier to remember! We have

$$\frac{d}{dx}sec(x) = \frac{d}{dx}(cos(x))^{-1} = (-1)(cos(x))^{-2} \frac{d}{dx}cos(x)$$

$$= -(cos(x))^{-2}(-sin(x)) = tan(x)sec(x).$$

Thus,

$$\frac{d}{dx}sec(x) = tan(x)sec(x).$$

$(csc(x))' = -cot(x)csc(x)$ Memorize!

(3)   To compute $\dfrac{d}{dx}csc(x)$ we write

$$\frac{d}{dx}csc(x) = \frac{d}{dx}(\sin(x))^{-1} = (-1)(\sin(x))^{-2}\frac{d}{dx}\sin(x)$$

$$= (-1)(\sin(x))^{-2}\cos(x) = -cot(x)csc(x).$$

We have shown that

$$\frac{d}{dx}csc(x) = -cot(x)csc(x).$$

$(tan(x))' = sec^2(x), (cot(x))' = -csc^2(x)$ Memorize!

(4)   Another important trigonometric function is the cotangent, $cot(x)$. For this function we find

$$\frac{d}{dx}cot(x) = \frac{d}{dx}\frac{\cos(x)}{\sin(x)} = \frac{(\cos(x))'\sin(x) - \cos(x)(\sin(x))'}{(\sin(x))^2} = -(\sin(x))^{-2}.$$

We have used the trigonometric identity $(\sin(x))^2 + (\cos(x))^2 = 1$ in deriving the above identity. The usual way this result is written is

$$\frac{d}{dx}cot(x) = -(csc(x))^2.$$

From BASIC TRIGONOMETRIC DERIVATIVES 2.10 we also have

$$\frac{d}{dx}tan(x) = (sec(x))^2.$$

Notation: $(sin(x))^2 = sin^2(x), \sqrt{x} = x^{1/2}$

(5)   Find $\dfrac{d}{dx}\sin^2(\sqrt{x})$. The notation $\sin^2(\sqrt{x})$ means $(\sin^2(\sqrt{x}))^2$ and is better written $(\sin(x^{1/2}))^2$. Let $h(x) = f(g(x))$ with $f(g) = g^2$ and $g(x) = \sin(x^{1/2})$. We have $h'(x) = 2\sin(x^{1/2})(\sin(x^{1/2}))'$. To compute $(\sin(x^{1/2})'$, we again use the CHAIN RULE applied to $w(x) = u(v(x))$ where $w(x) = \sin(x^{1/2})$,

$u(v) = \sin(v)$, and $v(x) = x^{1/2}$. We find $w'(x) = u'(v(x))v'(x) = \cos(x^{1/2})$ $(1/2)x^{-1/2}$. Our final conclusion is

$$\frac{d}{dx}\sin^2(\sqrt{x}) = x^{-1/2}\sin(x^{1/2})\cos(x^{1/2}).$$

The "Loglog" Function

(6)   Compute $\dfrac{d}{dx}\ln(\ln(x))$ and $\dfrac{d}{dx}\log_2(\log_2(x))$. Once again we use the CHAIN RULE applied to $f(g(x))$ where $f(g) = \ln(g)$ and $g(x) = \ln(x)$. The CHAIN RULE in the form $\dfrac{d}{dx}f(g(x)) = \dfrac{df}{dg}\dfrac{dg}{dx}$ becomes $\dfrac{d}{dx}\ln(\ln(x)) = (1/g)(1/x) = 1/(x\ln(x))$. If we replace $\ln(x)$ by $\log_2(x)$ and set $f(g) = \log_2(g)$, $g(x) = \log_2(x)$ then the CHAIN RULE $\dfrac{d}{dx}f(g(x)) = \dfrac{df}{dg}\dfrac{dg}{dx}$ becomes

$$\frac{d}{dx}\log_2(\log_2(x)) = (1/\ln(2)g)(1/\ln(2)x) = (\ln(2))^{-2}(x\log_2(x))^{-1}.$$

To summarize,

$$\frac{d}{dx}\ln(\ln(x)) = \frac{1}{x\ln(x)}$$

$$\frac{d}{dx}\log_2(\log_2(x)) = \frac{1}{(\ln(2))^2 x\log_2(x)}\ .$$

In general, we would have

$$\frac{d}{dx}\log_a(\log_a(x)) = \frac{1}{(\ln(a))^2 x\log_a(x)}\ .$$

Differentiating |x| And (ln|x|)' = 1/x   Memorize!

(7)   What is $\dfrac{d}{dx}\log_a(|x|)$? Most students dislike the function $|x|$. The best way to think of this function is as two functions pieced together

$$|x| = \begin{cases} x & \text{if } x \geqslant 0 \\ -x & \text{if } x < 0 \end{cases}.$$

The function $|x|$ has derivative function

$$\frac{d}{dx}|x| = \begin{cases} +1 & \text{if } x > 0 \\ -1 & \text{if } x < 0 \end{cases}.$$

The function $|x|$ does not have a derivative at $x = 0$ for the same reason that the function of FIGURE 1.2 does not have a derivative at the point C'. Even though $|x|$ does not have a derivative at $x = 0$, a composition $f(|x|)$ might have a derivative at $x = 0$. This is the case for $f(x) = x^2$, as $(|x|)^2 = x^2$ for all x and x^2 has derivative function $2x$ for all x. In our present example, $f(x) = \log_a(x)$. This function is not even defined for $x = 0$ so the composition $f(|x|) = \log_a(|x|)$ does not have a derivative at $x = 0$. For $x > 0$, $\log_a(|x|) = \log_a(x)$ and the derivative is $1/\ln(a)x$. For $x < 0$, $\log_a(|x|) = \log_a(-x)$. We can apply the CHAIN RULE $\dfrac{d}{dx}f(g(x)) = \dfrac{df}{dg}\dfrac{dg}{dx}$ with $f(g) = \log_a(g)$ and $g(x) = -x$. We obtain $\dfrac{d}{dx}\log(-x) = (\ln(a)g)^{-1}(-1) = 1/\ln(a)x$. Thus for all nonzero x we have

$$\frac{d}{dx}\log_a(|x|) = \frac{1}{\ln(a)x}.$$

This is an interesting example where the function to be differentiated, $\log_a(|x|)$, has a rather awkward description in terms of piecing together two different functions but its derivative has a simple form valid for all nonzero values of the argument, x. Memorize this rule!

x^x Is Very Different From x^n

(8) Here is a problem that invariably confuses the beginning student. Find $\dfrac{d}{dx}x^x$. The first thought is that $\dfrac{d}{dx}x^x$ is xx^{x-1} but this is not correct. Our formula $\dfrac{d}{dx}x^n = nx^{n-1}$ is based on the fact that n is constant. The right approach is to write $x^x = (e^{\ln(x)})^x = e^{x\ln(x)}$. Now apply the CHAIN RULE $\dfrac{d}{dx}f(g(x)) = \dfrac{df}{dg}\dfrac{dg}{dx}$ where $f(g) = e^g$ and $g(x) = x\ln(x)$. Doing this, we get

$$\frac{d}{dx}e^{x\ln(x)} = e^{x\ln(x)}\frac{d}{dx}(x\ln(x)).$$

We use the PRODUCT RULE to obtain

$$\frac{d}{dx}(x\ln(x)) = (1)\ln(x) + x(1/x) = \ln(x) + 1.$$

the final answer is thus

$$\frac{d}{dx}x^x = (\ln(x) + 1)x^x.$$

The Second Derivative $f''(x)$, $f^{(2)}(x)$, $\dfrac{d^2}{dx^2}f$

(9)   We should also practice taking higher order derivatives of logarithmic
and trigonometric functions. For example, what is $\dfrac{d^2}{dx^2}(\ln(x))^8$? We use the
CHAIN RULE

$$\frac{d}{dx}f(g(x)) = \frac{df}{dg}\frac{dg}{dx}$$

with $f(g) = (g)^8$ and $g(x) = \ln(x)$ to obtain

$$\frac{d^2}{dx^2}(\ln(x))^8 = \frac{d}{dx}\left(8(\ln(x))^7\frac{d}{dx}\ln(x)\right) = \frac{d}{dx}\frac{8(\ln(x))^7}{x}.$$

Now apply the QUOTIENT RULE to $8(\ln(x))^7/x$ to obtain the final answer

$$\frac{d^2}{dx^2}(\ln(x))^8 = 8\left(\frac{7(\ln(x))^6x^{-1} - (\ln(x))^7(1)}{x^2}\right).$$

This can also be written

$$\frac{d^2}{dx^2}(\ln(x))^8 = 8\frac{(\ln(x))^6}{x^2}(7/x - \ln(x)).$$

Higher Order Derivatives

(10)   We now compute $\dfrac{d^p}{dx^p}\sin(bx)$ where p is a nonnegative integer and
b is a real number. There are infinitely many such p so we shall give a formula
or rule for these derivatives in terms of p. To get a feeling for how this

computation goes, we compute the first few terms of the sequence of derivatives

$$\frac{d^0}{dx^0}\sin(bx), \ \frac{d^1}{dx^1}\sin(bx), \ \frac{d^2}{dx^2}\sin(bx), \ \ldots, \ \frac{d^p}{dx^p}\sin(bx), \ \ldots .$$

Here we mean that $\dfrac{d^0}{dx^0}\sin(bx) = \sin(bx)$ and $\dfrac{d^1}{dx^1}\sin(bx) = \dfrac{d}{dx}\sin(bx)$. Computing these terms we obtain the following sequence:

$$\sin(bx), \ b\cos(bx), \ -b^2\sin(bx), \ -b^3\cos(bx), \ b^4\sin(bx), \ b^5\cos(bx),$$

$$-b^6\sin(bx), \ -b^7\cos(bx), \ b^8\sin(bx), \ b^9\cos(bx), \ \text{etc.}$$

If you stare at this sequence for awhile, the pattern will become clear, but how do we describe it concisely? Here is the standard trick for giving a "formula" for the terms of this sequence. We say that for $j = 0, 1, 2, 3,$ $\ldots$

$$\frac{d^{2j}}{dx^{2j}}\sin(bx) = (-1)^j b^{2j}\sin(bx)$$

$$\frac{d^{2j+1}}{dx^{2j+1}}\sin(bx) = (-1)^j b^{2j+1}\cos(bx).$$

Although this terminology (that's all it is) looks impressive at first glance it is really a trivial idea. To see that it works try the various values of j to see that the sequence of derivatives that we computed above is obtained. A formal proof based on induction could be given of these formulas but we won't bother with that.

Know Your Trig Identities And Simplify First

(11)   In our first example, (1) above, we ran into problems because we failed to simplify an expression before differentiating it. This sort of oversight can become particularly embarrassing in the case of trigonometric functions. Consider the following difficult-looking problem:

$$\frac{d}{dx}\left(\frac{\sin^2(x) - \cos^2(x)}{\sin^2(x) + \cos^2(x)}\right) = \ ?$$

If you remember your trigonometric identities

$$\sin^2(x) + \cos^2(x) = 1$$

$$\sin^2(x) - \cos^2(x) = \cos(2x)$$

then the problem reduces to

$$\frac{d}{dx}\left(\frac{\sin^2(x) - \cos^2(x)}{\sin^2(x) + \cos^2(x)}\right) = \frac{d}{dx}\cos(2x) = -2\sin(2x).$$

Much easier if you know your basic trigonometric identities!

Extended Product Rule—Logarithmic Differentiation

(12)　You may have noticed by now that the PRODUCT RULE $(fg)' = f'g + fg'$ extends to three functions $(fgh)' = f'gh + fg'h + fgh'$ or more than three functions in the obvious way (each function in the product gets its turn to be differentiated). These more general rules follow easily from our basic PRODUCT RULE. For example, in the case of three functions, $(fgh)' = ((f)(gh))' = f'(gh) + f(gh)'$ by the PRODUCT RULE applied to the two functions f and gh. Now apply the PRODUCT RULE for two functions to $(gh)'$ to get $(fgh)' = f'(gh) + f(g'h + gh') = f'gh + fg'h + fgh'$. The general case of a product of n functions can be derived in the same way (formally, one would use induction).

There are certain types of problems where it is best to avoid using the extended PRODUCT RULE and use instead a method called "logarithmic differentiation." In spite of the fancy name, the idea is quite simple. If we have a function $h(x) = \ln(|g(x)|)$ then by the CHAIN RULE $\frac{d}{dx}f(g(x)) = \frac{df}{dg}\frac{dg}{dx}$ with $f(g) = \ln(|g|)$ we obtain $h'(x) = \frac{g'(x)}{g(x)}$. In other words, for any $g(x)$ we have

$$\frac{d}{dx}\ln(|g(x)|) = \frac{g'(x)}{g(x)}.$$

As an example, consider the product

$$g(x) = (4x + 5)^{9/4}(2x + 1)^{1/2}(3x + 2)^{1/3}(3x^2 + 1)^{1/6}.$$

We could find $g'(x)$ by the PRODUCT RULE, but another way is to first take the natural logarithm of the absolute value of both sides to get $\ln(|g(x)|)$ equal to

$$(9/4)\ln(|4x+5|)+(1/2)\ln(|2x+1|)+(1/3)\ln(|3x+2|)+(1/6)\ln(|3x^2+1|).$$

Now differentiate both sides to obtain

$$\frac{g'(x)}{g(x)} = \frac{9}{4x + 5} + \frac{1}{2x + 1} + \frac{1}{3x + 2} + \frac{x}{3x^2 + 1}.$$

The expression on the right side of the above equation is much simpler than what we would have obtained if we had differentiated g(x) by the PRODUCT RULE. Of course what we have obtained is not $g'(x)$ but $g'(x)/g(x)$. This is usually not much of a problem as we know g(x) and can multiply both sides of the equation by g(x) to obtain $g'(x)$. If we do that and then cancel common factors, we will obtain what we would have obtained by applying the PROD-UCT RULE in the first place (and there would have been no net computational advantage in using logarithmic differentiation). Logarithmic differentiation is more useful in finding the value of $g'(x)$ for a particular value of x. For example,

$$g'(1) = (9^{9/4})(3^{1/2})(5^{1/3})(4^{1/6})(1 + (1/3) + (1/5) + (1/4)).$$

There is one minor problem with logarithmic differentiation that we have not mentioned. Notice in our example that $g(-1/2) = 0$ and thus $g'(-1/2)/g(-1/2)$ is not defined. Generally, this happens only at a few points and is best dealt with in particular cases rather than by formulating some awkward general rule. How would you compute $g'(-1/2)$ in our particular case

$$g(x) = (4x + 5)^{9/4}(2x + 1)^{1/2}(3x + 2)^{1/3}(3x^2 + 1)^{1/6}?$$

Chain Rule With Limited Information

(13)   Here is another type of problem that one sees in calculus textbooks. Suppose $A(x) = B((y(x))^{-2})$, find $A'(1)$. As it stands, we don't have enough information to work this problem. If A, B, and y were known functions we could compute the function $A'(x)$ explicitly:

$$A'(x) = B'((y(x))^{-2})(-2)(y(x))^{-3}y'(x).$$

There may be some question in this notation about what B' means $\left(\text{it is not } \dfrac{dB}{dx}\right)$. In the differential notation we can be more specific. Let

$g(x) = (y(x))^{-2}$. Then $B'(g) = \dfrac{dB}{dg}$ and

$$\frac{dA}{dx} = \frac{dB}{dg}\frac{dg}{dx} = B'(g(x))(-2)(y(x))^{-3}y'(x).$$

To compute $A'(1)$ we at least need to know $y(1)$, $y'(1)$, and $B'(g(1))$. If you were writing a calculus book, you might ask the students the following ques-tion:

If $A(x) = B((y(x))^{-1})$, find $A'(1)$ where $y(1) = 2$, $y'(1) = 4$, and $B'(1/4) = 3$.

From the above formula, we have that $A'(1) = B'(g(1)) \, (-2) \, (y(1))^{-3} y'(1) = B'(1/4)(-2)(2^{-3})(4) = (3)(-2)(1/8)(4) = -3$. The 1/4 is there because $g(1) = (y(1))^{-2} = 2^{-2} = 1/4$. This type of problem may seem a bit artificial to you. How would one happen to know that $y(1) = 2$, $y'(1) = 4$, and $B'(1/4) = 3$ but not know enough about y and B to know A or at least a graph of A from which A' could be computed graphically? It is an unlikely situation that could only arise with graphical data in some very special situations. These problems are mostly just ways to test your understanding of the rules of differentiation.

Implicit Differentiation

(14)   If $L(x)$ is such that $L'(x) = x^{-2}$ then what is $\dfrac{d}{dx} L(x^3 + 2)$? This problem might seem a little strange at first glance because $L(x)$ is not known explicitly. If we set $g(x) = x^3 + 2$ then $L(g(x)) = L(x^3 + 2)$ and by the CHAIN RULE

$$\frac{d}{dx} L(x^3 + 2) = L'(g(x))g'(x) = \frac{1}{(x^3 + 2)^2} (3x^2).$$

Here we have used the fact that $L'(x) = 1/x$ so $L'(g) = 1/g$. This problem belongs to a general class called "implicit differentiation." We are given a relationship such as $yx^2 - y^3 + x^2 - 2 = 0$ where y is thought of as a function of x, but y is not explicitly known as a function of x. By differentiating, we obtain a relationship between y, y' and x. In this example, we get $y'x^2 + y(2x) - 3y^2 y' + 2x = 0$. Solving gives $y' = 2x(1 + y)/(3y^2 - x^2)$. If for a particular value of x, we know $y(x)$, then we can compute $y'(x)$. The original relationship between x and y can often be used to solve for y for that particular value of x.

Checking The Solution To A Differential Equation

(15)   Prove that $y = 3\sin(2t) + \cos(2t)$ satisfies the differential equation $y'' = -4y$. This problem, although it might sound hard if you are not familiar with the terminology, is very simple. It asks you to compute the second derivative y'' of the given expression and verify that it is the same as multiplying -4 times the expression. Thus we compute

$$y' = 3(\cos(2t))(2) - \sin(2t)(2)$$

$$y'' = (y')' = (6\cos(2t) - 2\sin(2t))' = -12\sin(2t) - 4\cos(2t)$$

The expression above is clearly the same as $-4y = -4(3\sin(2t) + \cos(2t))$.

We shall conclude this chapter with some exercises. In working these exercises you will find it very helpful to *memorize* certain basic facts about logarithmic and trigonometric functions. This will free your mind to think about the calculus rather than the precalculus aspects of these problems. For logarithmic functions we have BASIC FACTS 2.21

Memorize!

2.21 BASIC LOGARITHMIC FACTS

$$\log_a(xy) = \log_a(x) + \log_a(y)$$

$$\log_a(x/y) = \log_a(x) - \log_a(y)$$

$$\log_a(x^r) = r\log_a(x)$$

$$\log_a(b)\log_b(x) = \log_a(x)$$

$$\log_a(b) = \frac{1}{\log_b(a)} \, .$$

Everyone who has had a course in precalculus mathematics has seen the first three of the BASIC FACTS 2.21. The fourth identity, $\log_a(b)\log_b(x) = \log_a(x)$, is important for relating logarithms for different bases a and b. As we discussed in connection with FIGURE 2.13, for a fixed number x there is only one number z such that $a^z = x$. This number z is, by definition, $\log_a(x)$. Thus, to prove the fourth identity of BASIC FACTS 2.21, we need only show that for $z = \log_a(b)\log_b(x)$ we have $a^z = x$. We compute

$$a^{(\log_a(b)\log_b(x))} = (a^{(\log_a(b))})^{\log_b(x)} = b^{(\log_b(x))} = x.$$

That's all there is to it! The fifth identity of BASIC FACTS 2.21 follows immediately from the fourth by setting x = a and using the fact that $\log_a(a) = 1$. In memorizing such formulas it helps to talk to yourself a little bit about them! Regarding the fourth identity of BASIC FACTS 2.21, you might say "The two b's cancel each other." In regard to the fifth identity, you might say "The a and the b change roles." The idea is to aid your long-term memory with some verbal associations (no matter if they have precise mathematical meaning).

Finally, we list without proof some of the more frequently occurring tri-gonometric identities. Not all of these are required for the exercises at the end of this chapter, but hopefully you will work additional exercises at some time!

Memorize!

2.22 BASIC TRIGONOMETRIC FACTS

$$\tan(x) = \frac{\sin(x)}{\cos(x)} \qquad \cot(x) = \frac{\cos(x)}{\sin(x)} \qquad \sec(x) = (\cos(x))^{-1}$$

$$\csc(x) = (\sin(x))^{-1} \qquad (\sin(x))^2 + (\cos(x))^2 = 1 \qquad (\tan(x))^2 + 1 = (\sec(x))^2$$

$$1 + (\cot(x))^2 = (\csc(x))^2$$

$$\tan(x + y) = \frac{\tan(x) + \tan(y)}{1 - \tan(x)\tan(y)}$$

$$\sin(x + y) = \sin(x)\cos(y) + \cos(x)\sin(y)$$

$$\cos(x + y) = \cos(x)\cos(y) - \sin(x)\sin(y)$$

$$\sin(2x) = 2\sin(x)\cos(x)$$

$$\cos(2x) = (\cos(x))^2 - (\sin(x))^2 = 1 - 2(\sin(x))^2 = 2(\cos(x))^2 - 1$$

In the exercises that follow, be sure to pay attention to the values for which the various functions are defined. Some helpful formulas are given in Section 2.24 and the solutions are given in Section 2.25.

2.23 EXERCISES

(1) Differentiate the following:

(a) $\dfrac{x^2 - 1}{x + 1}$

(d) $\sin\left(\dfrac{3 - x^2}{x + 1}\right)$

(b) $\dfrac{1}{x^2 + 2x + 1}$

(e) $\cos\left(\dfrac{x^2 + 2x + 3}{x^3 + 6}\right)$

(c) $\left(\dfrac{x - 1}{x^{1/2} + 1}\right)^2$

(f) $\left(\dfrac{x - 1}{x + 1}\right)^3 (3x^2 + 1)^{1/2}$

(2) Differentiate the following:

(a) $\sin^2(t) + \cos^2(t)$

(c) $xe^{\sin(x)}$

(b) $(\sin(z)\cos(z))^{10}$

(d) $\ln(|\sin(x)|)$

(e) $(\log_a(|x|))^{1/3}$

(g) $\left(\dfrac{\ln(x) + 1}{\ln(x) + 2}\right)^3$

(f) $\dfrac{\log_a(|x|)}{\log_b(|x|)}$ $a \neq b$, $a > 1$, $b > 1$

(h) $\left(\dfrac{\tan(v) + 1}{\cot(v) + 2}\right)^5$

(3) (a) Find $\left[\dfrac{d^2}{dx^2}\csc(x)\right]_{x = \pi/4}$.

(e) If $f(x) = e^{|\ln(x)|}$ then $f'(x) = ?$

(b) Find $\dfrac{d^2}{dx^2}\sec(x)$.

(f) If $f(x) = \ln(e^{2x})$ then $f'(x) = ?$

(c) Find $\dfrac{d}{dx}(\cos^2(x) - \sin^2(x))^5$.

(g) Find $\dfrac{d}{dx}(\ln(x^9) - \ln(x^5))^2$.

(d) Find $\left[\dfrac{d}{dx}\ln(|\tan(x)|)\right]_{x = \pi/8}$.

(h) Find $\dfrac{d}{dx} 2^{\cot(x)}$.

(4) (a) $\dfrac{d}{dx}\log_2(\log_2(\log_2(x))) = ?$

(e) Find $\dfrac{d}{dx} x^{(1/x)}$.

(assume $x > 0$)

(b) For what values of x is $(\ln(x))^{1/2}$ defined and what is $\dfrac{d}{dx}(\ln(x))^{1/2}$?

(f) What is $\dfrac{d}{dx}\log_{10}(e^x)$?

(c) What is $\dfrac{d}{dx} x^{\ln(x)}$?

(assume $x > 0$)

(g) What is $\dfrac{d}{dx}\log_{10}(\ln(x))$?

(d) What is $\dfrac{d}{dx}(x + 1)^x$?

(assume $x > -1$)

Before giving the solutions to EXERCISES 2.23, we will summarize our logarithmic and trigonometric differentiation rules. These rules *should be memorized*. Recall that $\log_e(x)$ and $\ln(x)$ mean the same thing. In RULES 2.24, the second list of rules follows from the first by the CHAIN RULE: $f(g(x))' = f'(g(x))g'(x)$. Strictly speaking, we should not have to bother giving you the second set of rules. Some students have trouble remembering the factor $g'(x)$ when applying the CHAIN RULE. Perhaps the second set of rules will help impress on these students the need for this factor!

2.24   TRIG AND LOG DIFFERENTIATION RULES

$$\frac{d}{dx}\sin(x) = \cos(x)$$

$$\frac{d}{dx}\cos(x) = -\sin(x)$$

$$\frac{d}{dx}\sec(x) = \tan(x)\sec(x)$$

$$\frac{d}{dx}\csc(x) = -\cot(x)\csc(x)$$

$$\frac{d}{dx}\cot(x) = -(\csc(x))^2$$

$$\frac{d}{dx}\tan(x) = (\sec(x))^2$$

$$\frac{d}{dx}a^x = (\log_e(a))a^x$$

$$\frac{d}{dx}e^x = e^x$$

$$\frac{d}{dx}\log_a(x) = \frac{1}{(\log_e(a))x}$$

$$\frac{d}{dx}\log_e(x) = 1/x$$

$$\frac{d}{dx}\sin(g(x)) = \cos(g(x))g'(x)$$

$$\frac{d}{dx}\cos(g(x)) = -\sin(g(x))g'(x)$$

$$\frac{d}{dx}\sec(g(x)) = \tan(g(x))\sec(g(x))g'(x)$$

$$\frac{d}{dx}\csc(g(x)) = -\cot(g(x))\csc(g(x))g'(x)$$

$$\frac{d}{dx}\cot(g(x)) = -(\csc(g(x)))^2 g'(x)$$

$$\frac{d}{dx}\tan(g(x)) = (\sec(g(x)))^2 g'(x)$$

$$\frac{d}{dx}a^{g(x)} = (\log_e(a))a^{g(x)}g'(x)$$

$$\frac{d}{dx}e^{g(x)} = e^{g(x)}g'(x)$$

$$\frac{d}{dx}\log_a(g(x)) = \frac{1}{(\log_e(a))g(x)} g'(x)$$

$$\frac{d}{dx}\log_e(g(x)) = \frac{g'(x)}{g(x)}$$

Next we give the solutions to EXERCISE 2.23.

Study The Solution—Change The Problem—Rework It

2.25   SOLUTIONS TO EXERCISE 2.23

(1) **(a)** This looks like a problem that requires the QUOTIENT RULE. Notice, however, that the numerator factors into $(x - 1)(x + 1)$. The answer is 1. If you didn't notice this simplification don't feel too bad as you probably got some good practice with the QUOTIENT RULE.

(b) $\dfrac{d}{dx}(x^2 + 2x + 1)^{-1} = (-1)(x^2 + 2x + 1)^{-2}(2x + 2) = -2(x + 1)^{-3}$. Alternatively, you might factor before differentiating:

$$\frac{d}{dx}(x^2 + 2x + 1)^{-1} = \frac{d}{dx}(x + 1)^{-2} = -2(x + 1)^{-3}.$$

(c) Once again, we can apply the QUOTIENT RULE but notice that $x - 1 = (x^{1/2} - 1)(x^{1/2} + 1)$ so we need only compute

$$\frac{d}{dx}(x^{1/2} - 1)^2 = 2(x^{1/2} - 1)(1/2)x^{-1/2} = 1 - x^{-1/2}.$$

(d) We use the CHAIN RULE $\dfrac{d}{dx}f(g(x)) = \dfrac{df}{dg}\dfrac{dg}{dx}$ with $f(g) = \sin(g)$

and $g(x) = \left(\dfrac{3 - x^2}{x + 1}\right)$. We obtain

$$\cos\left(\frac{3 - x^2}{x + 1}\right) \frac{d}{dx}\left(\frac{3 - x^2}{x + 1}\right).$$

Using the QUOTIENT RULE we obtain

$$\frac{d}{dx}\left(\frac{3 - x^2}{x + 1}\right) = \frac{(-2x)(x+1) - (3-x^2)(1)}{(x + 1)^2}.$$

Doing a little algebra, the final answer may be written

$$-\cos\left(\frac{3 - x^2}{x + 1}\right)\left(\frac{x^2 + 2x + 3}{(x + 1)^2}\right).$$

(e) Using the CHAIN RULE we obtain first

$$-\sin\left(\frac{x^2 + 2x + 3}{x^3 + 6}\right) \frac{d}{dx}\left(\frac{x^2 + 2x + 3}{x^3 + 6}\right).$$

Now use the QUOTIENT RULE

$$\frac{d}{dx}\left(\frac{x^2 + 2x + 3}{x^3 + 6}\right) = \frac{(2x+2)(x^3+6) - (x^2+2x+3)3x^2}{(x^3 + 6)^2}.$$

The answer can be left the way it is or simplified slightly and be written

$$\left(\frac{x^4 + 4x^3 + 9x^2 - 12x - 12}{(x^3 + 6)^2}\right) \sin\left(\frac{x^2 + 2x + 3}{x^3 + 6}\right).$$

(f) Apply the PRODUCT RULE, $(f(x)g(x))' = f'(x)g(x) + f(x)g'(x)$, with $f(x) = \left(\dfrac{x - 1}{x + 1}\right)^3$ and $g(x) = (3x^2 + 1)^{1/2}$. For $f'(x)$ we compute

$$f'(x) = 3\left(\frac{x - 1}{x + 1}\right)^2 \left(\frac{(1)(x + 1) - (x - 1)(1)}{(x + 1)^2}\right).$$

For $g'(x)$ we compute

$$g'(x) = (1/2)(3x^2 + 1)^{-1/2}\, 6x.$$

After a little algebra, the final expression can be written

$$6\left(\frac{x - 1}{x + 1}\right)^2 \frac{(3x^2 + 1)^{1/2}}{(x + 1)^2} + \left(\frac{x - 1}{x + 1}\right)^3 \frac{3x}{(3x^2 + 1)^{1/2}}.$$

(2) **(a)** The notation $\sin^2(t)$ means $(\sin(t))^2$. From BASIC TRIGONO-METRIC FACTS 2.22 we have that $\sin^2(t) + \cos^2(t) = 1$. Thus

$$\frac{d}{dt}(\sin^2(t) + \cos^2(t)) = \frac{d}{dt}(1) = 0.$$

(b) Again, from BASIC TRIGONOMETRIC FACTS 2.22 we find

that $2\sin(z)\cos(z) = \sin(2z)$. Thus $\dfrac{d}{dz}((\sin(z)\cos(z))^{10} =$

$\dfrac{d}{dz}\left(\dfrac{\sin(2z)}{2}\right)^{10} = 2^{-9}.10\sin^9(2z)\cos(2z)$. If you didn't remember the

trigonometric identity you would compute $10(\sin(z)\cos(z))^9$

$\dfrac{d}{dz}(\sin(z)\cos(z)) = 10(\sin(z)\cos(z))^9(\cos^2(z) - \sin^2(z))$. This answer is

correct also, but not quite as nice! Can you show they're the same?

(c) $\dfrac{d}{dx}xe^{\sin(x)} = (1)e^{\sin(x)} + x\dfrac{d}{dx}e^{\sin(x)} = e^{\sin(x)} + xe^{\sin x}\cos(x) =$

$e^{\sin(x)}(1 + x\cos(x)).$

(d) Use the CHAIN RULE $\dfrac{d}{dx}f(g(x)) = \dfrac{df}{dg}\dfrac{dg}{dx}$ with $f(g) = \ln(|g|)$ and

$g(x) = \sin(x)$. We compute $\dfrac{df}{dg} = \dfrac{1}{g}$ and $\dfrac{dg}{dx} = \cos(x)$. Hence

$\dfrac{d}{dx}\ln(|\sin(x)|) = \dfrac{\cos(x)}{\sin(x)} = \cot(x).$

(e) Using the CHAIN RULE, we compute that

$$\frac{d}{dx}(\log_a(|x|)^{1/3} = (1/3)(\log_a(|x|))^{-2/3}\frac{d}{dx}\log_a(|x|)$$

The final answer can be written

$$\frac{1}{3\ln(a)x(\log_a(|x|))^{2/3}}.$$

(f) The answer is zero because $\dfrac{\log_a(|x|)}{\log_b(|x|)}$ is a constant function. To see

why, recall BASIC LOGARITHMIC FACTS 2.21, write $\log_a(|x|) =$

$\log_a(b)\log_b(|x|)$, and substitute this into the expression $\dfrac{\log_a(|x|)}{\log_b(|x|)}$. The

answer is $\log_a(b)$, which is a constant. That the ratio of the logarithm

functions for different bases is a constant is worth remembering.

(g) Use the CHAIN RULE to obtain

$$3 \left(\frac{\ln(x) + 1}{\ln(x) + 2}\right)^2 \frac{d}{dx}\left(\frac{\ln(x) + 1}{\ln(x) + 2}\right).$$

By the QUOTIENT RULE,

$$\frac{d}{dx}\left(\frac{\ln(x) + 1}{\ln(x) + 2}\right) = \frac{(1/x)(\ln(x) + 2) - (\ln(x) + 1)(1/x)}{(\ln(x) + 2)^2}.$$

After a little algebra, the final answer can be written

$$\frac{3(\ln(x) + 1)^2}{x(\ln(x) + 2)^4}.$$

In working this problem, one could notice that $\ln(x) + 1 = \ln(ex)$ and $\ln(x) + 2 = \ln(e^2x)$ and make these substitutions at the beginning. That does not simplify the calculations very much in this case.

(h)   We use the CHAIN RULE to compute

$$\frac{d}{dv}\left(\frac{\tan(v) + 1}{\cot(v) + 2}\right)^5 = 5\left(\frac{\tan(v) + 1}{\cot(v) + 2}\right)^4 \frac{d}{dv}\left(\frac{\tan(v) + 1}{\cot(v) + 2}\right).$$

Now use the quotient rule to compute

$$\frac{d}{dv}\left(\frac{\tan(v) + 1}{\cot(v) + 2}\right) = \frac{\sec^2(v)(\cot(v) + 2) - (\tan(v) + 1)\csc^2(v)}{(\cot(v) + 2)^2}.$$

These expressions can be simplified somewhat by various substitutions but it is not worth the trouble at this point.

(3)   **(a)**   The notation in this problem means *first* compute the second derivative of $\csc(x)$, and *then* evaluate the resulting function at $\pi/4$. To compute $\frac{d^2}{dx^2}\csc(x)$, we first compute $\frac{d}{dx}\csc(x) = -\csc(x)\cot(x)$. Next compute

$$\frac{d}{dx}(-\csc(x)\cot(x)) = (\csc(x)\cot(x))\cot(x) + (-\csc(x))(-\csc^2(x))$$

$$= \csc(x)\cot^2(x) + \csc^3(x).$$

Evaluated at $\pi/4$ this expression becomes $2^{1/2}(1)^2 + 2^{3/2}$, which is approximately 4.24.

(b)   We have

$$\frac{d}{dx}\sec(x) = \sec(x)\tan(x)$$

$$\frac{d}{dx}(\sec(x)\tan(x)) = \sec(x)\tan^2(x) + \sec^3(x).$$

(c) Computing without thinking we have

$$\frac{d}{dx}(\cos^2(x) - \sin^2(x))^5 = 5(\cos^2(x) - \sin^2(x))^4 \frac{d}{dx}(\cos^2(x) - \sin^2(x)).$$

Doing the indicated differentiation gives a messy but correct expression for the derivative. If we know our BASIC TRIGONOMETRIC FACTS 2.22, we recognize that $\cos^2(x) - \sin^2(x) = \cos(2x)$. The above expression is considerably simplified by this identity and becomes

$$\frac{d}{dx}(\cos(2x)^5 = 5(\cos(2x))^4 \frac{d}{dx}\cos(2x) = -10\cos^4(2x)\sin(2x).$$

(d) $\dfrac{d}{dx}\ln(|\tan(x)|) = (\tan(x))^{-1}\sec^2(x) = \cot(x)\sec^2(x) = (\sin(x)\cos(x))^{-1} = (\sin(2x))^{-1} = \csc(2x)$. Evaluating this expression at $\pi/8$ gives $2^{3/2}$.

(e) If you have forgotten what the graph of $\ln(x)$ looks like, refresh your memory by looking once again at FIGURE 2.13. From this graph it is clear that

$$|\ln(x)| = \begin{cases} \ln(x) & \text{if } x > 1 \\ -\ln(x) & \text{if } 0 < x < 1 \end{cases}.$$

From this observation it is clear that the graph of $|\ln(x)|$ has a sharp point or "cusp" at the point of the graph corresponding to $x = 1$ (like the point C' of FIGURE 1.2). This means that $|\ln(x)|$ has no derivative at $x = 1$ (i.e., at the point $(1,0)$ on the graph of $|\ln(x)|$). For points $0 < x < 1$, $\dfrac{d}{dx}|\ln(x)| = \dfrac{d}{dx}(-\ln(x)) = -1/x$. For points $x > 1$, $\dfrac{d}{dx}|\ln(x)| = \dfrac{d}{dx}\ln(x) = 1/x$. To summarize,

$$\frac{d}{dx}|\ln(x)| = \begin{cases} 1/x & \text{if } x > 1 \\ -1/x & \text{if } 0 < x < 1 \end{cases}.$$

We thus have that (simplify this further!)

$$\frac{d}{dx}e^{|\ln(x)|} = \begin{cases} e^{\ln(x)}(1/x) & \text{if } x > 1 \\ e^{-\ln(x)}(-1/x) & \text{if } 0 < x < 1 \end{cases}.$$

(f) By BASIC LOGARITHMIC FACTS 2.21, $\ln(e^{2x}) = 2x$. Thus, $f'(x) = 2$.

(g) By BASIC LOGARITHMIC FACTS 2.21, $\ln(x^9) - \ln(x^5) = \ln(x^9/x^5) = \ln(x^4) = 4\ln(x)$. Thus we have

$$\frac{d}{dx}(\ln(x^9) - \ln(x^5))^2 = \frac{d}{dx}(4\ln(x))^2 = 32 \ln(x)/x.$$

(h) We use the CHAIN RULE $\frac{d}{dx}f(g(x)) = \frac{df}{dg}\frac{dg}{dx}$ with $f(g) = 2^g$ and $g(x) = \cot(x)$. This gives

$$\frac{d}{dx} 2^{\cot(x)} = \ln(2)2^{\cot(x)}(-\csc^2(x)).$$

(4) **(a)** In EXAMPLES 2.20(6), we have already computed the derivative of the function $\log_2(\log_2(x))$. To use this result we shall apply the CHAIN RULE $\frac{d}{dx}f(g(x)) = \frac{df}{dg}\frac{dg}{dx}$ with $f(g) = \log_2(\log_2(g))$ and $g(x) = \log_2(x)$. From EXAMPLES 2.20(6) we have that

$$\frac{df}{dg} = \frac{1}{(\ln(2))^2 g\log_2(g)}.$$

We know that $\frac{dg}{dx} = \frac{1}{\ln(2)x}$. The final result is

$$\frac{d}{dx}\log_2(\log_2(\log_2(x))) = \frac{1}{(\ln(2))^3 x\log_2(x)\log_2(\log_2(x))}.$$

This result is valid if base 2 is replaced by base a.

(b) The function $(\ln(x))^{1/2}$ is defined (as a function from real numbers to real numbers) only when $\ln(x)$ is nonnegative. This means that $(\ln(x))^{1/2}$ is defined for x greater than or equal to 1. For $x > 1$ we have

$$\frac{d}{dx}(\ln(x))^{1/2} = (1/2)(\ln(x))^{-1/2}\frac{d}{dx}(\ln(x)) = \frac{1}{2x(\ln(x))^{1/2}}.$$

For $x = 1$, the function $(\ln(x))^{1/2}$ has no derivative. If you look at the expression above for the derivative when $x > 1$ you will see that it gets very large ("goes to infinity") as x gets close to 1.

(c) In EXAMPLES 2.20(8) we computed $\frac{d}{dx}x^x = (\ln(x) + 1)x^x$. The trick we used there was to replace x with $e^{\ln(x)}$ and differentiate $e^{x\ln(x)}$. We can use that same trick in this problem, writing $x^{\ln(x)} = (e^{\ln(x)})^{\ln(x)}$. Thus,

$$\frac{d}{dx}x^{\ln(x)} = \frac{d}{dx}e^{(\ln(x))^2} = e^{(\ln(x))^2}\frac{d}{dx}(\ln(x))^2 = \frac{2\ln(x)x^{\ln(x)}}{x}.$$

(d)   We use essentially the same trick as in **(c)**. Write $(x + 1)^x = e^{x\ln(x+1)}$. Using the CHAIN RULE $\frac{d}{dx}f(g(x)) = \frac{df}{dg}\frac{dg}{dx}$ with $f(g) = e^g$ and $g(x) = x\ln(x + 1)$ gives

$$\frac{d}{dx}(x+1)^x = e^{x\ln(x+1)}\frac{d}{dx}x\ln(x+1) = (x+1)^x\left(\ln(x+1)+\frac{x}{x+1}\right).$$

(e)   Write $x^{(1/x)} = e^{\ln(x)/x}$. Then

$$\frac{d}{dx}x^{(1/x)} = e^{\ln(x)/x}\frac{d}{dx}(\ln(x)/x) = x^{(1/x)}\left(\frac{1 - \ln(x)}{x^2}\right).$$

(f)   From BASIC LOGARITHMIC FACTS 2.21, we have that

$$\log_{10}(e^x) = \log_{10}(e)\log_e(e^x) = (\log_{10}(e))x.$$

Thus we compute easily

$$\frac{d}{dx}\log_{10}(e^x) = \log_{10}(e) = \frac{1}{\log_e(10)} = .4343 \text{ (approximately)}.$$

(g)   Use $\log_{10}(\ln(x)) = \log_{10}(e)\ln(\ln(x))$ to get $\frac{d}{dx}\log_{10}(\ln(x)) =$

$$\frac{\log_{10}(e)}{x\ln(x)} = \frac{1}{\ln(10)x\ln(x)}.$$

Some Students Memorize $(f(x)^{g(x)})' = f(x)^{g(x)}(g(x)\ln(f(x))'$

If you now feel that you have mastered EXERCISE 2.23, you can try to work the VARIATIONS on these exercises that follow. These variations are "parallel" to EXERCISE 2.23 in the sense that corresponding questions represent similar ideas. If you have trouble with a particular question, study the corresponding question in EXERCISE 2.23 and its solution in SOLUTIONS 2.25. Be sure to thoroughly memorize the differentiation rules of 2.24. When you feel ready, try the two practice exams following these VARIATIONS.

2.26   VARIATIONS ON EXERCISE 2.23

(1)   Differentiate the following:

(a) $\dfrac{x^2 - 4}{x - 2}$

(b) $\dfrac{1}{4x^2 + 4x + 1}$

(c) $\dfrac{x^{2/3} - 1}{x^{1/3} - 1}$

(e) $\sin\left(\dfrac{x^3 + 6}{x^2 + 2x + 3}\right)$

(d) $\cos\left(\dfrac{x + 1}{3 - x^2}\right)$

(f) $\left(\dfrac{x - 1}{x + 1}\right)^{1/2}(3x^2 + 1)^3$

(2) Differentiate the following:

(a) $\dfrac{1 - \cos^2(t)}{\sin^2(t)}$

(e) $(\log_a|x^{1/3}|)^{1/3}$

(b) $(\cos^2(t) - \sin^2(t))^{10}$

(f) $\dfrac{\log_a|x^2|}{\log_b|x^3|}$ $\begin{array}{l} a > 1 \\ b > 1 \\ a \neq b \end{array}$

(c) $(\sin x)e^x$

(g) $\left(\dfrac{\ln(x^2) + 1}{\ln(x^2) + 2}\right)^3$

(d) $\ln|\tan^2(x)|$

(h) $\left(\dfrac{\cot(r) + 2}{\tan(r) + 2}\right)^5$

(3) **(a)** Find $\left[\dfrac{d^2}{dx^2}\sec(x)\right]_{x = \pi/4}$.

(e) If $f(x) = 2^{|\log_2(x)|}$ then $f'(x) = ?$

(b) $\dfrac{d^2}{dx^2}\csc(x) = ?$

(f) If $f(x) = \ln(e^{ex})$ then $f'(x) = ?$

(c) $\dfrac{d}{dx}(1 + \tan^2(x)) = ?$

(g) Find $\dfrac{d}{dx}(\ln(x^9) - \ln(x^5))^2$.

(d) $\left[\dfrac{d}{dx}\ln(|\tan(x)|)\right]_{x = \pi/8} = ?$

(h) Find $\dfrac{d^2}{dx^2}\cot^2(x)$.

(4) **(a)** $\dfrac{d}{dx}\log_{10}(\log_{10}(\log_{10}(x))) = ?$

(d) What is $\dfrac{d}{dx}(x + 1)^{(x + 1)}$? (assume $x > 0$)

(b) Where is $(\ln(x))^{1/4}$ defined and what is its derivative?

(e) Find $\dfrac{d}{dx}(x + 1)^{\frac{1}{x + 1}}$. (assume $x > -1$)

(c) What is $\dfrac{d}{dx}x^{\log_2(x)}$? (assume $x > 0$)

(f) What is $\dfrac{d}{dx}\log_2(e^x)$?

(g)   What is $\dfrac{d}{dx}\log_{10}(\log_2(x))$?

2.27   VARIATIONS ON EXERCISE 2.23

(1)   Differentiate the following:

(a) $\dfrac{x^4 - 25}{x^2 + 5}$

(d) $\sin\left(\dfrac{x^4 + 2x^2 + 1}{x^2 + 2}\right)$

(b) $\dfrac{1}{2x^{1/2} + 3}$

(e) $\cos\left(\dfrac{x^{1/2} + 1}{2x^2}\right)$

(c) $\left(\dfrac{x^{1/2} - 1}{x^{1/4} + 1}\right)^2$

(f) $\left(\dfrac{x^2 - 1}{x + 1}\right)^2 (4x^{3/2} + 1)^{1/6}$

(2)   Differentiate the following:

(a) $4\cos^2(2t) + 4\sin^2(2t)$

(e) $\log_a(|x^3 + 1|)$

(b) $\sin^2(6x) - \cos^2(6x)$

(f) $\dfrac{\log_a(|3x + 1|)}{\log_b(|3x + 1|)}$, $a, b > 1$ $a \neq b$

(c) $(x^2 + 3)e^{x\tan x}$

(g) $\left(\dfrac{\ln(\sin x) + 3}{\ln(\cos x) + 1}\right)^3$

(d) $\ln(|\sin^2 2x|)$

(h) $\left(\dfrac{(\cot x)^3 + 3x}{\cot(x^3 + 3x)}\right)^2$

(3)   (a)   Find $\left[\dfrac{d^2}{dx^2}\csc(x + 2)\right]_{x = \frac{\pi}{4} - 2}$.

(e)   If $f(x) = 10^{|\log_{10}(x^2 + 1)|}$ then $f'(x) = ?$

(b)   Find $\dfrac{d^2}{dx^2}\sec(x + 2)$.

(f)   If $f(x) = \log_7 7^{(\sin x)e^x}$ then $f'(x) = ?$

(c)   Find $((5\cos(2x)\sin(2x))^{1/2})'$.

(g)   Find $\dfrac{d}{dx}[(\ln(x^{3/2}) + \ln(x^{1/2})\quad \ln(x^{5/2})]$.

(d)   Find $\left[\dfrac{d}{dx}\ln(|\tan^2 2x|)\right]_{x = \frac{\pi}{8}}$.

(h)   Find $\dfrac{d^3}{dx^3}(\tan x)(\sin x)$.

(4) **(a)** $\dfrac{d}{dx}\ln(\ln(\ln(\ln(3x^3+2)))) = ?$ **(e)** Find $\dfrac{d}{dx}(x+1)^{1/x^2}$. (assume

$x \neq 0$).

(b) Where is $(\ln(\sqrt{2x+1}))^{3/4}$
defined and what is its   **(f)** What is $\dfrac{d}{dx}\log_{10}(e^{\sqrt{3x^2+2}})$?
derivative?

(c) What is $\dfrac{d}{dx}(x^{1/2}+1)^{\ln x}$?
                                             (g) What is $\dfrac{d}{dx}\ln(\log_7 e^{x^2})$?
(assume $x > 0$)

(d) What is $\dfrac{d}{dx}(x^{-1/2}+4)^{6x^2}$?

(assume $x \neq 0$)

2.28 VARIATIONS ON EXERCISE 2.23

(1) Differentiate the following:

(a) $\dfrac{x^2 - 16}{x + 4}$ **(d)** $\sin\left(\dfrac{3x^2 + 2}{x + 1}\right)$

(b) $\dfrac{1}{3x^2 + 2x + 1}$ **(e)** $\cos\left(\dfrac{4x^2 + 2}{16x^4 + 8x + 2}\right)$

(c) $\left(\dfrac{x^3 - 1}{x^{3/2} - 1}\right)$ **(f)** $\left(\dfrac{x-2}{x+2}\right)^{3/2}(4x^3+2x+1)^{1/5}$

(2) Differentiate the following:

(a) $\dfrac{\cos^2 t}{1 - \sin^2 t}$ **(e)** $(\log_a(|x^2|))^{1/2}$

(b) $[\sin(2z)\cos(2z)]^9$ **(f)** $\dfrac{\log_a(|x^2+1|)}{\log_b(|x-3|)}$, $a,b > 1$
                                                                    $a \neq b$

(c) $x^3 e^{\cos x}$ **(g)** $\left(\dfrac{\ln(x^{1/2}) + 1}{\ln(x^{1/5}) + 2}\right)^{2/7}$

(d) $\ln(|\cos 3x^4|)$ **(h)** $\left(\dfrac{\tan(2x) + 1}{\cot(x^2) + 2}\right)^{3/2}$

(3) **(a)** Find   **(b)** Find $\dfrac{d^2}{dx^2}(\csc(2x+1))$.

$\left[\dfrac{d^2}{dx^2}\sec(2x + 1)\right]_{x = \frac{\pi}{8} - \frac{1}{2}}$.

(c) Find $((1 + \cot^2(3x))^{1/2})'$.

(f) If $f(x) = \log_5 5^{(x^2+1)}$ then $f'(x) = ?$

(d) Find $\dfrac{d}{dx}\ln(|\cot 3x|)\Big|_{x=\frac{\pi}{21}}$.

(g) Find
$$\dfrac{d}{dx}(\log_5(x^{7/8}) + \log_5(x^{8/7}))^{12}.$$

(e) If $f(x) = 3^{|\log_3(x+1)|}$ then $f'(x) = ?$

(h) Find $\dfrac{d^3}{dx^3}\tan^2(x^3)$.

(4) (a) $\dfrac{d}{dx}2^{(3^{(4^x)})} = ?$

(e) $\dfrac{d}{dx}x^{1/4x} = ? \; x > 0.$

(b) Where is $(\ln(3x))^{1/3}$ defined and what is its derivative?

(f) $\dfrac{d}{dx}\log_8(e^{4x^2}) = ?$

(c) $\dfrac{d}{dx}x^{\ln^2(x)} = ? \; x > 0.$

(g) $\dfrac{d}{dx}\ln(\log_{10}x) = ? \; x > 0.$

(d) $\dfrac{d}{dx}(4x^2+3)^{3x} = ?$

2.29 VARIATIONS ON EXERCISE 2.23

(1) Differentiate the following:

(a) $\dfrac{3x^2 - 12}{x + 4}$

(d) $\ln(1 + 2\sqrt{x})$

(b) $\cos\left(\dfrac{1 - x^2}{1 + x^2}\right)$

(e) $\left(\dfrac{3x+1}{2-4x}\right)^2 \left(\dfrac{2x^2+1}{5+3x}\right)^{1/3}$

(c) $\left(\dfrac{\sin x + x}{1 + \cos x}\right)^3$

(f) $\dfrac{2}{3x^2 + 4x + 1}$

(2) Differentiate the following:

(a) $\cos(2x) + \sin^2 x - 1$

(e) $x \cdot (\tan(3x))^4$

(b) $e^{\sin x} \cdot e^{\cos x}$

(f) $e^{(1 + \ln x)}$

(c) $(\sec x + \tan^2 x)^{4/5}$

(g) $\left(\dfrac{\ln x + 4}{\ln x - 4}\right)^5$

(d) $(\ln x^2)^{1/2}$

(3) **(a)** Find

$\dfrac{d}{dx}(3\ln x + \ln x^{1/4} - \ln x^2)^8.$

(e) Find $\dfrac{d}{dx}\cos^2(3x)\sin x \Big|_{x=\frac{\pi}{3}}.$

(b) Find $\dfrac{d^2}{dx^2}\sec(3x).$

(f) Find $\dfrac{d}{dx}e^{x^2} \cdot e^{3x} \cdot e^{\sin x}.$

(c) Find $\dfrac{d}{dx}\ln|\csc x + \cot x|.$

(g) Find $\dfrac{d}{dx}\tan(e^x).$

(d) Find $\dfrac{d}{dx}\ln(e^{|x|}).$

(h) Find $\dfrac{d}{dx}(5\cos x\sin x)^3.$

(4) **(a)** Find $\dfrac{d}{dx}(1/x)^x.$

(d) Find $\dfrac{d}{dx}(\sin x)^{\ln x}.$

(b) Find $\dfrac{d}{dx}\ln(\sin(e^x)).$

(e) Find

$\dfrac{d}{dx}\left(\dfrac{x^2+x+1}{2x+4}\right)^3\left(\dfrac{x^2+1}{2-3x}\right)^{-4}.$

(c) Find $\dfrac{d}{dx}(\ln x)^x.$

2.30   VARIATIONS ON EXERCISE 2.23

(1) Differentiate the following:

(a) $(3x+1)^2(x^2+4x+1)^3$

(d) $\dfrac{x^3+1}{x+1}$

(b) $\sin^2(x^3-1)$

(e) $\ln\left(\dfrac{x^2+\sin x}{2+\cos 3x}\right)$

(c) $\sqrt{\sin x} \cdot (x^3+1/x)$

(f) $\left(\dfrac{2}{1+x}\right)^{3/2}$

(2) Differentiate the following:

(a) $e^{(\tan x - \sec x)}$

(d) $\sin(\ln(x^{3/2}))$

(b) $(\ln x)(e^x)$

(e) $\left(\dfrac{\tan x + 4\sin x + 1}{\cos(2x) - 3\cos x + x}\right)^4$

(c) $\ln\sqrt{x}$

(f) $\left(\dfrac{2x+1}{x^3-4}\right)^2\sin(3x)$

(g) $\sin(e^x)$ **(h)** $\tan^2\left(\dfrac{-3x + 4}{2x + 1}\right)$

(3)   Find the following derivatives:

(a)   Find $\dfrac{d}{dx}(\sec^2 x + \tan x)\Big|_{x=\frac{\pi}{3}}$.   **(e)**   Find $\dfrac{d}{dx}\ln|\sec x + \tan x|$.

(b)   Find $\dfrac{d}{dx}\ln(3 + 4x^2)\Big|_{x=1}$.   **(f)**   Find $\dfrac{d}{dx}\sin^5(\cos x)$.

(c)   Find $\dfrac{d^2}{dx^2}\ln|x|$.

(g)   Find
$\dfrac{d}{dx}\left(\dfrac{x^2\sin x - 3x + 1}{\ln x + e^x}\right)^2$.

(d)   Find $\dfrac{d}{dx}x \cdot e^{\cos x}\Big|_{x=\pi}$.

(4)   Differentiate the following:

(a) $(1/x)^{2x}$ **(e)** $\ln(\sin(e^x))$

(b) $\ln(\sqrt{e^{3x}})$ **(f)** $\tan(\log_5 x^2)$

(c) $e^x \cdot x^x$ **(g)** $x \cdot (\ln x)^{1/2}$

(d) $\ln\left|\dfrac{1}{\ln|x|}\right|$

2.31   VARIATIONS ON EXERCISE 2.23

(1)   Differentiate the following:

(a) $\dfrac{x^3 + 2x - 1}{x^2 + 1}$ **(d)** $\tan(x^2(x + 1))$

(b) $\dfrac{3}{x^2 + x}$ **(e)** $\sqrt{\dfrac{x^2 + 1}{3x}} \cdot \sin(x^2)$

(c) $\ln\left(\dfrac{x - 1}{x + 1}\right)$ **(f)** $\left(\dfrac{3x^2 - 2x + 1}{-x^4 + 5x + 2}\right)^3$

(2)   Differentiate the following:

(a) $\sin^2 t + \cos^2(2t)$ **(b)** $(\cos^2 x - \sin^2 x)^4$

(c) $\ln|\cos x|$

(d) $e^{(\sin x + \cos x)}$

(e) $(\tan x + 3\sec(2x))^3$

(f) $\dfrac{\sin^2 x - 2\tan x}{\ln|x^3| + \sec(4x)}$

(g) $x^2 e^{\cos x}$

(h) $\ln(1 + e^x)$

(3)   Find the following derivatives:

(a)   Find $\dfrac{d^2}{dx^2}(\ln x)$.

(b)   Find $\dfrac{d}{dx}\sec^2 x$.

(c)   Find $\dfrac{d}{dx}e^{\tan x}\bigg|_{x=\frac{\pi}{4}}$.

(d)   Find

$\dfrac{d}{dx}(\ln x^5 - \ln x^4 + 3\ln x)$.

(e)   Find

$\dfrac{d}{dx}\ln((x-1)(x+2)^2/(x^2+1))$.

(f)   Find $\dfrac{d}{dx}\ln(e^{2x^2})$.

(g)   Find $\dfrac{d}{dx}\ln(e^{2x^2} + 1)$.

(h)   Find $\dfrac{d}{dx}\tan(e^{\cos x})$.

(4)   Find the following derivatives:

(a)   Find $\dfrac{d}{dx}(\sin(\sin x))$.

(b)   Find $\dfrac{d}{dx}e^{(e^x)}$.

(c)   Find $\dfrac{d}{dx}(\ln x)^x$.

(d)   Find $\dfrac{d}{dx}e^{\log_{10} x}$.

(e)   Find $\dfrac{d}{dx}\sin(\ln(\cos x))$.

(f)   Find $\dfrac{d}{dx}x^x$.

2.32   VARIATIONS ON EXERCISE 2.23

(1)   Differentiate the following:

(a) $\dfrac{1 - 3x^2}{2 + \cos x}$

(b) $\left(\dfrac{2}{1 - 3x + 4x^2}\right)^3$

(c) $\ln\left(\dfrac{1 - x^2}{1 + x^2}\right)$

(d) $\left(\dfrac{3x + 1}{4x}\right)\cos(x^2 + 5)$

(e) $e^{(3x\sin(4x))}$

(f) $\sqrt{\dfrac{2 + 3x^2}{6 - x^3}}$

(2) Differentiate the following:

    (a) $\sin^2 x - \cos^2 x$

    (b) $(\sin^3 x + 4x)/(\tan^2 x + 1)$

    (c) $\ln|2 + 4x\sin x|$

    (d) $\sqrt{\ln(2x)}$

    (e) $e^{2\tan^2 x}$

    (f) $(\tan x \sec x)^5$

    (g) $\left(\dfrac{2 + \ln x^2}{3 + e^{5x}}\right)^4$

    (h) $\cos(\tan x)$

(3) **(a)** Find $\dfrac{d^2}{dx^2}(\tan x)\Big|_{x=\frac{\pi}{3}}$.

    (b) Find $\dfrac{d^2}{dx^2}e^{(x^2)}$.

    (c) Find $f'(4)$, if $f(x) = ((3x^2 + 4x + 1)/(2x^2 + x - 1))^3$.

    (d) Find $\dfrac{d}{dx}|x^3|$.

    (e) Find $\dfrac{d^2}{dx^2}(\sec^2 x)$.

    (f) Find $\dfrac{d}{dx}(\ln(xe^x))$.

    (g) Find $\dfrac{d}{dx}\cos(e^{\sin x})$.

(4) **(a)** Find $\dfrac{d}{dx}(1/x)^{\ln x}$.

    (b) Find $\dfrac{d}{dx}(x + 2)^3(3x - 4)^4(5x - 2)^{-3}$ at $x = 1$.

    (c) Find $\dfrac{d}{dx}(1 + x^2)^{2x+3}$.

    (d) Find $\dfrac{d}{dx}2^{\ln x}$.

    (e) Find $\dfrac{d}{dx}\sqrt{e^x}$.

    (f) Find $\dfrac{d}{dx}\ln(\sin(\sqrt{1 + x^2}))$.

2.33 PRACTICE EXAM 1

If you are feeling confident about Chapters 1 and 2, try these practice exams. You should be able to work each of these in one hour, closed book.

(1) **(a)** Find $h(x) = f(g(x))$ where $f(t) = 2t^4 + 3$ and $g(x) = (-x - 1)^{-2}$.

    (b) Find $h(z) = f(g(z))$ where $f(g) = \ln(g)$ and $g(z) = \sin(z)$. For what values of z is $h(z)$ defined?

(2) Find functions $f(g)$ such that $h(x) = f(g(x))$ in each of the following cases.

(a) $h(x) = (x - 2)^3 + x^2 - 4x - 1$ and $g(x) = x - 2$.

(b) $h(x) = \dfrac{x^2 - 6x}{(x - 3)^3}$ and $g(x) = x - 3$.

(3)   (a)   If $D(x) = (A(x) + B(x) + C(x))^2$ find $D'(2)$ if $A(2) = B(2) = 1$, $C(2) = 3$, $A'(2) = B'(2) = 1$, and $C'(2) = 2$.

(b)   Find the equation of the line tangent to $y = 2x^3 + x^2 + 5$ at $x = 1$.

(4)   Differentiate the following:

(a) $\left(\dfrac{x - 1}{x^{1/2} + 1}\right)^3$

(b) $\sin\left(\dfrac{x^2 + 1}{x^3 + 1}\right)$

(5)   Differentiate the following:

(a) $t^{(\sin^2(t) + \cos^2(t))}$

(b) $x^2 e^{\cos(x)}$

(c)   Find $\left[\dfrac{d^2}{dx^2}\sin(2x)\right]_{x = \pi/4}$.

(6)   (a)   Differentiate $\ln(|\cos(x)|)$.

(b)   If $f(x) = \ln(e^{5x})$ then $f'(x) = $?

(c)   Find $\dfrac{d}{dx}(\ln(x^3) - \ln(x^7))^3$.

(7)   (a)   What is $\dfrac{d}{dx}(2x + 3)^{\ln(x)}$?

(b)   What is $\dfrac{d}{dx}(2x + 3)^x$?

END OF PRACTICE EXAM 1

2.34   PRACTICE EXAM 2

If you had trouble with PRACTICE EXAM 1, find the problems similar to the ones you had difficulty with in EXERCISE 1.16 and EXERCISE 2.23. Study the solutions to these problems before trying PRACTICE EXAM 2.

(1)   (a)   Find $h(x) = f(g(x))$ where $f(t) = 2t^{-4} + 3$ and $g(x) = (-x - 1)^{-2}$.

(b) Find $h(z) = f(g(z))$ where $f(g) = \ln \left(\dfrac{g - 3}{g + 3} \right)$ and $g(z) = \sqrt{z}$.

For what values of z is $h(z)$ defined?

(c) Find $h(k(y))$ and $k(h(y))$ where $h(y) = 4^y$ and $k(y) = \log_4(y)$. For what values of y are these functions defined?

(2) Find functions $f(g)$ such that $h(x) = f(g(x))$ in each of the following cases.

(a) $h(x) = (x + 2)^3 + x^2 + 4x$ and $g(x) = x + 2$.

(b) $h(x) = \dfrac{x^2 + 6x}{(x + 3)^5}$ and $g(x) = x + 3$.

(3) **(a)** If $D(x) = (A(x) + B(x) + C(x))^4$ find $D'(2)$ if $A(2) = B(2) = 3$, $C(2) = 1$, $A'(2) = B'(2) = 1/3$, and $C'(2) = 2$.

(b) Find the equation of the line tangent to $y = x^3 + 2x^2 + 1$ at $x = 1$.

(c) Find the equation of the line normal to the curve $y = x^{2/3}$ at $x = 8$.

(4) Differentiate the following:

(a) $\ln \left(\dfrac{x - 1}{x^{1/2} + 1} \right)$

(b) $e^{\sin \left(\frac{x^2 + 1}{x^3 + 1} \right)}$

(5) Differentiate the following:

(a) $t^{\log_2(\sin^2(t) + \cos^2(t))}$

(b) $x^2 e^{\tan(x)}$

(c) Find $\left[\dfrac{d^2}{dx^2} \sin(2x) \right]_{x = \pi/4}$.

(6) **(a)** Differentiate $\ln(|\sec(x)|)$.

(b) If $f(x) = \ln(e^{8x})$ then $f'(x) = ?$

(c) Find $\dfrac{d}{dx}(\ln(x^3) - \ln(x^7))^3$

(7) **(a)** What is $\dfrac{d}{dx}(2x + 3)^{(\ln(x) + 1)}$?

(b) What is $\dfrac{d}{dx}(2x + 3)^{x+1}$?

END OF PRACTICE EXAM 2

2.35   INVERSE TRIGONOMETRIC AND HYPERBOLIC FUNCTIONS

The inverse trigonometric functions (called arcsin, arccos, or $\sin^{-1}$, $\cos^{-1}$, etc.) and hyperbolic functions (sinh, cosh, etc.) will be the last class of functions that we shall add to our basic list of functions whose derivatives we shall memorize. For the beginning calculus student, the former class, the inverse trig functions, are the most important. The hyperbolic functions are little more than a notational contrivance from the point of view of the beginning student. The hyperbolic functions have many properties that are remarkably like the trig functions. The reason for this will be clear to any student who takes more mathematical analysis beyond the first course in calculus.

A More Careful Look At Compositional Inverses

We now take a more careful look at the idea of compositional inverses of functions. We saw this idea applied in 1.17(1)f,g in connection with the solutions to EXERCISE 1.16, in the proof of LEMMA 2.2, and in the proof of THEOREM 2.18. We are now going to make a more systematic use of this idea. It will be helpful to be a little more precise about the notion of a function. If a mathematician says ''I have a function in this envelope!'' he or she means that the following information is contained in the envelope:

(1)   A set D is specified. D is called the *domain* of the function.

(2)   A set R is specified. R is called the *range* of the function.

(3)   A rule is specified which assigns to each element of the domain D an element of the range R. (Exactly one element is assigned to each element of D. When an element of R is assigned to an element of D it is not removed from R. It may be assigned to other elements of D as well.)

Here are some examples of functions:

Notice The Notation For Intervals: [a,b], (a,b], [a,b), (a,b)

FUNCTION 1

(1)   D is the set $(0,1] = \{x : 0 < x \leqslant 1\}$.

(2)   R is the set $(0,1]$ also.

(3)   To each element x in D assign $\sqrt{x}$.

FUNCTION 2

(1)   D is $(0,1]$ just as in FUNCTION 1.

(2)   R is (0,2].

(3)   To each element x in D assign $\sqrt{x}$. This is the same rule as in FUNCTION 1.

When Are Two Functions Equal?

Notice that the only difference between FUNCTION 1 and FUNCTION 2 is that in FUNCTION 1 the range is (0,1] and in FUNCTION 2 the range is (0,2]. The graphs of these two functions are shown in FIGURE 2.36. Technically, two functions are equal if they have the same domain, the same range, and the same rule of assignment. Thus, FUNCTION 1 and FUNCTION 2 are different in the sense of this definition as they have different ranges. This may seem a bit like nit-picking to you, in which case you are right. In calculus, we usually just take the range of a function to be the set of real numbers (which we also denote by R). If we had done this for FUNCTION 1 and FUNCTION 2, they would be the same function.

Image Of f Is The Set Of All f(x), x In The Domain Of f

If a function has domain D and range R, then a mathematician would say that the function "maps D to R." The IMAGE of a function is the set of all values in R that are assigned to some value in D. FUNCTION 1 and FUNCTION 2 have the same IMAGE. In both cases, the IMAGE is the set (0,1]. A function is "surjective" or "onto" if its IMAGE is the same as its range R. In this sense, FUNCTION 1 above is onto, but FUNCTION 2 is not onto. Let's look at a couple of additional functions:

FUNCTION 3

(1)   D is the set (0,1].

(2)   R is the real numbers.

(3)   To each element x in D assign $\sqrt{x}$.

FUNCTION 4

(1)   D is the set (0,2].

(2)   R is the set of real numbers.

(3)   To each element x in D assign $\sqrt{x}$.

Graph Of f Is The Set Of All (x,f(x)) x In Domain Of f

The graphs of FUNCTION 3 and 4 are shown in FIGURE 2.36. The graph of a function is, by definition, the set of all pairs (x,y) where x is in the

FIGURE 2.36   Four Functions

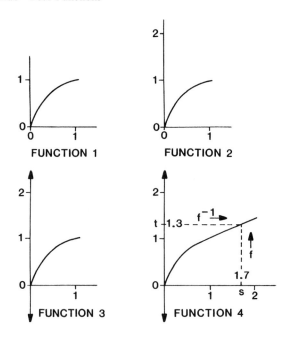

FUNCTION 1

FUNCTION 2

FUNCTION 3

FUNCTION 4

domain D of the function and y is the value of R assigned to x by the function. The function $\sqrt{x}$ is defined (as a real valued function) for all nonnegative real numbers x. Sometimes one hears, "The domain of definition of $\sqrt{x}$ is the set of all nonnegative real numbers." It is not just nit-picking to be careful about specifying the domains of the functions we talk about in calculus. As in the case of FUNCTION 3 and 4, we may not always take the domain of a function to be *all* values for which the expression defining the function is valid (i.e., the "domain of definition").

"One-To-One" Or "Injective"—It's The Same

The functions graphed in FIGURE 2.36 also have another important property. Notice that, for these functions, no value in the range is assigned to two or more domain values. To see a situation where this property is not valid, look at FIGURE 2.37(a). For this function, the value 1 in the range is assigned to 1/4, 1, and 7/4. A function which has the property that no range value is assigned to more than one domain value is called "one-to-one" or "injective." Another way to say the same thing is "f is one-to-one if s $\neq$ t implies f(s) $\neq$ f(t)" or, equivalently, "f is one-to-one if f(s) = f(t) implies s = t."

FIGURE 2.37 A Function That Is Not One-To-One

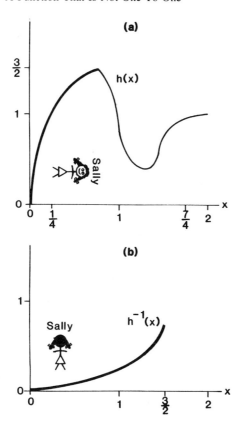

You should think carefully about why these various definitions define the same thing.

"Functional" Or "Compositional" Inverses—Same Thing

Look again at FIGURE 2.36. Note that FUNCTIONS 1, 2, 3, and 4 are one-to-one. Any one-to-one function f has another function associated with it called its "functional inverse" or, simply, "inverse" f^{-1}. To each number in the IMAGE of f, we associate the number s which has $f(s) = t$. We call this unique number s, $f^{-1}(t)$. If, for example, we take f to be FUNCTION 4 of FIGURE 2.36, then $t = 1.3$ is associated with $s = 1.7$ because $f(1.7) = (1.7)^{1/2} = 1.3$. Thus $f^{-1}(1.3) = 1.7$. In the case of a one-to-one function f, we may take the domain of f^{-1} to be the IMAGE of f and the range of f^{-1} to be the real numbers. Thus for $f =$ FUNCTION 4, the domain of f^{-1}

is the interval $[0,2^{1/2}] = \{x : 0 \leqslant x \leqslant 2^{1/2}\}$. For values of x in this interval, $f^{-1}(x) = x^2$. Thus, $f^{-1}(1.3) = (1.3)^2 = 1.7$. Be careful with the notation for functional inverses! The notation $f^{-1}(1.3)$ is not the same as $(f(1.3))^{-1}$. The latter means "compute f(1.3) and take its reciprocal." Thus we compute $f(1.3) = 1.14$ and $1/1.14$ is 0.88, so $(f(1.3))^{-1} = 0.88$, which is not the same as $f^{-1}(1.3) = (1.3)^2 = 1.7$. Functional inverses are also called "compositional inverses."

$$f(f^{-1}(x)) = x, \quad f^{-1}(f(x)) = x$$

The general situation is illustrated in FIGURE 2.38. In this figure, a one-to-one function f is defined by giving its graph. The domain of f is $[-1/4, 1]$ and the range is the real numbers. FIGURE 2.38(b) is an exact copy of FIGURE 2.38(a) except that the vertices of the rectangle are interpreted in terms of f in (a) and in terms of f^{-1} in (b). In (a), $P = (0, f(s))$ and in (b), $P = (0, t)$. P is a fixed point in the plane so this means that $(0, f(s)) = (0, t)$ or, in particular, $f(s) = t$. Similarly, $Q = (s, 0) = (f^{-1}(t), 0)$ so $f^{-1}(t) = s$. If in the identity $f(s) = t$ we substitute $s = f^{-1}(t)$, we obtain

$$f(f^{-1}(t)) = t \text{ for any t in the domain of } f^{-1}.$$

Likewise, if in the identity $f^{-1}(t) = s$ we substitute $t = f(s)$ we obtain

$$f^{-1}(f(s)) = s \text{ for any s in the domain of f.}$$

These observations are summarized in FIGURE 2.38.

FIGURE 2.38   Basic Properties of Compositional Inverses

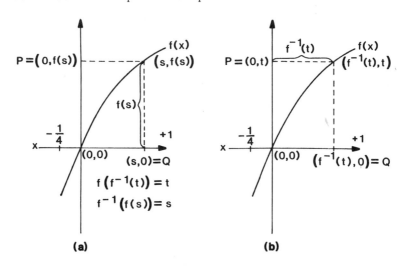

(a)   (b)

If It's Not One-To-One, Restrict The Domain!

Unfortunately, many of the most important functions in calculus are not one-to-one. A typical situation is shown in FIGURE 2.37(a) and (b). The function h(x) shown in FIGURE 2.37(a) is defined graphically and has domain [0,2]. Clearly, h is not one-to-one. Note, however, that if we restrict the function h to the interval [0,3/4] then this function is one-to-one. The graph of this new function is shown as the darker portion of the graph of h in FIGURE 2.37(a). Shown also is our old friend SALLY of FIGURE 2.13. Again, SALLY is slightly behind the page. FIGURE 2.37(b) shows the darkened portion of the graph as viewed by SALLY. This new graph is called h^{-1} *even though it is the inverse of the restriction of h to the interval* $[0,3/4]$ *and not the inverse of the original* *h*. The original h has no inverse as it is not one-to-one. There is nothing magical about the choice of the interval [0,3/4]. The interval [3/4,5/4] could have been chosen as well. The reader should sketch the graph of h^{-1} based on the restriction of h to [3/4,5/4].

Compositional Inverses Of Trig Functions

We now use the idea of FIGURE 2.37 to define the compositional inverses of the trigonometric functions. These compositional inverses are shown in FIGURE 2.40. The first function shown is the compositional inverse of the cosine. This function is defined by restricting the cosine to the interval [0,π]. The inverse function is denoted by $\cos^{-1}(x)$ or arccos(x). The first notation is common but very poor. It is standard practice to write $\cos^n(x)$ for $(\cos(x))^n$. Thus $\cos^{-1}(x) = 1/\cos(x)$ in this notation. We shall use arccos(x) in this book, but you should be forwarned of the alternative notation $\cos^{-1}$ for arccos. Usually, it will be clear from the context whether $\cos^{-1}(x)$ is intended to mean arccos(x) or $(\cos(x))^{-1}$.

The Chain Rule And Compositional Inverses

There is a standard technique for finding the derivative of the compositional inverse h^{-1} given the derivative h'. We have that $h(h^{-1}(x)) = x$ for all x in the domain of h^{-1}. By the CHAIN RULE, $1 = (x)' = (h(h^{-1}(x)))' = h'(h^{-1}(x))(h^{-1}(x))'$. Thus, we have

2.39   BASIC RULE

$$(h^{-1}(x))' = \frac{1}{h'(h^{-1}(x))} .$$

Applying this rule to arccos(x) gives

$$(\arccos(x))' = \frac{1}{-\sin(\arccos(x))} \cdot$$

Look now at the small triangle near the graph of arccos(x) in FIGURE 2.40. From this triangle, it is evident that $y = \arccos(x) = \arcsin((1 - x^2)^{1/2})$. Thus, $\sin(\arccos(x)) = \sin(\arcsin((1 - x^2)^{1/2})) = (1 - x^2)^{1/2}$. This gives

$$(\arccos(x))' = \frac{-1}{(1 - x^2)^{1/2}} \cdot$$

Using the small triangles in FIGURE 2.40 together with BASIC RULE 2.39 we get the following rules for differentiating the compositional inverses of the trigonometric functions.

FIGURE 2.40   Basic Inverse Trigonometric Functions

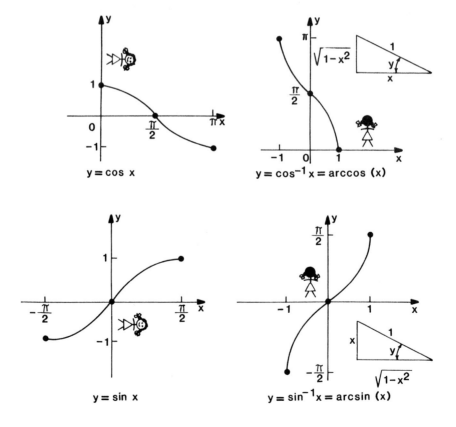

FIGURE 2.40   Continued

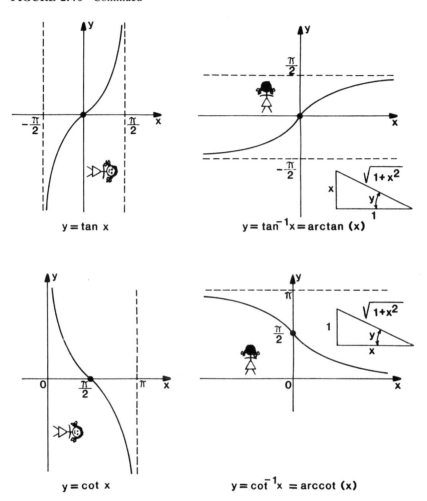

$y = \tan x$

$y = \tan^{-1}x = \arctan\ (x)$

$y = \cot x$

$y = \cot^{-1}x\ = \text{arccot}\ (x)$

Memorize!

2.41   DIFFERENTIATION RULES FOR INVERSE TRIG FUNCTIONS

$$(\arcsin(x))' \ = \ \frac{1}{(1 \ - \ x^2)^{1/2}} \qquad (\arccos(x))' \ = \ \frac{-1}{(1 \ - \ x^2)^{1/2}}$$

$$(\arctan(x))' \ = \ \frac{1}{1 \ + \ x^2} \qquad (\text{arccot}(x))' \ = \ \frac{-1}{1 \ + \ x^2}$$

$$(\text{arcsec}(x))' = \frac{1}{x(x^2 - 1)^{1/2}} \qquad (\text{arccsc}(x))' = \frac{-1}{x(x^2 - 1)^{1/2}}$$

The latter two rules of DIFFERENTIATION RULES 2.41 are the rules of differentiation for the compositional inverses of the secant and cosecant. The graphs of these functions are shown in FIGURE 2.43.

2.42 EXERCISES

(1) Derive all of the DIFFERENTIATION RULES 2.41 that we did not derive in the text.

(2) In EXERCISES 2.23 and the VARIATIONS on these exercises, replace sin, cos, tan, cot, sec, and csc by arcsin, arccos, arctan, arccot, arcsec, and

FIGURE 2.43 Inverse secant and cosecant

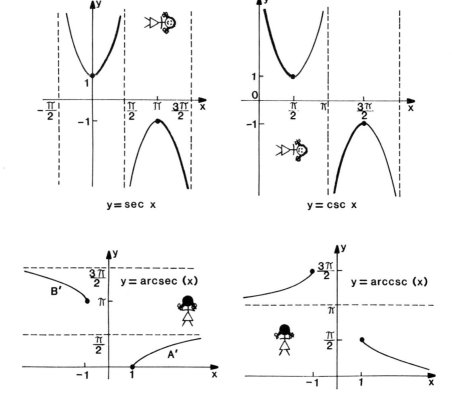

arccsc respectively. Differentiate the resulting functions and compare your answers with the answers to the original problems.

The Hyperbolic Functions

Our final class of functions is the class of "hyperbolic" functions. These functions will not be of any great importance to us in the sequel. They do occur naturally in some applied problems involving calculus (hanging cables and ocean waves are two classical examples). We give a brief description of these functions and some of their properties at this point to guard the reader against an anxiety attack in the event they crop up in some later course.

2.44 DEFINITION (hyperbolic functions)

$$\sinh(x) = \frac{e^x - e^{-x}}{2}$$

$$\cosh(x) = \frac{e^x + e^{-x}}{2}$$

$$\tanh(x) = \frac{e^x - e^{-x}}{e^x + e^{-x}} = \frac{\sinh(x)}{\cosh(x)}$$

In addition to these basic functions we have

$$\coth(x) = \frac{1}{\tanh(x)} \quad \sech(x) = \frac{1}{\cosh(x)} \quad \csch(x) = \frac{1}{\sinh(x)}.$$

The function $\sinh(x)$ is pronounced "sinch of x" and $\cosh(x)$ is pronounced "kohsch of x." Alternatively, one may just say "hyperbolic sin of x" for $\sinh(x)$" or "hyperbolic cosine of x" for $\cosh(x)$. This seems to be the form for $\tanh(x)$, $\coth(x)$, $\sech(x)$, and $\csch(x)$ where one says, "hyperbolic tangent of x," hyperbolic cotangent of x," etc. The differentiation rules for these functions are trivial to derive:

2.45 DIFFERENTIATION RULES FOR HYPERBOLIC FUNCTIONS

$$\frac{d}{dx}\sinh(x) = \cosh(x) \qquad \frac{d}{dx}\cosh(x) = \sinh(x)$$

$$\frac{d}{dx}\tanh(x) = \sech^2(x) \qquad \frac{d}{dx}\coth(x) = -\csch^2(x)$$

$$\frac{d}{dx}\text{sech}(x) = -\text{sech}(x)\tanh(x) \qquad \frac{d}{dx}\text{csch}(x) = -\text{csch}(x)\coth(x)$$

Graphs of the hyperbolic functions are shown in FIGURE 2.46.

Several questions usually occur to the student at this point. One is "Why are these functions called 'hyperbolic'?" If you were to graph all points in the plane of the form $(x,y) = (\cosh(t),\sinh(t))$ as t ranges over all real numbers, you would obtain the graph of a curve called a *hyperbola*. The points (x,y) on this curve satisfy $x^2 - y^2 = 1$ because $\cosh^2(t) - \sinh^2(t) = 1$. A second, more practical, question is usually "How much of this stuff should we memorize?" If you have a good memory for this sort of stuff then memorize it all, as the more you know the better. There is a danger for the beginning student of getting confused between the differentiation rules for the trigonometric functions and the much less important hyperbolic functions. If you feel this might be a problem for you, then memorize just the definitions of the hyperbolic functions (DEFINITION 2.44) and the rules for differentiating the sinh, cosh, and tanh. Note that these rules are the same as for the sin, cos, and tan, except that $(\cos(x))' = -\sin(x)$ but $(\cosh(x))' = +\sinh(x)$.

Just as with the trigonometric functions, one can define the compositional inverses of the hyperbolic functions. These functions can be expressed directly

FIGURE 2.46   Graphs of the Hyperbolic Functions

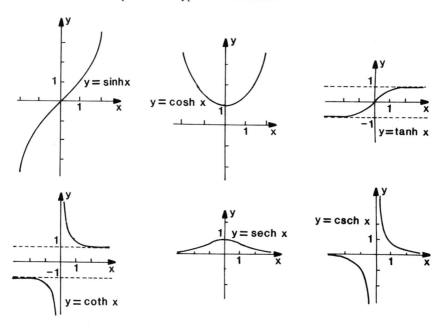

in terms of the function ln (log base e). Because of this, the derivatives of these functions can be computed directly by differentiating ln and applying the CHAIN RULE. We conclude by stating these results.

2.47 INVERSE HYPERBOLIC FUNCTIONS

$$\text{arcsinh}(x) = \ln(x + (1 + x^2)^{1/2})$$

$$\text{arccosh}(x) = \ln(x + (x^2 - 1)^{1/2}) \ (x \geq 1)$$

$$\text{arctanh}(x) = (1/2)\ln\frac{1 + x}{1 - x} \ (x^2 < 1)$$

$$\text{arccoth}(x) = (1/2)\ln\frac{x + 1}{x - 1} \ (x^2 > 1)$$

$$\text{arcsech}(x) = \ln\left(\frac{1 + (1 - x^2)^{1/2}}{x}\right) \ (0 < x \leq 1)$$

$$\text{arccosh}(x) = \ln\left(\frac{1}{x} + \frac{(1 + x^2)^{1/2}}{|x|}\right) \ (x \neq 0)$$

2.48 DIFFERENTIATION RULES FOR INVERSE HYPERBOLIC FUNCTIONS

$$\frac{d}{dx}\text{arcsinh}(x) = \frac{1}{(1 + x^2)^{1/2}}$$

$$\frac{d}{dx}\text{arccosh}(x) = \frac{1}{(x^2 - 1)^{1/2}} \ (x > 1)$$

$$\frac{d}{dx}\text{arctanh}(x) = \frac{1}{1 - x^2} \ (x^2 < 1)$$

$$\frac{d}{dx}\text{arccoth}(x) = \frac{1}{1 - x^2} \ (x^2 > 1)$$

$$\frac{d}{dx}\text{arcsech}(x) = \frac{-1}{x(1 - x^2)^{1/2}} \ (0 < x < 1)$$

$$\frac{d}{dx}\text{arccsch}(x) = \frac{-1}{|x|(1 + x^2)^{1/2}} \ (x \neq 0)$$

Are you tempted to say that arctanh(x) and arccoth(x) have the same derivative? If so, reread the discussion of functions prior to FIGURE 2.36, noting that the domain for the derivative of arctanh(x) is all x with $x^2 < 1$ and the domain of the derivative of arccoth(x) is all x with $x^2 > 1$.

Chapter 3

APPLICATIONS OF DERIVATIVES

In this chapter we shall take a look at various applications of derivatives. We are, however, going to be a little more critical of these "applications" than the standard calculus textbook. To begin with, we look at the problem of graphing functions. Let's take as our first example the function $y(x) = x - (1/3)x^3$. We want to sketch the graph of this function. In what sense is this an application of derivatives? It seems to be a simple problem for a hand calculator or a computer and not really a problem for calculus as advertised in the calculus textbooks. We shall see that calculus really plays a very minor role in these graphing problems. The best strategy is to avoid calculus as much as possible in these problems and only use derivatives to "fine tune" your results. This fine tuning can be important in some problems and there we will see calculus concepts coming into play in interesting ways.

The Inventors Of Calculus Didn't Have Computers—You Do!

In PROGRAMS 3.2, we see two programs in BASIC to compute values of the function $y = x - (1/3)x^3$. PROGRAM 3.2(a) gives a sequence of points on the graph of y. Statement 10 in this program records what the function is. This is a good idea in case you have been doing several functions and have a number of these tables of values lying around on your desk. Statement 20 generates the values of the dependent variable between -2 and $+2$ in steps of .2. In statement 30, "(" prints a left parenthesis, X prints the value of the dependent variable, "," prints a comma, X − (X^3)/3 prints the value of the function, and ")" prints the closing parenthesis. The

98

BASIC program of PROGRAM 3.2(a) was executed on a good but very inexpensive computer and does not involve any formatting. Any computer that implements BASIC should be able to do at least this well. PROGRAM 3.2(b) is done in a little bit more sophisticated version of BASIC which allows the PRINT USING statement for formatted output. Computers are so common in our society at this point that any student can gain access to one or buy one for very little expense. The language BASIC can be learned to the point required by PROGRAMS 3.2 in less than 30 minutes. We assume in our discussion of graphing of functions that the reader has taken the trouble to adventure at least this far into the world of computers. The function y = x − (1/3)x³ and its derivative y′ = 1 − x² have been graphed in FIGURE 3.1.

FIGURE 3.1 Graphs of $y = x - (1/3)x^3$ and $y' = 1 - x^2$

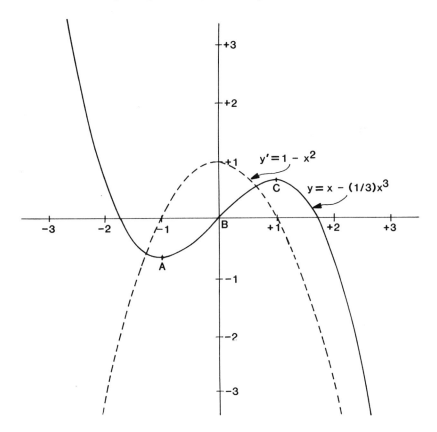

3.2 PROGRAMS FOR GRAPHING $x - (1/3)x^3$

(a)

```
10 PRINT "(X,X – (X^3)/3)"
20 FOR X = – 2 TO 2 STEP .2
30 PRINT "("X","X – (X^3)/3")"
40 NEXT X
```

$(X,X - (X^3)/3)$

(– 2 , .666666667)
(– 1.8 , .144)
(– 1.6 , – .234666667)
(– 1.4 , – .485333333)
(– 1.2 , – .624)
(– .999999999 , – .666666666)
(– .799999999 , – .629333333)
(– .599999999 , – .528)
(– .399999999 , – .378666666)
(– .199999999 , – .197333333)
(5.82076609E – 10 , 5.82076609E – 10)
(.200000001 , .197333334)
(.400000001 , .378666667)
(.600000001 , .528)
(.800000001 , .629333333)
(1 , .666666666)
(1.2 , .623999999)
(1.4 , .485333331)
(1.6 , .234666663)
(1.8 , – .144000005)

(b)

```
10 PRINT "      X";"      X – (X^3)/3"
20 FOR X = – 2 TO 2 STEP .2
30 PRINT USING " + ##.## ";X,X – (X^3)/3
40 NEXT X
```

X	X – (X^3)/3
– 2.00	+ 0.67
– 1.80	+ 0.14
– 1.60	– 0.23
– 1.40	– 0.49
– 1.20	– 0.62
– 1.00	– 0.67
– 0.80	– 0.63

X	X − (X^3)/3
− 0.60	− 0.53
− 0.40	− 0.38
− 0.20	− 0.20
+ 0.00	+ 0.00
+ 0.20	+ 0.20
+ 0.40	+ 0.38
+ 0.60	+ 0.53
+ 0.80	+ 0.63
+ 1.00	+ 0.67
+ 1.20	+ 0.62
+ 1.40	+ 0.49
+ 1.60	+ 0.23
+ 1.80	− 0.14

Local Minima And Maxima

Take a look now at FIGURE 3.1. Three points, $A = (-1, -2/3)$, $B = (0, 0)$, and $C = (1, 2/3)$ are shown on the graph of $y(x) = x - (1/3)x^3$. The point A is called a "local minimum" of y, the point B is called a "point of inflection" of y, and the point C is called a "local maximum" of y. We shall now discuss these terms and what they mean in terms of derivatives. Notice that the point $A = (-1, -2/3)$, where the value of y is $-2/3$, is not the smallest value that the function $y(x) = x - (1/3)x^3$ ever assumes. For all values of $x > 2$, $y(x) < -2/3$. We know that because $y(2) = -2/3$ and it is apparent from the graph of y that y decreases as x increases for values of $x > 1$. How can we be *absolutely sure* that y decreases for all values of $x > 1$? We might worry that for x large, say $x > 100$, the function starts to increase again! First of all, a little common sense will convince us that such a situation will not occur. The term x^3 gets large much faster than the term x so that once the term $(1/3)x^3$ begins to get larger than x for positive x, the whole expression $x - (1/3)x^3$ continues to decrease. At this point, if you still need reassuring you can use a little calculus. The derivative function $y'(x) = 1 - x^2$ is shown in FIGURE 3.1. Note that for $x > 1$, $1 - x^2 < 0$. A function with a negative derivative at a point is decreasing. Thus, $y(x) = x - (1/3)x^3$ is decreasing for all values of $x > 1$.

Even though there are infinitely many values of x for which $y(x) < y(-1) = -2/3$, we say that y has a "local minimum at $x = -1$" or "local minimum at $A = (-1, -2/3)$" because the value $y(-1) = -2/3$ is minimal in the vicinity of A. In other words, for some interval of values of x, centered at $x = -1$, the value $y(-1) = -2/3$ is the minimal value of f. Similarly, the function y has a "local maximum" at $x = +1$ (at the point C).

Critical Points

How did we know that the local minimum of y was *exactly* at x $= -1$? Just looking at the graph of y in FIGURE 3.1, we really can't be sure that the local minimum of y near x $= -1$ doesn't actually occur at x $= -1.001$, say. Here calculus can help. We compute $y'(x) = 1 - x^2$. Setting $y'(x) = 0$ and solving for x we get x $= +1$ or x $= -1$. The values of x where $y'(x) = 0$ are called the "critical points" of y(x). Since $y'(-1.001)$ is not zero, -1.001 is not a critical point of y(x). It should be obvious to you at this point that if x is not a critical point of y then x is not a local maximum or minimum. Thus, x $= -1$ is the only critical point in its immediate vicinity and hence x $= -1$ must be exactly the local minimum. In general, for differentiable functions y(x), the local maxima and minima will be found among the critical points.

Weren't we lucky that we were easily able to solve for the values of x such that $y'(x) = 0$! That is, we were easily able to find the critical points of y(x) by solving the simple quadratic equation $1 - x^2 = 0$. Since every student learns to solve a quadratic $ax^2 + bx + c = 0$ in high school algebra, calculus problems are frequently concocted so that $y'(x)$ is a quadratic equation. In the real world, things are not usually this nice. If $y'(x)$ is a polynomial of degree 3 or higher or some horrible trigonometric or logarithmic function, then it may not be so easy to locate the critical points other than by some graphical or numerical method involving a computer. As we have already remarked, the other critical point, x $= +1$, is a local maximum. What about the mysterious point B of FIGURE 3.1?

Points Of Inflection

To understand what is special about B, the so-called "point of inflection," it is necessary to look at the graph of y' shown also in FIGURE 3.1. Note that for values of x < 0, y' has been increasing. At x $= 0$, y' starts to decrease. A value of x where y' changes from increasing to decreasing or from decreasing to increasing is called a "point of inflection" of y. One also says "(x, y(x)) is a point of inflection of y." Thus, we might say "B $=$ (0, 0) is a point of inflection of y $= x - (1/3)x^3$" or "0 is a point of inflection of y $= x - (1/3)x^3$." By definition, a point of inflection x of y(x) will be a local maximum or minimum of y'. If y' itself has a derivative, y" (the second derivative of y), then $y''(x) = 0$ at a point of inflection. This can sometimes be a help in pinning down the exact location of a point of inflection. As with critical points, the usefulness of the "second derivative tests" has been exaggerated by most calculus texts.

We now consider some additional examples of graphing functions.

3.3 EXAMPLE Graph the function $f(x) = 1 + 2x + 3x^2 + 4x^3 + 5x^4$ and find the critical points.

Use Calculus To Fine-Tune Graphical Information

Following the precept to avoid calculus if at all possible, we wrote the two BASIC programs of PROGRAM SEQUENCE 3.4(a) and (b). PROGRAM 3.4(a) gives a general feeling for the graph of the function $1 + 2x + 3x^2 + 4x^3 + 5x^4$. Note that this function gets very large for large negative and positive values of x; in fact already at $x = -3$ the function value is 319 and for $x = +3$ the function value is 547. PROGRAM 3.4(a) seems to indicate that f(x) has a single minimum in the vicinity of $x = -.3$. In PROGRAM 3.4(b) we take a more careful look at f(x) in the vicinity of $x = -.3$. It seems that $-.44$ is quite close to the value of x where f(x) assumes its minimum value. We could continue in this manner without using any calculus to try and locate the value of x where f(x) assumes its minimum. The reader should attempt this on his or her computer. You will discover in attempting this that it becomes quite difficult to gain additional accuracy in locating the minimum. A more sensitive way to locate the minimum is to pass to the derivative $f'(x) = 2 + 6x + 12x^2 + 20x^3$ and try to find where $f'(x) = 0$ in the vicinity of $-.44$. There are fairly sophisticated polynomial root finding subroutines available on many computers but a straightforward approach works fine for us. PROGRAMS 3.4(c), (d), and (e) search directly for the root of $f'(x)$ near .44 and end up with $-.43708$ where f(x) takes on the value of 0.547439. This is, to within the accuracy of interest to us, the minimum value of f(x). It is both an "absolute minimum" and a local minimum. By writing a program to evaluate $f'(x)$ from -3 to $+3$, such as PROGRAM 3.4(a) does for f(x), you can easily convince yourself that there are no other roots of $f'(x)$ than the one we have located near $-.43708$.

What about points of inflection of f(x)? The second derivative is $f''(x) = 6 + 24x + 60x^2$. By graphing this function, you can easily see that $f''(x)$ has no real roots. Thus there are no points of inflection for f(x). A common calculus problem is to specify a function and interval on which the function is to be considered. The student is asked to find the local maxima and minima of the function on that interval and to also find the absolute maxima and minima. Consider, for example, our function $f(x) = 1 + 2x + 3x^2 + 4x^3 + 5x^4$ and the interval $-3 \le x \le +3$. This interval is also written $[-3, +3]$. From our above discussion, we have that there are no local maxima of f(x) in $[-3, +3]$. The maximum value is unique (i.e., there is only one x in $[-3, +3]$ such that f(x) is maximal) and this occurs at x = +3 where f(+3) = 547. There is one local minima at x = -.43708 where f(x) = .547439. This is also the absolute minima. At this point the reader should take some reasonable interval, say $[-1, +1]$, and actually draw the graph of f(x).

3.4 PROGRAM SEQUENCE FOR $f(x) = 1 + 2x + 3x^2 + 4x^3 + 5x^4$

(a)

```
10 FOR X = -3 TO 3 STEP .3
20 PRINT X,1 + 2*X + 3*X^2 + 4*X^3 + 5*X^4
30 NEXT X
```

READY.

X	$1 + 2x + 3x^2 + 4x^3 + 5x^4$
−3	319
−2.7	204.458501
−2.4	124.072
−2.1	70.2265003
−1.8	36.2800001
−1.5	16.5625
−1.2	6.37600003
− .900000001	1.99450001
− .600000001	.664000002
− .300000001 ←	.6025
−9.31322575E-10	.999999998
.299999999	2.0185
.599999999	4.79199999
.899999999	11.4265
1.2	25
1.5	49.5625
1.8	90.136
2.1	152.7145
2.4	244.264
2.7	372.7225
3	547

(b)

```
10 FOR X = -.45 TO -.35 STEP .01
20 PRINT X,1 + 2*X + 3*X^2 + 4*X^3 + 5*X^4
30 NEXT X
```

READY.

X	$1 + 2x + 3x^2 + 4x^3 + 5x^4$
−.45	.54803125
−.44 ←	.5474688

x	$1 + 2x + 3x^2 + 4x^3 + 5x^4$
−.43	.54761205
−.42	.5484328
−.41	.54990405
−.4	.552
−.39	.55469605
−.38	.5579688
−.37	.56179605
−.36	.5661568

(c)

```
10 FOR X = − .44 TO − .43 STEP .001
20 PRINT X,2 + 6*X + 12*X^2 + 20*X^3
30 NEXT X
```

READY.

x	$2 + 6x + 12x^2 + 20x^3$
−.44	− .0204800006
−.439	− .0134383801
−.438	− 6.42544015E-03
−.437 ←	5.58940577E-04
−.436	7.51488047E-03
−.435	.0144425011
−.434	.021341921
−.433	.0282132618
−.432	.0350566421
−.431	.0418721824

(d)

```
10 FOR X = − .438 TO − .437 STEP .0001
20 PRINT X,2 + 6*X + 12*X^2 + 20*X^3
30 NEXT X
```

READY.

x	$2 + 6x + 12x^2 + 20x^3$
−.438	− 6.42543909E-03
−.4379	− 5.7257183E-03
−.4378	− 5.02628303E-03
−.4377	− 4.32713272E-03
−.4376	− 3.62826793E-03
−.4375	− 2.92969014E-03
−.4374	− 2.23139368E-03
−.4373	− 1.53338422E-03
−.4372	− 8.35658761E-04
−.4371 ←	− 1.38218893E-04
−.437	5.58937114E-04

(e)

```
10 FOR X = -.4371 TO -.437 STEP .00001
20 PRINT X,2+6*X+12*X^2+20*X^3
30 NEXT X
```

READY.

x	$2 + 6x + 12x^2 + 20x^3$
-.4371	-1.38217398E-04
-.43709	-6.84889564E-05
-.43708 ←	1.23628342E-06
-.43707	7.0958904E-05
-.43706	1.40678323E-04
-.43705	2.10395389E-04
-.43704	2.80108838E-04
-.43703	3.49820082E-04
-.43702	4.19528591E-04
-.43701	4.89234011E-04
-.437000001	5.58936088E-04

Parameterized Families

In our next example, we see a situation where calculus is a major help beyond straight computational techniques. Instead of considering one function, we consider a "family of functions." Let $f_a(x) = a(x - a)^2$ where a is a real number. For each actual value of a we can write down a specific function of this type. For $a = 1$ we obtain $f_1(x) = x(x - 1)^2$ and for $a = 2$ we obtain $f_2(x) = 2x(x - 2)^2$. For $a = -3$ we obtain $f_{-3}(x) = -3x(x + 3)^2$. Graphs of $f_1(x)$ and $f_2(x)$ are shown in FIGURE 3.6. The "variable" a that is required to specify a function $f_a(x)$ is called a "parameter" of the family of functions. The parameter "ranges over all real numbers."

3.5 EXAMPLE Consider the parameterized family of functions $f_a(x) = ax(x - a)^2$ where the parameter a ranges over all real numbers. Find the local maxima, local minima, and points of inflection of the functions $f_a(x)$ *in terms of the parameter a when $a \neq 0$.*

Using the product rule, we compute that $f_a'(x) = a(x - a)^2 + (ax)2(x - a) = a(x - a)(3x - a)$. The second derivative is $a(6x - 4a)$. The critical points are, by definition, all points where $f_a'(x) = 0$. There are two such points, $x = a$ and $x = a/3$. We are lucky that it is so easy to solve the equation $f_a'(x) = 0$. It is easy to construct other examples where it is very difficult or impossible to solve the equation explicitly in terms of the param-

FIGURE 3.6 The Parameterized Family f_a

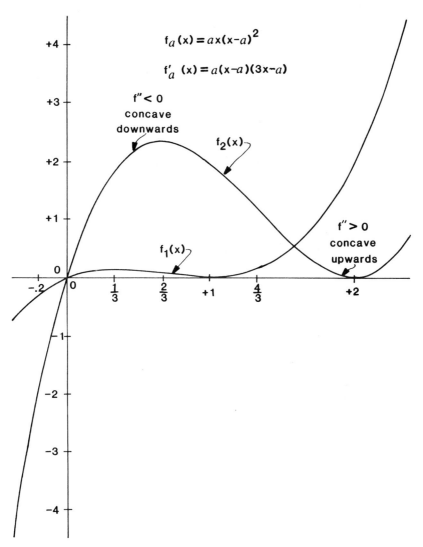

$$f_a(x) = ax(x-a)^2$$

$$f'_a(x) = a(x-a)(3x-a)$$

f″ < 0
concave
downwards

$f_2(x)$

f″ > 0
concave
upwards

$f_1(x)$

eter. We should graph a few of the functions $f_a(x)$ from our family of functions. We have graphed $f_1(x)$ and $f_2(x)$ in FIGURE 3.6. The reader should graph $f_a(x)$ for some negative value of a such as $a = -1$ or $a = -2$. Looking at the graphs of FIGURE 3.6 , it is apparent that $x = a/3$ corresponds to a local maximum and $x = a$ corresponds to a local minima. Neither points are absolute maxima or minima. The second derivative, $a(6x - 4a)$ is zero when $x = 2a/3$. Thus, $f_1(x)$ has a point of inflection at $x = 2/3$ and $f_2(x)$ has a

point of inflection at x = 4/3. It is clear from FIGURE 3.6 that the points
of inflection are about where calculus techniques tell us they are. It would
be awkward to locate these points of inflection without calculus and *very
awkward* to locate them as functions of the parameter *a* without calculus.

Second Derivative Test

There is another important fact that we can learn from FIGURE 3.6. The
function $f_2(x)$ has a local maximum at 2/3. Look at values of $f'_2(x)$ for values
of x starting a little bit less than 2/3 and ranging to values a little bit more
than 2/3. Note that $f'_2(x)$ is decreasing as x ranges through these values. This
means that the second derivative, $f''_2(2/3)$, is negative. The general rules are
as follows:

If t is a local maximum of a function f(x) then $f''(t) \le 0$.

If t is a local minimum of a function f(x) then $f''(t) \ge 0$.

An equivalent and more useful statement of these rules is given in SECOND
DERIVATIVE TEST 3.7.

3.7 SECOND DERIVATIVE TEST   Let f(x) be a function and suppose
that at x = t, $f'(t) = 0$. If $f''(t) < 0$ then f has a local maximum at t. If
$f''(t) > 0$ then f has a local minimum at t.

At $a/3$, $f''_a(a/3) = -2a^2$ is negative for all values of *a* and hence $a/3$ is
a local maximum. At *a*, $f''_a(a) = 2a^2$ is positive for all *a* and hence *a* is a
local minimum. These facts are obvious from the graphs of FIGURE 3.6.
For *any point* x, if $f''(x) < 0$ then f is said to be "concave downwards" at
x and if $f''(x) > 0$ then f is said to be "concave upwards" at x. Referring
to FIGURE 3.6, f_2 is concave downward at x = .5 and concave upwards at
x = 1.8.

We now review the basic principles of graphing functions.

3.8   REVIEW OF GRAPHING FUNCTIONS

Compute Enough Values To Know The General Shape

(1)   Given a function f(x) to be graphed, write a simple BASIC program
(or program in some other computer language) such as those of PROGRAMS
3.1 and use it to generate points on the graph of f(x). Sketch a graph of the
function to get a good feeling for its shape. In the case where you are dealing
with a parameterized family of functions, draw several graphs for represen-

tative choices of the parameter (as was done in FIGURE 3.6). When you have had some experience with this sort of problem you may not have to draw the graph but simply look at the list of points on the graph (i.e., the output of your program) to get all of the information you need at this stage.

Study The Critical Points

(2)   Next try to locate the local and absolute maxima and minima. The interval over which the function is defined will be important here as some local maxima and minima may not be included in the interval. Also, an absolute maximum and/or minimum may occur at the endpoints of the interval. If a local maximum or minimum of a function $f(x)$ occurs at a point x where $f'(x)$ is defined, then $f'(x) = 0$ at this point x. This fact can be a help in locating these local maxima and minima. The points x where $f'(x) = 0$ are called the *critical points* of f. Just because $f'(x) = 0$ it does not necessarily follow that x is a local maximum or minimum of $f(x)$. For example, take $f(x) = x^3$ and $x = 0$. If you are dealing with a parameterized family of functions, try to express the critical points, local maxima, and local minima in terms of the parameter or parameters as was done in EXAMPLE 3.3. This may not always be possible as it may be very difficult to solve the equation $f'(x) = 0$ in terms of the parameter. Even when no explicit solution can be given, there are sometimes more sophisticated techniques that can give useful information about the roots of $f'(x) = 0$ in terms of the parameters.

Find The Points Of Inflection

(3)   After steps (1) and (2) you should have a very good idea of what the graph of $f(x)$ looks like. You should now notice where the function is concave upwards and concave downwards. A function is concave upwards at x if the second derivative $f''(x) > 0$ and concave downwards at x if $f''(x) < 0$. Points x where $f(x)$ changes from being concave upwards to concave downwards or from concave downwards to concave upwards are called *points of inflection of f*. If x is a point of inflection of f and $f''(x)$ is defined, then $f''(x) = 0$. This fact can help in locating points of inflection precisely. Just because $f''(x) = 0$ doesn't mean that you have a point of inflection. For example, look at $f(x) = x^4$ and $x = 0$. If x is such that $f'(x) = 0$ and f is concave downwards at x (i.e., $f''(x) < 0$) then x is a local maximum of f. If $f'(x) = 0$ and f is concave upwards at x then x is a local minimum of f. These facts are the basis for SECOND DERIVATIVE TEST 3.7.

Taylor Polynomials, L'Hopital's Rule

We shall conclude our discussion of graphing functions by giving a number of exercises. Before doing so, however, we introduce some additional ideas that will be helpful in solving these problems. These ideas, which are standard topics in calculus courses, are called "limits," "Taylor polynomials," and "L'Hopital's Rule." Here we emphasize intuitive ideas, returning for a more careful treatment in Chapter 5. If you take a more advanced course in calculus or in mathematical analysis you will see a "rigorous" treatment of these topics.

To illustrate these topics we shall study the following two examples:

$$s(x) = \frac{e^x - e^{-x}}{\sin(x)} \quad \text{and} \quad a(x) = \frac{\ln(x)}{|x - 1|}.$$

Values of $s(x)$ are computed in PROGRAM SEQUENCE 3.10(a) and (b). Values of $a(x)$ are computed in PROGRAM SEQUENCE 3.10(c) and (d). The first thing to notice about these functions is that, at $x = 0$ for $s(x)$ and at $x = 1$ for $a(x)$, the numerators and denominators are zero. This is why statement 10 in each of the four programs of PROGRAM SEQUENCE 3.10 is constructed so that the values $x = 0$, for $s(x)$, and $x = 1$, for $a(x)$, are not generated. The function $s(x)$ is a very crazy function which undergoes wild oscillations as x gets larger and larger ("goes to infinity") or smaller and smaller ("goes to minus infinity"). These oscillations are caused by the oscillations of $\sin(x)$ in the denominator.

At the moment, we are concerned only about the behavior of $s(x)$ near x = 0 and $a(x)$ near x = 1. In PROGRAM SEQUENCE 3.10(a) we have generated some values of $s(x)$ in the interval $[-3, +3]$. Notice that as x approaches 0 through negative values, the values of $s(x)$ approach 2. Likewise, as x approaches 0 through positive values, the values of $s(x)$ approach 2. This should be compared with the behavior of $a(x)$ near x = 1 as shown in PROGRAM SEQUENCE 3.10(c). As x approaches 1 through values less than 1, the values of $a(x)$ approach -1. As x approaches 1 through values greater than 1, the values of $a(x)$ approach $+1$. Alternatively, we say "the limit of $s(x)$ as x approaches 0 from below is 2" or "the limit of $a(x)$ as x approaches 1 from above is $+1$." Symbolically, these statements are made as in LIMIT EXPRESSIONS 3.9.

3.9 LIMIT EXPRESSIONS

$$\lim_{x \to 0-} s(x) = +2 \quad \text{and} \quad \lim_{x \to 0+} s(x) = +2$$

$$\lim_{x \to 1^-} a(x) = -1 \quad \text{and} \quad \lim_{x \to 1^+} a(x) = +1$$

$$\text{where } s(x) = \frac{e^x - e^{-x}}{\sin(x)} \quad \text{and} \quad a(x) = \frac{\ln(x)}{|x - 1|}$$

3.10   PROGRAM SEQUENCE

(a)

```
10 FOR X = -3 TO +3 STEP .21
20 PRINT X,(EXP(X) - EXP( - X))/SIN(X)
30 NEXT X
```

x			s(x)
−3			141.976678
−2.79			47.0961274
−2.58			24.6394442
−2.37			15.2075866
−2.16			10.2910545
−1.95			7.41303783
−1.74			5.60182164
−1.53			4.40530662
−1.32			3.58855362
−1.11			3.01976488
− .9			2.62091526
− .69			2.3441513
− .48			2.15967082
− .27			2.04919586
− .0600000005	←	→	2.00240144
.149999999	←	→	2.01505641
.359999999			2.08829661
.569999999			2.22881798
.779999999			2.45004504
.989999999			2.77462404
1.2			3.23905162
1.41			3.90214945
1.62			4.86107476
1.83			6.28337298
2.04			8.47666511
2.25			12.0584241
2.46			18.4425582
2.67			31.6322641
2.88			68.6651573

(b)

```
10 FOR X = -.0055 TO .0055 STEP .001
20 PRINT X,(EXP(X) - EXP( - X))/SIN(X)
30 NEXT X
```

x	s(x)
-5.5E-03	2.00002015
-4.5E-03	2.00001349
-3.5E-03	2.00000814
-2.5E-03	2.00000412
-1.5E-03	2.00000144
-5E-04 ← →	2.00000027
5E-04 ← →	1.99999983
1.5E-03	2.0000016
2.5E-03	2.00000422
3.5E-03	2.00000822
4.5E-03	2.00001345
5.5E-03	2.00002012

(c)

```
10 FOR X = +0.15 TO 2 STEP 0.1
20 PRINT X,LOG(X)/ABS(X - 1)
30 NEXT X
```

x	a(x)
.15	-2.23190587
.25	-1.84839248
.35	-1.61511096
.45	-1.45183218
.55	-1.32852667
.65	-1.23080833
.75	-1.15072829
.85	-1.08345953
.95 ← →	-1.02586588
1.05 ← →	.975803296
1.15	.931746283
1.25	.892574207
1.35	.857441693
1.45	.825696793
1.55	.796827147
1.65	.77042352
1.75	.746154384
1.85	.723747811
1.95	.702978287

(d)

```
10 FOR X = +0.99 TO +1.01 STEP 0.0011
20 PRINT X,LOG(X)/ABS(X – 1)
30 NEXT X
```

x	a(x)
.99	– 1.00503356
.9911	– 1.00447657
.9922	– 1.00392038
.9933	– 1.00336501
.9944	– 1.00281047
.9955	– 1.00225676
.9966	– 1.00170385
.9977	– 1.00115177
.9988	– 1.0006005
.9999 ←	→ – 1.00005084
1.001 ←	→ .99950097
1.0021	.998951765
1.0032	.998403555
1.0043	.997856251
1.0054	.997309778
1.0065	.996764087
1.0076	.99621922
1.0087	.995675124
1.0098	.995131837

In our discussion of LIMIT EXPRESSIONS 3.9 we did not define the concept of a limit precisely. A precise definition of a limit is given in CHAPTER 5 (DEFINITION 5.66). The intuitive idea of a limit as expressed in our previous discussion is important, however. The validity of the assertions of LIMIT EXPRESSIONS 3.9 is, for the moment, based solely on the computer data of PROGRAM SEQUENCE 3.10. From PROGRAM SEQUENCE 3.10(a) we see that $s(-.06) = 2.002$ and $s(+.15) = 2.02$. In PROGRAM SEQUENCE 3.10(b) we take a closer look at what is happening to $s(x)$ near $x = 0$. There we look at the values of $s(x)$ for x ranging from $-.0055$ (or $-5.5E\text{-}03$ in scientific notation) to $+.0055$. We find that $s(-.0005) = 2.00000027$ and $s(+.0005) = 1.99999983$. If you try these computations on your computer, you might get slightly different answers due to different methods of computing the functions $\sin(x)$ and e^x. The message will be the same, however, that

$$\lim_{x \to 0-} s(x) = +2 \quad \text{and} \quad \lim_{x \to 0+} s(x) = +2.$$

The numerator of s(x) is $e^x - e^{-x}$ and the denominator is sin(x). Both of these functions are zero when x = 0. For this reason we say that s(x) is "not defined" for x = 0. If we want to define s(x) for x = 0 the natural choice is s(0) = +2 since the limits from above and below are both equal to +2 at x = 0. Formally, we could write

$$s(x) = \begin{cases} \dfrac{e^x - e^{-x}}{\sin(x)} & \text{if } x \neq 0 \\ +2 & \text{if } x = 0 \end{cases}$$

The following definition will be useful in discussing this expanded definition of s(x):

3.11 DEFINITION   Let f(x) be a function. Suppose that for x = t we have

$$\lim_{x \to t-} f(x) = r \quad \text{and} \quad \lim_{x \to t+} f(x) = r.$$

Then we say that "the limit of f(x) as x approaches t is r" and write

$$\lim_{x \to t} f(x) = r.$$

The idea of DEFINITION 3.11 is that if both the limit from above and the limit from below exist at a point t and are equal to the same number r then that value r is simply called *the limit* at t. Thus for our function s(x),

$$\lim_{x \to 0} s(x) = +2.$$

This leads to the definition of continuous functions.

Continuous Functions

3.12 DEFINITION   Let f(x) be a function. Suppose that at x = t, f(t) is defined and

$$\lim_{x \to t} f(x) = f(t).$$

Then we say that f(x) is *continuous* at x = t.

The intuitive idea of DEFINITION 3.12 is that the graph of a continuous function has no jumps or tears in it. In other words, you can draw the graph without taking your pencil off the paper. To better understand this concept, we now take a closer look at the function a(x) which, as we have already noted informally in connection with LIMIT EXPRESSIONS 3.9, is *not continuous* (or "discontinuous") at x = 1.

PROGRAM SEQUENCE 3.10(c) gives values of $a(x) = (\ln(x))/|x - 1|$ for various values of x between x = 0.15 and x = 2. The value x = 1 is not included here as both the numerator and denominator of a(x) vanish at x = 0. By inspecting these values, one becomes strongly suspicious that

$$\lim_{x \to 1-} a(x) = -1 \quad \text{and} \quad \lim_{x \to 1+} a(x) = +1.$$

This suspicion is confirmed by PROGRAM SEQUENCE 3.10(d), where we take a closer look at values of a(x) near x = 1. This means that the graph of a(x) takes a sudden jump as x goes past the point x = 1. We could define a(1) to be something, just as we assigned a value to s(0) above. If we define a(1) = -1 then a(x) would be "continuous from below" or "left continuous" at x = 0. If we define a(1) = +1 then a(x) would be "continuous from above" or "right continuous" at x = 1. No assignment of a value to a(1) can make the function a(x) continuous at x = 1 in the sense of DEFINITION 3.12, however.

Limit As x Approaches Infinity

In discussing limits, $\lim_{x \to t} f(x)$, we have assumed that t is a real number. We can also discuss $\lim_{x \to +\infty} f(x)$ where the symbol $+\infty$ stands for "plus infinity." We can also look at $\lim_{x \to -\infty} f(x)$ where $-\infty$ stands for "minus infinity." To say that

$$\lim_{x \to +\infty} f(x) = L$$

means that as x gets larger and larger the values of f(x) get closer and closer to L. As an example, consider $f(x) = (2x + x^{-1})/(4x + x^{-1})$. A BASIC program to evaluate the limit of f(x) as x gets larger and larger is given in PROGRAM 3.13. This program contains an infinite loop and execution must be aborted by the user. It is evident that

$$\lim_{x \to +\infty} f(x) = .5.$$

It is obvious from the definition of f(x) that f(x) tends to .5 as x tends to plus infinity. The term x^{-1} goes to zero as x gets large. Thus for large values of x, f(x) is essentially equal to 2x/4x, which is .5.

3.13   PROGRAM FOR AN INFINITE LIMIT

```
10 X = 1000
20 PRINT (2*X + X^ − 1)/(4*X + X^ − 1)
30 X = X + 1000
40 GOTO 20
```

READY.

```
.500000125
.500000016
.500000007
.500000004
.500000002
.500000002
.500000001
.500000001
.500000001
.500000001
.5
.5
.5
.5
.5
.5
.5
.5
.5
.5
.5
.5
```

If the reader will think about the type of calculations used to evaluate limits, namely those of PROGRAM SEQUENCE 3.10 and PROGRAM 3.13, it will be obvious that the following rules for computing limits are valid:

3.14   RULES FOR LIMITS   For any real number r and any functions $f(x)$ and $g(x)$

(1) $\displaystyle\lim_{x \to t} rf(x) = r(\lim_{x \to t} f(x))$

(2) $\displaystyle\lim_{x \to t} (f(x) + g(x)) = \lim_{x \to t} f(x) + \lim_{x \to t} g(x)$

(3) $\displaystyle\lim_{x \to t} (f(x)g(x)) = (\lim_{x \to t} f(x))(\lim_{x \to t} g(x))$

$$(4) \quad \lim_{x \to t} \frac{f(x)}{g(x)} = \frac{\displaystyle\lim_{x \to t} f(x)}{\displaystyle\lim_{x \to t} g(x)} \cdot$$

In rule (4) we must have $\lim_{x \to t} g(x)$ nonzero. These rules apply if $\lim_{x \to t}$ is replaced by $\lim_{x \to t-}$ or $\lim_{x \to t+}$ and if t is plus or minus infinity.

You will find that you will apply the RULES FOR LIMITS 3.14 without even thinking about it. Notice how much simpler RULES 3.14(3) and (4) are than the corresponding rules for derivatives!

Indeterminant Forms

If we try to apply RULE 3.14(4) to either of the functions s(x) or a(x) of LIMIT EXPRESSIONS 3.9 we end up with no useful information. For example, apply RULE 3.14(4) to $a(x) = f(x)/g(x)$ where $f(x) = e^x - e^{-x}$ and $g(x) = |x - 1|$ (take t = 1):

$$\lim_{x \to 1-} \frac{f(x)}{g(x)} = \frac{\displaystyle\lim_{x \to 1-} f(x)}{\displaystyle\lim_{x \to 1-} g(x)} = \frac{0}{0} \cdot$$

The expression 0/0 is not defined. In words, we say that the limit from below of $f(x)/g(x)$ is "of the form 0/0." For this example, the limit from above of $f(x)$ is also of the form 0/0. Another example of this type of difficulty is obtained by taking $f(x)$ to be x^2 and $g(x)$ to be e^x. Again trying to apply RULE 3.14(4), now with $t = \infty$, we obtain

$$\lim_{x \to \infty} \frac{f(x)}{g(x)} = \frac{\displaystyle\lim_{x \to \infty} f(x)}{\displaystyle\lim_{x \to \infty} g(x)} = \frac{\infty}{\infty} \cdot$$

In the above case, we say that the limit as x approaches infinity of $f(x)/g(x)$ is of the form ∞/∞. We need not have $t = \infty$ to obtain a limit of the form ∞/∞. For example, let $f(x) = 1 + 2(x - 2)^{-3}$ and let $g(x) = 1 + 3(x - 2)^{-3}$. Then the limit as x approaches 2 of $f(x)/g(x)$ is of the form ∞/∞.

We now discuss some methods to deal with limits of the form 0/0 or ∞/∞. More examples will be given in the exercises. One simple way that works with ratios of polynomials (so called "rational functions") is to use some algebraic tricks to make the limiting value obvious when using rules such as 3.14(4). As a trivial example, consider $f(x)/g(x)$ where $f(x) = x^3$ and $g(x) = x^2$. Using 3.14(4), this is of the form 0/0 as x approaches 0 and of the form ∞/∞ as x approaches infinity. But, of course, we would just write

$f(x)/g(x) = x$. Then it is obvious that $f(x)/g(x)$ approaches 0 as x approaches 0 and approaches infinity (i.e., gets arbitrarily large) as x approaches infinity. A slightly more interesting example is the case $f(x) = 1 + 2(x - 2)^{-3}$ and $g(x) = 1 + 3(x - 2)^{-3}$. In this case we multiply numerator and denominator by $(x - 2)^3$ to obtain

$$\frac{f(x)}{g(x)} = \frac{(x - 2)^3 + 2}{(x - 2)^3 + 3}.$$

It is obvious from this expression for $f(x)/g(x)$ that the limit as x approaches 2 is 2/3. This type of trick works for ratios of polynomials but what about ratios such as those for $s(x)$ and $a(x)$ of LIMIT EXPRESSIONS 3.9? We have $s(x)$ as a ratio of $e^x - e^{-x}$ and $\sin(x)$ and neither of these functions are polynomials. A mathematician named Taylor (he died in 1731) discovered a way to replace such functions as $e^x - e^{-x}$ and $\sin(x)$ by polynomials. When this is done we can then use algebraic tricks to find limits. Before discussing Taylor's method, we discuss a useful related method known as L'Hopital's rule.

L'Hopital's Rule

3.15   L'HOPITAL'S RULE   For reasonable functions $f(x)$ and $g(x)$, if

$$\lim_{x \to t} \frac{f(x)}{g(x)} = \frac{0}{0} \text{ or } \frac{\infty}{\infty}$$

then

$$\lim_{x \to t} \frac{f(x)}{g(x)} = \lim_{x \to t} \frac{f'(x)}{g'(x)}.$$

Instead of $\lim_{x \to t}$ we could use $\lim_{x \to t-}$ or $\lim_{x \to t+}$.

You should, of course, wonder what is meant by "reasonable" in L'HOPITAL'S RULE 3.15. To learn more, read the article by A. E. Taylor referenced in Appendix 2. We won't bother with a precise definition of "reasonable" at this point. Our point of view is that we can use L'HOPITAL'S RULE to help us guess the limit, but we shall be a bit wary of our answer and, if concerned, check it against other evidence (such as a computer generation of points on the graph of $f(x)/g(x)$). We shall give an example of L'HOPITAL'S RULE failing, but first let's apply the rule to our functions $s(x)$ and $a(x)$ of LIMIT EXPRESSIONS 3.9.

Consider our function $s(x) = f(x)/g(x)$ where $f(x) = e^x - e^{-x}$ and $g(x) = \sin(x)$. Viewed in this way, $s(x)$ has the form $0/0$ as x approaches 0. Thus we can apply L'HOPITAL'S RULE with $f'(x) = e^x + e^{-x}$ and $g'(x) = \cos(x)$. By L'HOPITAL'S RULE

$$\lim_{x \to 0} \frac{e^x - e^{-x}}{\sin(x)} = \lim_{x \to 0} \frac{e^x + e^{-x}}{\cos(x)} = \frac{2}{1} = 2.$$

This result certainly checks with PROGRAM SEQUENCE 3.10(a) and (b), so $f(x) = e^x - e^{-x}$ and $g(x) = \sin(x)$ must be reasonable functions (it's true, they are!).

Now let's take a look at $a(x) = f(x)/g(x)$ where $f(x) = \ln(x)$ and $g(x) = |x - 1|$. With this choice of $f(x)$ and $g(x)$, $a(x)$ has the form $0/0$ as x approaches 1. We apply L'HOPITAL'S RULE with $f'(x) = 1/x$ and

$$g'(x) = \begin{cases} -1 \text{ if } x < 1 \\ +1 \text{ if } x > 1 \end{cases}.$$

By L'HOPITAL'S RULE, we find that for the limit from below

$$\lim_{x \to 1-} \frac{\ln(x)}{|x - 1|} = \lim_{x \to 1-} \frac{1/x}{-1} = \frac{+1}{-1} = -1$$

and for the limit from above

$$\lim_{x \to 1+} \frac{\ln(x)}{|x - 1|} = \lim_{x \to 1+} \frac{1/x}{+1} = \frac{+1}{+1} = +1.$$

These results check with what we found in PROGRAM SEQUENCE 3.10(c) and (d).

You May Have To Apply L'Hopital's Rule Many Times

As yet another example of L'HOPITAL'S RULE, consider $f(x)/g(x)$ where $f(x) = x^2$ and $g(x) = e^x$. As x approaches plus infinity, the limit is of the form ∞/∞. By L'HOPITAL'S RULE, we have

$$\lim_{x \to \infty} \frac{f(x)}{g(x)} = \lim_{x \to \infty} \frac{f'(x)}{g'(x)} = \lim_{x \to \infty} \frac{2x}{e^x}.$$

But, the latter limit is also of the form ∞/∞. We can apply L'HOPITAL'S RULE to $2x/e^x$ to get

$$\lim_{x \to \infty} \frac{2x}{e^x} = \lim_{x \to \infty} \frac{2}{e^x} = 0.$$

We thus conclude that

$$\lim_{x \to \infty} \frac{x^2}{e^x} = 0.$$

To reach this conclusion we had to apply L'HOPITAL'S RULE twice. If we had tried to find the limit as x approaches infinity of x^3/e^x using L'HOPITAL'S RULE we would end up applying the rule 3 times to conclude

$$\lim_{x \to \infty} \frac{x^3}{e^x} = 0.$$

In fact, by applying L'HOPITAL'S RULE n times to x^n/e^x (for n a fixed positive integer) we would find

$$\lim_{x \to \infty} \frac{x^n}{e^x} = 0.$$

Results like these will be considered in the exercises. We conclude our brief discussion of L'HOPITAL'S RULE with an example of the failure of L'HOPITAL'S RULE. This example can be skipped without affecting your ability to work the exercises that follow. Mathematically inclined students will rightfully insist on such an example.

There Are "Unreasonable" Functions

Consider $f(x)/g(x)$ where $f(x) = x^2\sin(1/x)$ and $g(x) = \sin(x)$. As x approaches zero, this expression becomes of the form 0/0. That $g(x) = \sin(x)$ approaches 0 as x approaches 0 is obvious. The function $f(x)$ is a bit crazy, however! Some values of $\sin(1/x)$ near $x = 0$ are shown in PROGRAM SEQUENCE 3.16(a). Of course, all values of $\sin(1/x)$ will be between -1 and $+1$, but notice how wildly the values of $\sin(1/x)$ oscillate in PROGRAM SEQUENCE 3.16(a). This is because as x approaches 0, $1/x$ goes to infinity, forcing $\sin(1/x)$ to oscillate infinitely often between -1 and $+1$. When we take the product $x^2\sin(1/x)$ these oscillations are forced to be smaller and smaller by the fact that x^2 goes to zero. Thus $f(x) = x^2\sin(1/x)$ approaches 0 as x approaches 0.

3.16   PROGRAM SEQUENCE FOR L'HOPITAL'S FAILURE

(a)

```
10 FOR X = - .01 TO + .01 STEP .0015
20 PRINT X,SIN(1/X)
30 NEXT X
```

READY.

x	$\sin(1/x)$
− .01	.506365668
− 8.5E-03	.986799078
− 7E-03	.996362208
− 5.5E-03	.384062462
− 4E-03	.970528058
− 2.5E-03	.850919195
− 1E-03	− .826878546
4.99999998E-04	.93003695
2E-03	− .467772183
3.5E-03	.169818632
5E-03	− .87329726
6.5E-03	.0917568965
8E-03	− .616040486
9.5E-03	− .999803908

(b)

```
10 FOR X = - .01 TO + .01 STEP .0015
20 PRINT X,((X^2)*SIN(1/X))/SIN(X)
30 NEXT X
```

READY.

x	$(x^2\sin(1/x)/\sin(x)$
− .01	− 5.06374108E-03
− 8.5E-03	− 8.38789317E-03
− 7E-03	− 6.97459243E-03
− 5.5E-03	− 2.11235419E-03
− 4E-03	− 3.8821226E-03
− 2.5E-03	− 2.12730021E-03
− 1E-03	8.26878689E-04
4.99999998E-04	4.650E-04
2E-03	− 9.35544989E-04
3.5E-03	5.94366426E-04

x	$(x^2\sin(1/x)/\sin(x))$
5E-03	−4.3665045E-03
6.5E-03	5.96424027E-04
8E-03	−4.92837645E-03
9.5E-03	−9.49828001E-03

(c)

```
10 FOR X = −.01 TO +.01 STEP .0015
20 PRINT X,COS(1/X)
30 NEXT X
```

READY.

x	$\cos(1/x)$
− .01	.862318856
−8.5E-03	− .161949312
−7E-03	− .0852194656
−5.5E-03	.923307113
−4E-03	.240988149
−2.5E-03	− .525296685
−1E-03	.562380848
4.99999998E-04	− .367466
2E-03	− .883849161
3.5E-03	− .985475317
5E-03	.487187742
6.5E-03	− .995781433
8E-03	.787714491
9.5E-03	.019802684

(d)

```
10 FOR X = −.01 TO +.01 STEP .0015
20 PRINT X,(2*X*SIN(1/X) − COS(1/X))/COS(X)
30 NEXT X
```

READY.

x	$(2x\sin(1/x) - \cos(1/x))/\cos(x)$
− .01	−.872489794
−8.5E-03	.145178972
−7E-03	.0712721408
−5.5E-03	−.92754583
−4E-03	−.248754363
−2.5E-03	.521043718
−1E-03	−.560727371

x	$(2x\sin(1/x) - \cos(1/x))/\cos(x)$
4.99999998E-04	.368396096
2E-03	.881979837
3.5E-03	.986670091
5E-03	$-.495926914$
6.5E-03	.996995335
8E-03	$-.797596662$
9.5E-03	$-.0388007091$

Now look at PROGRAM SEQUENCE 3.16(b) where we have generated values of $f(x)/g(x)$. These values seem to indicate that

$$\lim_{x \to 0} \frac{f(x)}{g(x)} = 0.$$

It is in fact true that the above limit is 0. But, if we try to apply L'HOPITAL'S RULE we compute $g'(x) = \cos(x)$ and $f'(x) = 2x\sin(1/x) - \cos(1/x)$. The function $\cos(1/x)$ that occurs in $f'(x)$ is another "bad actor" and oscillates wildly near zero as is shown in PROGRAM SEQUENCE 3.16(c). These oscillations cause $f'(x)/g'(x)$ to oscillate wildly also near zero (see PROGRAM SEQUENCE 3.16(d)) so that $\lim_{x \to 0} \dfrac{f'(x)}{g'(x)}$ does not exist. Thus, even though $f(x)/g(x)$ behaves well near $x = 0$ and has a limit there, L'HOPITAL'S RULE is no help in finding it. Basically, this is how all failures of L'HOPITAL'S RULE occur. If you are interested, you should look up the precise statement of L'HOPITAL'S RULE (see Appendix 2). You will see that the existence of the limit $f'(x)/g'(x)$ is the key assumption necessary for the rule to succeed. The correctly stated L'HOPITAL'S RULE never fails.

We have already remarked above that functions such as $\sin(x)$, $\cos(x)$, e^x, and $\ln(x)$ can be approximated by polynomials (called "Taylor Polynomials"). In TAYLOR POLYNOMIALS 3.17, we give such a list.

Taylor Polynomials

3.17   TAYLOR POLYNOMIALS   For x near zero we have

(1) $e^x = 1 + x + \dfrac{x^2}{2}$

(2) $\ln(1 + x) = x - \dfrac{x^2}{2} + \dfrac{x^3}{3}$

(3) $\sin(x) = x - \dfrac{x^3}{6}$

(4) $\cos(x) = 1 - \dfrac{x^2}{2}$

(5) $\tan(x) = x + \dfrac{x^3}{3}$

We are being a bit sloppy in TAYLOR POLYNOMIALS 3.17 because none of these statements are true equalities. They are only approximations. The reader should now take a close look at PROGRAM SEQUENCE 3.18 which shows how good these approximations are for various values of x near x = 0. Of course, when a computer evaluates a function like sin(x), it is programmed to make a certain approximation selected by an expert in the subject of numerical analysis (hopefully!). Thus, what we are doing is comparing our simple Taylor polynomials with these approximations. As you can see, the Taylor polynomials of PROGRAM SEQUENCE 3.18 are pretty good over the ranges chosen. You should try out different sequences of x near 0 on your own computer. For each of the functions f(x) of PROGRAM SEQUENCE 3.18, the general rule is to form the polynomial

$$p(x) = f(0) + f^{(1)}(0)x + \frac{f^{(2)}(0)}{2}x^2 + \frac{f^{(3)}(0)}{6}x^3.$$

In the above formula for p(x), the notation $f^{(n)}(0)$ means the nth derivative of f(x) evaluated at 0. In the case of e^x in TAYLOR POLYNOMIALS 3.17, we have left off the term involving x^3. One can compute Taylor polynomials of any degree for the functions f(x) above. In general, the coefficient of x^n is $f^{(n)}(0)/n!$ where $n! = n(n - 1)(n - 2) \ldots (3)(2)(1)$ is the product of the first n integers. The higher the degree of the polynomial the better it approximates f(x) near x = 0. Of course, anybody can write down a polynomial and claim it "approximates f(x)." The question, of course, is *how well* does it approximate f(x)? We have answered this question by a simple computer test. A more theoretical study of the errors involved in approximating with Taylor polynomials can be found in most calculus books. Sometimes the Taylor polynomials such as we have been considering (approximating f(x) near x = 0) are called Maclaurin polynomials or "series."

3.18   PROGRAM SEQUENCE FOR TAYLOR POLYNOMIALS

(a)

```
10 FOR X = -.18 TO .18 STEP .04
20 PRINT X,EXP(X),1 + X + (X^2)/2
30 NEXT X
```

x	e^x	$1 + x + x^2/2$
−.18	.835270212	.8362
−.14	.869358235	.8698
−.1	.904837418	.905
−.06	.941764534	.9418
−.02	.980198673	.9802
.02	1.02020134	1.0202
.06	1.06183655	1.0618
.1	1.10517092	1.105
.14	1.1502738	1.1498

(b)

```
10 FOR X = −.18 TO .18 STEP .04
20 PRINT X,LOG(1 + X)",X − (X^2)/2 + (X^3)/3
30 NEXT X
```

x	$\ln(1 + x)$	$x - x^2/2 + x^3/3$
−.18	−.198450939	−.198144
−.14	−.15082289	−.150714667
−.1	−.105360516	−.105333333
−.06	−.0618754037	−.061872
−.02	−.020202707	−.0202026667
.02	.0198026279	.0198026667
.06	.0582689078	.058272
.1	.0953101797	.0953333334
.14	.131028262	.131114667

(c)

```
10 FOR X = −.18 TO .18 STEP .04
20 PRINT X,SIN(X),X − (X^3)/6
30 NEXT X
```

x	$\sin(x)$	$x - x^3/6$
−.18	−.179029573	−.179028
−.14	−.139543115	−.139542667
−.1	−.0998334167	−.0998333333
.06	−.0599640065	−.059964
−.02	−.0199986667	−.0199986667
.02	.0199986667	.0199986667
.06	.0599640065	.059964
.1	.0998334167	.0998333334
.14	.139543115	.139542667

(d)

```
10 FOR X = −.18 TO .18 STEP .04
20 PRINT X,COS(X),1 − (X^2)/2
30 NEXT X
```

x	cos(x)	$1 - x^2/2$
−.18	.983843693	.9838
−.14	.990215996	.9902
−.1	.995004165	.995
−.06	.99820054	.9982
−.02	.999800007	.9998
.02	.999800007	.9998
.06	.99820054	.9982
.1	.995004165	.995
.14	.990215996	.9902

(e)

```
10 FOR X = −.18 TO .18 STEP .04
20 PRINT X,TAN(X),X + (X^3)/3
30 NEXT X
```

READY.

x	tan(x)	$x + x^3/3$
−.18	−.181969529	−.181944
−.14	−.140921895	−.140914667
−.1	−.100334672	−.100333333
−.06	−.0600721038	−.060072
−.02	−.0200026671	−.0200026667
.02	.0200026671	.0200026667
.06	.0600721038	.060072
.1	.100334672	.100333333
.14	.140921895	.140914667

We now give some exercises on computing limits and graphing functions. The reader should memorize TAYLOR POLYNOMIALS 3.17. For additional hints and examples look at SOLUTIONS 3.20.

3.19 EXERCISES

(1) In this exercise we ask the student to "play calculus instructor" and make up some exercises on limits along the guidelines suggested.

(a)   One way to make hard-looking limit problems is to factor a polynomial and form the quotient of the polynomial with one or more of its factors. For example, divide $(x - 2)(x + 1)$ by $x - 2$ and take the limit as x approaches 2. In symbols,

$$\lim_{x \to 2} \frac{(x - 2)(x + 1)}{(x - 2)}.$$

This limit is of the form ∞/∞ as both the numerator and denominator evaluated at 0 become 0. The obvious thing to do in this case is to cancel $x - 2$ from the numerator and denominator (i.e., divide numerator and denominator by $x - 2$) to obtain

$$\lim_{x \to 2} (x + 1) = 3.$$

So far this seems simple enough. Suppose you give this problem to a friend but, before doing so, multiply $(x - 2)$ times $(x + 1)$ to get $x^2 - x - 2$. The problem becomes

$$\lim_{x \to 2} \frac{x^2 - x - 2}{x - 2} = ?$$

If your friend hasn't seen the first version of the problem, he or she will have to divide the numerator by the denominator to evaluate the limit. Make up two problems along these lines and have your classmates do the same. Exchange problems and practice taking limits of this sort. Make these problems so that the numerator divides the denominator and solve them by polynomial division.

(b)   If you worked part (a) above, you may have found that it is easy to construct hard problems that require some awkward polynomial arithmetic to solve. You might, for example, start with

$$\lim_{x \to 1} \frac{(x^3 - x^2 - x - 1)(x^2 - 1)}{(x^2 - 1)}.$$

Now take the product $(x^3 - x^2 - x - 1)(x^2 - 1) = x^5 - x^4 - 2x^3 + x + 1$ and give your friend the problem

$$\lim_{x \to 1} \frac{x^5 - x^4 - 2x^3 + x + 1}{x^2 - 1} = ?$$

One could use polynomial division to work this problem (divide numerator into denominator and evaluate the resulting polynomial at $x =$

1). An easier way is to use L'HOPITAL'S RULE 3.15. In this case, $f(x) = x^5 - x^4 - 2x^3 + x + 1$ and $g(x) = x^2 - 1$. Thus, $f'(x) = 5x^4 - 4x^3 - 6x^2 + 1$ and $g'(x) = 2x$. We then see that

$$\lim_{x \to 1} \frac{f'(x)}{g'(x)} = \lim_{x \to 1} \frac{5x^4 - 4x^3 - 6x^2 + 1}{2x} = -2.$$

Note in this last limit the denominator evaluated at $x = 1$ is not zero so the limit is obtained by substituting $x = 1$ and evaluating the resulting expression. Notice that, instead of dividing by $x^2 - 1$ in this problem, we could have divided by some polynomial times $x^2 - 1$. For example, we could have divided by $(x + 1)(x^2 - 1) = x^3 + x^2 - x - 1$. Using the method of polynomial division of (a) above would not work on this problem, but L'HOPITAL'S RULE still works fine. Make up five problems along these lines and share them with your classmates. Some of these problems may require several applications of L'HOPI-TAL'S RULE.

(2)   This exercise is concerned with limits that can be evaluated using the TAYLOR POLYNOMIALS 3.17. First of all, you should recall that these polynomials are only valid approximations for small values of x (x near zero). For example, $\cos(x) = 1 - x^2/2$ cannot be a good approximation for large x as the right-hand side becomes very large and negative for all large values of x while the $\cos(x)$ function varies between -1 and $+1$ for all values of x. Another important observation for these problems is that in evaluating a limit $p(x)/q(x)$ as x goes to zero where $p(x)$ and $q(x)$ are polynomials, it is only necessary to consider the terms of lowest degree in $p(x)$ and $q(x)$. Thus

$$\lim_{x \to 0} \frac{2x + 3x^2 + 4x^3}{3x + 5x^3} = \lim_{x \to 0} \frac{2x}{3x} = 2/3.$$

Suppose we try the following limit problem:

$$\lim_{x \to 0} \frac{\sin(x) - x}{\tan(x) - x} = ?$$

Using TAYLOR POLYNOMIALS 3.17, we write $\sin(x) - x = -x^3/6$ and $\tan(x) - x = x^3/3$. These identities are only approximate, but are very good for small values of x. As we are looking at $\lim_{x \to 0}$, the values of x will be small. Thus we have

$$\lim_{x \to 0} \frac{\sin(x) - x}{\tan(x) - x} = \lim_{x \to 0} \frac{-x^3/6}{x^3/3} = -1/2.$$

Make up three problems of this type and exchange them with your classmates.

We now work some more exercises designed to give you practice with limits and graphing functions. In all graphing problems, you are encouraged to write a program to generate data. Calculus can then be used to fine-tune your results. In the limit problems try several methods when possible.

3.20 EXERCISES

(1) Find the following limits:

(a) $\displaystyle \lim_{x \to -2} \frac{x^2 + 5x + 6}{x + 2}$

(b) $\displaystyle \lim_{x \to 9} \frac{x^2 - 81}{2(x - 9)}$

(2) Find the following limits:

(a) $\displaystyle \lim_{x \to 0} \frac{\sin x}{x}$

(b) $\displaystyle \lim_{x \to 0} \frac{\sin 2x}{\sin x}$

(3) Find the following limits:

(a) $\displaystyle \lim_{x \to \infty} \frac{x + \sin x}{2x}$

(b) $\displaystyle \lim_{x \to \infty} \frac{\ln x}{x^a}, \, a > 0$

(4) Graph and find all local maxima, minima, and points of inflection for the following functions:

(a) $y = -2x^2 + 3x + 5$

(b) $y = 12(1 - x)/x^2$

(5) Graph (for selected values of α) and find all local maxima, minima, and points of inflection in terms of the parameter $\alpha > 0$.

(a) $f_\alpha(x) = 2\alpha x(x - 3\alpha)$

(b) $f_\alpha(x) = 1/x^2 - \alpha/x$.

3.21 VARIATIONS ON EXERCISE 3.20

(1) Find the following limits:

(a) $\displaystyle \lim_{x \to 0} \frac{2x^2 + 3x}{x - 1}$

(b) $\lim\limits_{x \to 4} \dfrac{\sqrt{x} - 2}{x - 4}$

(2) Find the following limits:

    (a) $\lim\limits_{x \to 0} \dfrac{e^x - 1}{\tan x}$

    (b) $\lim\limits_{x \to 1} \dfrac{2\ln x}{\sin(\pi x)}$

(3) Find the following limits:

    (a) $\lim\limits_{x \to 0} \dfrac{\ln(1 + x)}{\sin x}$

    (b) $\lim\limits_{x \to 0} x^a \ln x, \ a > 0$

(4) Graph and find all local maxima, minima, and points of inflection for the following functions:

    (a) $y = e^x - x$

    (b) $y = (x^2 + 1)/x$

(5) Graph (for selected values of α) and find all local maxima, minima, and points of inflection in terms of the parameter $\alpha > 0$.

    (a) $f_\alpha(x) = e^{\alpha x} - e^{-\alpha x}$

    (b) $f_\alpha(x) = x^3 - 3\alpha x^2 + 3x + 2$

3.22 VARIATIONS ON EXERCISE 3.20

(1) Find the following limits:

    (a) $\lim\limits_{x \to 3} \dfrac{3x^3 + 4x^2 - x + 1}{2x^2 + 4x - 5}$

    (b) $\lim\limits_{x \to -1} \dfrac{x^2 + 1}{x^2 - 1}$

(2) Find the following limits:

    (a) $\lim\limits_{x \to 0} \dfrac{\ln(\tan x)}{x^3}$

    (b) $\lim\limits_{x \to 0} \dfrac{\sin x}{2\tan \sqrt{x}}$

(3)   Find the following limits:

(a) $\displaystyle\lim_{x\to 0} \frac{ex - e\sqrt{x}}{\sqrt{x}}$

(b) $\displaystyle\lim_{x\to \pi/2} \frac{\sin(\cos x)}{\cos x}$

(4)   Graph and find all local maxima, minima, and points of inflection for the following functions:

(a) $y = x\ln|x|$

(b) $y = (x^2 + 1)/(x^2 - 1)$

(5)   Graph (for selected values of α) and find all local maxima, minima, and points of inflection in terms of the parameter $\alpha > 0$.

(a) $f_\alpha(x) = x - \alpha/x$

(b) $f_\alpha(x) = x^\alpha - x^{\alpha/2}$, $x > 0$

3.23   VARIATIONS ON EXERCISE 3.20

(1)   Find the following limits:

(a) $\displaystyle\lim_{x\to 0} \frac{\sqrt{x}}{\sqrt[3]{x}}$

(b) $\displaystyle\lim_{x\to 1} \frac{x^3 - 3x^2 + 3x - 1}{x - 1}$

(2)   Find the following limits:

(a) $\displaystyle\lim_{x\to \pi/2} \frac{\pi e^{\cos x} - 2x}{\cos x}$

(b) $\displaystyle\lim_{x\to 0} \frac{\ln|\sec x + \tan x|}{\tan x}$

(3)   Find the following limits:

(a) $\displaystyle\lim_{x\to 0} \frac{\cos x - 1}{x}$

(b) $\displaystyle\lim_{x\to e} \frac{\ln(\ln|x|)}{\ln x - 1}$

(4)   Graph and find all local maxima, minima, and points of inflection for the following functions:

(a) $y = x + \sin(x)$

(b) $y = \ln|x^3 - 3x|$

(5) Graph (for selected values of α) and find all local maxima, minima, and points of inflection in terms of the parameter $\alpha > 0$.

(a) $f_\alpha(x) = x\ln|\alpha x|$

(b) $f_\alpha(x) = (x^2 + \alpha)/(x - \alpha)$

Now The Classical Max–Min And Related Rate Problems

3.24 CLASSICAL APPLICATIONS OF DIFFERENTIATION We now look at some typical classical applications of differentiation. Remember, when you work these problems, think about whether or not you really need calculus to solve them. Would drawing a graph do just as well? With this attitude, you will make more intelligent use of calculus and gain more insight into the real nature of these exercises.

3.25 EXERCISES

(1) A cylindrical beer can is to be made with 50 in^2 of aluminum. What should the dimension of the can be in order to maximize the volume of the can? Sketch the optimum can.

(2) A person is at point A on the bank of a straight river and wants to get to point B on the opposite shore as quickly as possible. If he can run 10 miles per hour and swim 5 miles per hour, if the river is 1 mile wide, and point B is 6 miles down the river from point A, what path should he take to get there as quickly as possible? This is a slowly flowing river, so ignore the motion of the water.

(3) A 10 foot wide hallway leads into a 5 foot wide hallway. What is the longest length of straight board that can be pushed along the floor around the corner joining the two hallways?

(4) A 6 foot tall man is walking away from a 20 foot high lamppost at the rate of 3 feet per second. When he is 10 feet from the lamppost, at what rate is the length of his shadow changing?

(5) If an object weighs w lbs. on the surface of the earth, then it weighs $w(1 + .00025r)^{-2}$ lbs. when it is r miles above the surface of the earth. What is the rate of decrease of weight of a 10,000 lb. rocket 500 miles above the earth and traveling away at a velocity of 3 miles per second.

(6) A toy rocket is fired vertically with an initial velocity of 200 feet per second. Find the maximum height of the rocket and find its velocity when it strikes the ground.

(7) The parametric equations of the curve traveled by a particle are x = 2 − 3cos(t) and y = 3 + 2sin(t), where t is the time in seconds. Find the rate at which the tangent to the curve is changing as a function of t. Sketch the curve, showing some points corresponding to specific values of t.

(8) The parametric equations of the curve traveled by a particle are x = −1 + cos(t) and y = 1 + 2sin(t). Find the rate at which the particle is moving along the curve as a function of time.

We now give the solutions to EXERCISE 3.25. As previously, you should try to work EXERCISE 3.25 without looking at the solutions. If you get stuck, take a peek at the solution and try again. If you still get stuck, study the solution carefully and then make up and work your own variation to the problem, however minor it may be.

3.26 SOLUTIONS TO EXERCISE 3.25

(1) The surface area is fixed and given as 50 in². The surface area is twice the area of the base ($2\pi r^2$) plus the area of the lateral surface ($2\pi rh$). Thus, if S denotes the surface area, $S = 2\pi r^2 + 2\pi rh = 50$ in². There are two unknowns, r and h. Solving for h we get, $h = (25/\pi r) - r$. We wish to maximize the volume, $V = \pi r^2 h$. We haven't discussed maximizing functions of two variables (that comes in more advanced courses). By substituting the expression for h in terms of r into the expression for V, we obtain $V(r) = 25r - \pi r^3$. This expresses V as a function of only one variable. So far, no calculus! Here are a couple of basic programs that tell us that $V(r)$ is maximized for r = 1.65 and h = 3.30.

(a)

```
10 FOR R=0 TO 3 STEP .2
20 PRINT R,25*R−3.14*R^3
30 NEXT R
```

r	V(r)
0	0
.2	4.97488
.4	9.79904
.6	14.32176

r	V(r)
.8	18.39232
1	21.86
1.2	24.57408
1.4	26.38384
1.6	→ 27.13856
1.8	26.68752
2	24.88
2.2	21.56528
2.4	16.59263
2.600001	9.811341
2.800001	1.070694

(b)

```
10 FOR R = 1.2 TO 2 STEP .05
20 PRINT R,25*R − 3.14*R^3
30 NEXT R
```

r	V(r)
1.2	24.57408
1.25	25.11719
1.3	25.60142
1.35	26.02442
1.4	26.38384
1.45	26.67732
1.5	26.9025
1.55	27.05703
1.6	27.13856
1.65	→ 27.14473
1.7	27.07318
1.75	26.92157
1.8	26.68752
1.849999	26.3687
1.899999	25.96275
1.949999	25.4673
1.999999	24.88001

We still haven't used any calculus! To use calculus, we should compute $\frac{d}{dr}V(r) = V'(r) = 25 - 3\pi r^2$. This gives, by setting $V'(r) = 0$ and solving for r, $r = (25/3\pi)^{1/2} = 1.63$ with $h = 3.26$. It seems that the optimal beer can has its diameter the same as its height. I've never seen a beer can of this shape. In fact, looking through my pantry, I found only one can of this shape.

It was a can of nuts. Hopefully, this is not a commentary on the usefulness of calculus!

(2)   Figure 3.27 is a sketch of the situation:

From FIGURE 3.27, we see that the time, T(r), to get from A to B, given that the person runs along the bank a distance of r miles before starting to swim straight towards B, is given by

$$T(r) = \frac{r}{10} + \frac{(1 + (6 - r)^2)^{1/2}}{5}.$$

FIGURE 3.27   A Straight, Slow River

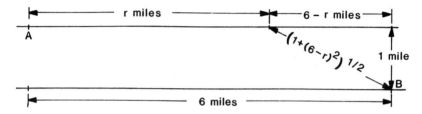

Here are a couple of simple BASIC programs that generate values of T(r). It is evident from these programs that r = 5.42 miles (approximately) gives the minimum time.

(a)

```
10 FOR R=0 TO 6 STEP .2
20 PRINT R,(R/10)+((1+(6−R)^2)^.5 )/5
30 NEXT R
```

r	T(r)	r	T(r)
0	1.216552	3.200001	.9146427
.2	1.197115	3.400001	.8971355
.4	1.177717	3.600001	.8799999
.6	1.158362	3.800001	.8633218
.8	1.139056	4.000001	.8472135
1	1.119804	4.200001	.8318251
1.2	1.100612	4.4	.8173592
1.4	1.081488	4.6	.804093
1.6	1.062441	4.8	.79241
1.8	1.043481	5	.7828428
2	1.024621	5.2	.776125
2.2	1.005875	→ 5.399999	→ .7732381
2.4	.9872616	5.599999	.7754066
2.600001	.9688017	5.799999	.7839607
2.800001	.9505221	5.999999	.7999998
3.000001	.9324554		

(b)

```
10 FOR R = 5.3 TO 5.5 STEP .02
20 PRINT R,(R/10) + ((1 + (6 − R)^2)^.5 )/5
30 NEXT R
```

r	T(r)
5.3	.7741311
5.32	.7738595
5.34	.7736331
5.36	.7734531
5.38	.7733211
5.4	.7732381
→ 5.42	→ .7732055
5.44	.7732248
5.46	.7732972
5.48	.773424
5.5	.7736068

This requires no calculus. To use calculus on this problem, we compute $T'(r)$ to get

$$T'(r) = \frac{1}{10} - \frac{6 - r}{5(1 + (6 - r)^2)^{1/2}} \cdot$$

Setting $T'(r) = 0$ gives $(1 + (6 - r)^2)^{1/2} = 2(6 - r)$. Squaring both sides and solving for $6 - r$ gives $6 - r = (1/3)^{1/2}$ or $r = 5.423$. This agrees with the BASIC program given above.

(3) Note in FIGURE 3.28 that the "longest board" is the shortest line segment touching the interior corner and opposite sides of the two hallways. Evidently, the length of such a line segment is

$$L(t) = \frac{10}{\cos(t)} + \frac{5}{\sin(t)} \cdot$$

Here are a couple of BASIC programs that compute $L(t)$ for selected values of t, measured in radians. Note that the minimum segment occurs when $t = .67$ radians or 38.4 degrees. The largest board length is 20.8097 feet.

FIGURE 3.28   Board Sliding Around Corner

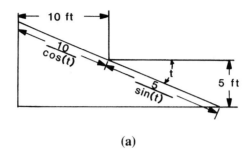

(a)

```
10 FOR T = .5 TO 1.3 STEP .05
20 PRINT T,(10/COS(T)) + (5/SIN(T))
30 NEXT T
```

T	L(T)
.5	21.82409
.55	21.29582
.6	20.97144
→ .6500001	→ 20.82341
.7000001	20.83595
.7500001	21.00227
.8000001	21.32328
.8500001	21.80721
.9000001	22.47029
.9500001	23.33841
1	24.45014
1.05	25.86183
1.1	27.65641
1.15	29.95842
1.2	32.96161
1.25	36.98235
1.3	42.57242

(b)

```
10 FOR T = .6 TO .7 STEP .005
20 LPRINT T,(10/COS(T)) + (5/SIN(T))
30 NEXT T
```

T	L(T)
.6	20.97144
.605	20.94904
.61	20.92838
.615	20.90943

T	L(T)
.62	20.89218
.625	20.8766
.63	20.86269
.635	20.85043
.64	20.83981
.645	20.8308
.65	20.82341
.655	20.81761
.66	20.8134
.665	20.81077
→ .67	→ 20.80972
.675	20.81022
.68	20.81228
.685	20.81588
.69	20.82103
.695	20.82772
.7	20.83594

To use calculus on this problem, we compute $L'(t)$ and solve for t such that $L'(t) = 0$. We calculate that

$$L'(t) = \frac{10\sin(t)}{\cos^2(t)} - \frac{5\cos(t)}{\sin^2(t)} .$$

Thus, $L'(t) = 0$ when $10\sin^3(t) = 5\cos^3(t)$. In other words, $\tan^3(t) = 1/2$ or $\tan(t) = 0.79$. But, $\arctan(.79) = .67$ radians or 38.4 degrees, which is no surprise as we knew that already from the BASIC program above.

(4) From FIGURE 3.29 we see by similar triangles that

$$\frac{s}{6} = \frac{s + d}{20} .$$

FIGURE 3.29 Man Walking Away from Lamppost

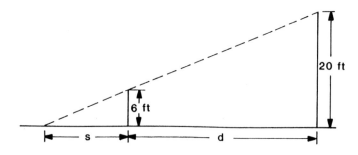

From this relationship, we conclude that s $=$ (3/7)d. The man is walking at 3 feet per second, so d $=$ 3t is a function of time. Thus, s(t) $=$ (9/7)t and s$'$(t) $=$ 9/7 feet per second. Are you surprised that the rate of change of the shadow is a constant? Does it seem intuitive to you that the rate of change of the shadow should be the same if the man is very close to the lamppost or very far away?

(5) Let W(r) denote the given function that expresses weight as a function of r. By the CHAIN RULE, $\dfrac{dW}{dt} = \dfrac{dW}{dr}\,\dfrac{dr}{dt}$ gives the rate of change of weight as a function of time. But $\dfrac{dW}{dr} = $ w(-2)(1 $+$ .00025r)$^{-3}$(.00025) $= -$w(5 $\times$ 10^{-4})(1 $+$ (2.5 $\times$ 10^{-4})r)$^{-3}$. Multiplying this expression, with r $=$ 500 miles, times $\dfrac{dr}{dt} = $ 3 miles per second gives $-$10.5 lbs. per second as the rate of decrease of weight.

(6) This problem requires a little bit of knowledge of elementary physics. Let s(t) denote the distance of an object, as a function of time, from the surface of the earth or, alternatively, from any other fixed point of reference (50 miles up, for example). Usually, we measure s(t) as positive going up from the point of reference and negative going down. This is not always the case, however. Let's assume for now that up is positive. The velocity, v(t), of the object as a function of time is defined to be $\dfrac{d}{dt}$s(t) $=$ s$'$(t). If the object is moving up, its velocity is positive and moving down its velocity is negative. The acceleration, a(t), of the object is, by definition, $\dfrac{d}{dt}$v(t) $=$ v$'$(t). The "speed" of an object at a time t is the absolute value of the velocity v(t). In our model, an object has positive acceleration if it is heading toward the surface of the earth with decreasing speed or heading away from the earth with increasing speed. Otherwise (away with decreasing speed or toward with increasing speed), the acceleration is negative. Long ago, people discovered that an object dropped freely, without any push, fell toward the earth with increasing speed. Such an object has negative acceleration based on our assumption that s(t) is positive directed away from the earth's surface.

But there was a surprise about objects falling toward the surface of the earth awaiting the ancients! The acceleration didn't seem to depend on the "weight" of the object. A small coin would accelerate at the same rate as a heavier coin. This acceleration is about $-$32 feet per second per second or

-32 ft/sec^2. In metric units, the acceleration at the earth's surface is -9.8 meters per second per second. Two basic laws of physics lie behind this observation. The first is the famous F $=$ ma formula, which says that the force acting on an object is the product of the object's mass and its acceleration. The second is the law of gravitation, which states that the force between two objects of mass M and m is given by F $=$ GMm/r^2 where r is the distance between the two objects and G is a constant. Just how this distance should be measured is itself a good calculus problem. For spheres of uniform density, the distance is measured between their centers. If m is the mass of a small marble that you are about to drop from one foot above your desk, then F $=$ (GM/r^2)m where M is the mass of the earth and r is the distance from the center of the earth to the center of the marble. This means that the acceleration acting on the marble is GM/r^2. If you now drop the marble from one foot above the floor, the distance r is a few feet less so, technically, the acceleration is a little less. The difference between taking r in the expression GM/r^2 to be the distance from the center of the earth to the top of the desk or the center of the earth to the floor is so minute that GM/r^2 may be treated as constant. For all points near the surface of the earth, GM/r^2 is usually treated as a constant, 32 ft/sec^2. Notice that the acceleration GM/r^2 does not depend on m. This explains why heavy objects fall as fast as light ones. We are ignoring air resistance here. Obviously, a feather or a bubble will not fall as fast as a dime in the air currents in your room.

To get back to the problem at hand, problem (6), the rocket, we assume, is launched from the ground, which we take to be at time zero and height zero. Thus, s(0) $=$ 0. The acceleration is -32 ft/sec^2. This means that the velocity must be v(t) $=$ -32t $+$ v$_0$, where v$_0$ $=$ 200 ft/sec is the initial velocity, v(0). The distance s(t) must satisfy s$'$(t) $=$ v(t) and s(0) $=$ 0 and hence must look like s(t) $=$ -16t^2 $+$ 200t. We picture the rocket shooting straight upwards until the time t when v(t) $=$ 0. This occurs when -32t $+$ 200 $=$ 0 or t $=$ 25/4 seconds. The height of the rocket at this time is s(25/4) $=$ 625 ft. As we are ignoring air resistance, you will guess, if you have had a little physics, that the velocity when the rocket hits the ground will be exactly -200 ft/sec. The minus is because the falling object is headed toward the earth. You should show, by starting with the fact that the acceleration is -32 ft/sec^2, that any object dropped from 625 feet will strike the ground at -200 ft/sec.

(7) FIGURE 3.30 shows the curve and a basic program to generate points on the curve.

FIGURE 3.30 Rotation of Tangent to Ellipse

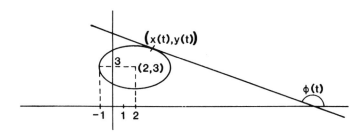

```
10 PI = 3.14159
20 FOR T = 0 TO 2*PI STEP 2*PI/20
30 PRINT T,2 − 3*COS(T),3 + 2*SIN(T)
40 NEXT T
```

t	x(t)	y(t)
0	− 1	3
.314159	− .8531702	3.618034
.628318	− .427052	4.17557
.942477	.2366424	4.618034
1.256636	1.072946	4.902113
1.570795	1.999996	5
1.884954	2.927047	4.902114
2.199113	3.763352	4.618037
2.513272	4.427048	4.175574
2.827431	4.853168	3.618039
3:14159	5	3.000006
3.455749	4.853173	2.381972
3.769908	4.427058	1.824435
4.084067	3.763365	1.38197
4.398226	2.927062	1.097889
4.712385	2.000013	.9999998
5.026544	1.072962	1.097884
5.340703	.2366562	1.38196
5.654862	− .4270418	1.824421
5.969021	− .8531644	2.381955
6.28318	− 1	2.999988

We are asked to find "the rate at which the tangent is changing as a function of t." There are two ways that the tangent is changing. The point of contact of the tangent line with the curve changes with t in the obvious way. The slope of the tangent line also changes. The slope of the tangent line is $\tan(\phi(t))$ for some angle $\phi(t)$ which changes with t. The most interesting measure of change of the tangent line is $\dfrac{d}{dt}\phi(t) = \phi'(t)$, which measures the rate of rotation of the tangent line with respect to t. To compute $\phi'(t)$, we first compute $\dfrac{dx}{dt} = 3\sin(t)$ and $\dfrac{dy}{dt} = 2\cos(t)$. By the CHAIN RULE, we have $\dfrac{dy}{dt} = \dfrac{dy}{dx}\dfrac{dx}{dt}$ or $2\cos(t) = \dfrac{dy}{dx}3\sin(t)$. Solving, we obtain $\dfrac{dy}{dx} = (2/3)\cot(t)$. Thus, $\phi(t) = \arctan((2/3)\cot(t))$ and

$$\frac{d\phi}{dt} = \frac{-6\csc^2(t)}{9 + 4\cot^2(t)}.$$

This gives the rate of rotation of the tangent with respect to t (see DIFFERENTIATION RULES 2.41 for $\dfrac{d}{dx}\arctan x$).

(8) The curve is an ellipse. You should write a short program to generate points on this ellipse and sketch the curve as was done in problem (7). The distance at time t that a point on the curve has traveled along the curve from a fixed reference point, say $(x(0), y(0))$, is called the "arclength at time t" and is denoted by L(t). The basic fact we need about arclength is that $dL = ((dx)^2 + (dy)^2)^{1/2}$. We discuss this idea more fully in CHAPTER 4. As for now, you should look ahead to FIGURE 4.53 to understand this statement about differentials of arclength. In particular, we have

$$\frac{dL}{dt} = ((\frac{dx}{dt})^2 + (\frac{dy}{dt})^2)^{1/2} = (\sin^2(t) + 4\cos^2(t))^{1/2} = (1 + 3\cos^2(t))^{1/2}.$$

We now look at some VARIATIONS on EXERCISE 3.25. Our first VARIATIONS involve exactly the same problems but stated more generally in terms of parameters. Here you will find computational methods less of a help and calculus much more important as a tool in finding solutions in terms of parameters. Even in dealing with problems formulated in terms of parameters, it is a good idea to assign values to the parameters and do some computational studies. We have already done this in SOLUTIONS TO EXERCISE 3.25, so in VARIATIONS 3.31 you can concentrate on the calculus.

3.31 VARIATIONS ON EXERCISE 3.25

(1) A cylindrical beer can is to be made with s in^2 of aluminum. What should the dimensions of the can be, as a function of s, in order to maximize the volume of the can?

(2)   A person is at point A on the bank of a straight river and wants to get to a point B on the opposite shore as quickly as possible. If he can run ρ miles per hour and swim σ miles per hour, if the river is ω miles wide, and point B is β miles down the river from point A, what path should he take to get there as quickly as possible? This is a slowly flowing river, so ignore the motion of the water.

(3)   A β foot wide hallway leads into a λ foot wide hallway. What is the longest length of a straight board that can be pushed along the floor around the corner joining the two hallways?

(4)   A man τ feet tall is walking away from a λ foot high lamppost at the rate of ρ feet per second. When he is δ feet from the lamppost, at what rate is the length of his shadow changing?

(5)   If an object weighs ω lbs. on the surface of the earth, then it weighs $\omega(1 + .00025r)^{-2}$ lbs. when it is r miles above the surface of the earth. What is the rate of decrease of weight of an R lb. rocket M miles above the surface of the earth and traveling away at a velocity of v miles per second?

(6)   A toy rocket is fired vertically with an initial velocity of V_0 feet per second. Find the maximum height of the rocket and find its velocity when it strikes the ground.

(7)   The parametric equations of a curve traveled by a particle are x = a + bcos(t) and y = c + dsin(t), where t is the time in seconds. Find the rate at which the tangent to the curve is changing as a function of t,a,b,c, and d.

(8)   The parametric equations of the curve traveled by a particle are x = a + bcos(t) and y = c + dsin(t). Find the rate at which the particle is moving along the curve as a function of t,a,b,c, and d.

3.32   VARIATIONS ON EXERCISE 3.25

(1)   A cylindrical beer can has radius r and height h and is made of aluminum. Suppose the cost per in^2 of aluminum is 1 cent. Let V denote the volume of the can. In terms of V and r, give a formula for the cost of the least expensive can.

(2)   Consider the parabola $y = (x - 1)^2 + 4$. Find the point on it such that the distance from this point to the origin is minimal.

(3) A sport field has the pictured geometric form. If the perimeter of the field is 1 km, what should x and y be such that the surface area is maximal? What if the field is a rectangle with only one half circle attached?

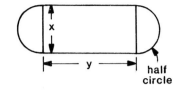

(4) A 6 foot tall man stands at the point A. A light source is located at the point C and moves upwards at the rate of 3 feet per second. When the light source is 12 feet over the earth at the point D, at what rate is the length of the man's shadow changing?

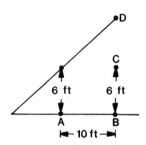

(5) Find two positive numbers whose sum is 20 such that their product is maximal. Find two positive numbers whose product is 20 such that their sum is maximal.

(6) A toy rocket, 10 meters from the ground, is fired at an angle of 60° with an initial velocity of 200 feet per second. Find the maximum height of the rocket, its velocity when it strikes the ground, and the distance from A to B.

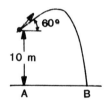

(7) A stone is dropped into water and causes concentric circles propagating at a rate of 5 m/s. After 10 seconds, at what rate is the area enclosed by the first concentric circle expanding?

(8) The parametric equations of the curve traveled by a particle are $x = t + 1$ and $y = -t^2 + 8t + 9$ where $t \geq 0$ is the time. At what time is the absolute value of the tangential velocity (rate of change of arclength) minimal?

Now go back over VARIATIONS 3.32 and introduce parameters into the problems, just as was done in VARIATIONS 3.31 relative to EXERCISES 3.25. For example, in 3.32(4), the man can be T feet tall, the light source can be moving upwards with a velocity of v feet per second, etc. Do the same to the remaining VARIATIONS ON EXERCISES 3.25 after you have solved the problems as stated.

3.33   VARIATIONS ON EXERCISE 3.25

(1)   A container with a square bottom is made with S in^2 of wood. What should the dimensions of the container be in order to maximize its volume?

(2)   Suppose a natural gas supplier is located at A and plans a pipeline to a location at B. If the cost for the pipeline in the river is four times that on the earth, find the point C such that the cost for the pipeline is minimal. How can this problem and 3.25(2) be combined into one "general idea"? (The pipe will go straight from A to C along land and straight from C to B under the river.)

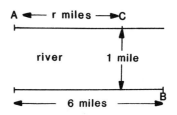

(3)   We are considering a half elliptical hallway 1 meter wide and want to slide a straight board along the floor entering at A and leaving at B. What is the longest board that can be brought through this hall? Prove your answer is correct.

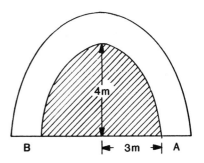

(4)   A man standing at the point A observes a car moving at the rate of 50 km/h. When the car is 200 m at the point B, at what rate is the distance between the man and the car changing?

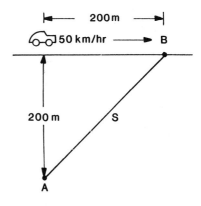

(5) We consider a right triangle with a dimensions 3, 4, and 5 m. Find the dimensions of the largest rectangle included in it as pictured.

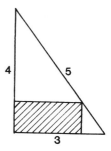

(6) A 6 foot tall man is walking away from a light at a rate of $\dfrac{70}{15+x}$ ft/sec where x is the horizontal distance from the light. The light starts at height of 20 ft when x = 0 and is moving upwards at a rate of 10 ft/sec. How fast is the man's shadow changing when x = 20 ft?

(7) The parametric equations of the curve traveled by a particle are x = $-1 + 3\cos^2 t$ and y = $1 - 3\sin(2t)$. Find the rate at which the particle is moving along the curve as a function of time. Sketch the curve.

(8) A light source located at A is reflected through a mirror and is caught at the point B. Find the point C such that the sum of the distances AC and CB is minimal. Show that for this C, the angle α equals the angle β. Does this result depend on the distances AP, PQ, and BQ? Explain.

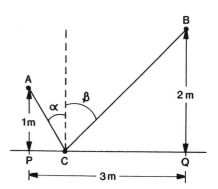

3.34   VARIATIONS ON EXERCISE 3.25

(1) A cylindrical can has a fixed volume of v in^3 and is made of aluminum. The cost for the top and bottom is half of the cost of the other parts of the can (which costs more because of decorations for advertising). Find the height h and the radius r of the can such that the cost for aluminum is minimal.

(2)   A pond is the shape of an equi-
laterial triangle, 40 meters on each
side. A man at A, the midpoint of
one side, wants to get to B, the midpoint of another side. If he can run
8m/sec and swim 2m/sec what route
should he follow to minimize the
time to get to B? Replace 8m/sec by
v_r and 2m/sec by v_w and analyze the
strategy. Could there ever be an advantage not to run or swim in a
straight line?

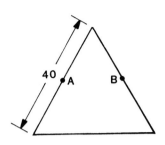

(3)   Given the curve y = 1/x, find the shortest distance between the point
(0, 1) and this curve.

(4)   An ellipse with semi-major axis
a = 3 meters and semi-minor axis
b = 2 meters starts to expand. If a
expands at a rate of 8m/sec and b
at a rate of 4m/sec, what is the rate
of increase of the area enclosed by
the ellipse when b = 12 meters?
The area of an ellipse is πab.

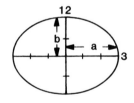

(5)   Find the dimensions of a rectangle included in a half circle of
radius r such that the area of the
rectangle is maximal.

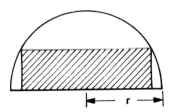

(6) A circular island of radius 30m is surrounded by a canal of radius 10m. A man at A wants to get to B as quickly as possible. Once on the island, he must run to B along the shore of the canal (not in a straight line). If he can run 6m/sec and swim 3m/sec (he's wearing swim fins!), what is his best route to get from A to B? Replace 6m/sec by v_r and 3m/sec by v_w and analyze the strategy. Is it ever to his advantage to run first, then swim? Is it ever to his advantage to not swim in a straight line? Don't be afraid to use a computer to get a feeling for this problem. What happens if he is allowed to run in a straight line on the island?

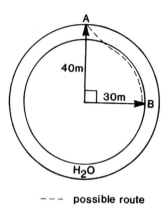

- - - **possible route**

(7) The parametric equations of the curve traveled by a particle are $x = \sin^2 t$ and $y = \cos^2 t$. Find the rate at which the particle is moving along the curve as a function of time t. Sketch the curve for $0 \le t \le \pi/2$.

(8) We consider a family of straight lines in the xy-plane passing through the point (2,2). Suppose they intersect the x-axis at A and the y-axis at B; find the slope of the line such that the shaded area is minimal. A and B depend on the slope of the line.

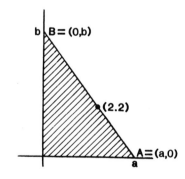

Chapter 4

INTEGRALS

Imagine that some students have taken a test on differentiation. The professor has taken the exam papers home. After dinner, she sits down to grade the exam and, looking at the first answer on the first student's exam, she sees

$$2x\cos(x^2).$$

It is then that she realizes, with great disgust, that she left the exam at the office and has forgotten what the first problem on the exam was! She sees that all of the other students got the same answer to this problem, so she assumes it is right. The exam problem was of the type "What is the derivative of $F(x)$?" But what was $F(x)$? This latter question is what the study of "integration" is all about in calculus. The function $F(x)$ is called the "integral" or "antiderivative" of $2x\cos(x^2)$. If you have had a lot of practice at differentiating functions, especially using the chain rule, you might guess $F(x) = \sin(x^2)$. This is correct as $(\sin(x^2))' = \cos(x^2)(x^2)' = 2x\cos(x^2)$. There is a good deal of such guesswork in finding integrals. It is easy to write down functions $f(x)$ for which no nice expression for the integral is known. An example is e^{x^2}.

The symbol "$\dfrac{d}{dx}$" is used to mean "the derivative of." Thus, $\dfrac{d}{dx}\sin(x)$ means "the derivative of $\sin(x)$." Correspondingly, we use the strange symbol "$\int$" to mean "the integral of." Thus, we write $\int 2x\cos(x^2) = \sin(x^2)$ to mean "the integral of $2x\cos(x^2)$ is $\sin(x^2)$." This idea is summarized in FIGURE 4.1.

FIGURE 4.1   The Derivative vs. the Integral

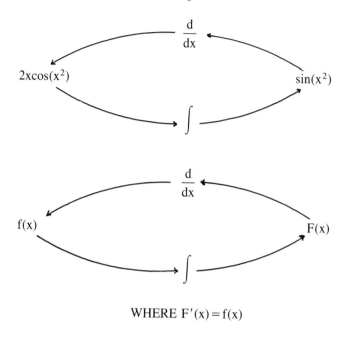

WHERE $F'(x) = f(x)$

The variable in our examples so far has been x. But, if f(t) is a function of t then $\int f(t)$ is a function F(t) such that $\frac{d}{dt}F(t) = f(t)$. But what if we walk into an abandoned classroom and see "$\int 2 = ?$" on the blackboard. What was the variable? If it was x, then the answer is 2x. If the variable was t, then the answer is 2t. To avoid this and related confusions, the notation for integrals or antiderivatives is usually $\int f(x)dx$ or $\int f(t)dt$. Thus, if we had seen "$\int 2dt = ?$" then the answer would have been 2t. If we had seen "$\int 2dx = ?$" then the answer would have been 2x.

F(x) And F(x) + C Have The Same Derivative

There is another simple but important observation about integrals that needs to be stated. As we have noted, $\int 2x\cos(x^2) = \sin(x^2)$, which means that $\frac{d}{dx}\sin(x^2) = 2x\cos(x^2)$. But, of course, $\frac{d}{dx}(\sin(x^2) + 10) = 2x\cos(x^2)$ also. In fact, $\frac{d}{dx}(\sin(x^2) + C) = 2x\cos(x^2)$ for any constant function C. This fact is sometimes incorporated into the notation for integrals by writing

$$\int 2x\cos(x^2)dx = \sin(x^2) + C.$$

This notation is intended to remind us that there are infinitely many functions with derivative $2x\cos(x^2)$ and they all differ by a constant function. Once this observation has been made and we understand what we are talking about, it is quite all right to write simply

$$\int 2x\cos(x^2)dx = \sin(x^2).$$

We understand in this latter notation that, in fact, $\sin(x^2)$ is a representative from an infinite class of antiderivatives for $2x\cos(x^2)$ and all of the rest are obtained by adding a constant function to $\sin(x^2)$.

If The Zero Function Is Its Derivative, The Function Is Constant

The fact that all antiderivatives of a given function f(x) differ by a constant is a more subtle idea than it seems at first glance. Suppose we have two functions, F(x) and G(x), such that $F'(x) = G'(x) = f(x)$. Let $H(x) = F(x) - G(x)$. Then $H'(x) = F'(x) - G'(x) = f(x) - f(x) = 0$. To claim that F(x) and G(x) *differ by a constant function* is the same as claiming that H(x) *is a constant function*. This means that the statement that "any two antiderivatives F(x) and G(x) differ by a constant" is the same as the statement that "any function H(x) with derivative function the zero function must be a constant function" (something like $H(x) = 2$ so $F(x) = G(x) + 2$). This latter statement has strong intuitive appeal. Suppose $H'(x) = 0$ for all x. Let's try to draw the graph of such an H(x). Suppose $H(0) = 2$, for example. Put your pencil at the point (0, 2) and try to imagine what the graph is like near this point. If, in going right or left, you draw the graph with the slightest bit of slope up or down you will construct points on the graph where $H'(x)$ is not 0. You're stuck at $H(x) = 2$ and must draw the graph of the constant function 2. For more advanced courses in mathematical analysis it is essential that this intuitive idea be given a precise analytical formulation. The intuitive idea, however, will suffice for our studies of calculus.

Constant Functions Can Wear Many Disguises

There is another complication that will occur in connection with the ideas associated with the previous paragraph. Suppose John decides that

$$\int 2\sin(2x)dx = -\cos(2x)$$

and suppose that Mary decides that

$$\int 2\sin(2x)dx = 2\sin^2(x).$$

If they are both right (and they are in this case) then $2\sin^2(x)$ and $-\cos(2x)$ must differ by a constant (i.e., $2\sin^2(x) - (-\cos(2x)) = 2\sin^2(x) + \cos(2x)$ is a constant function). If you know your basic trigonometric identities, then you will recognize that this is true and, in fact, $2\sin^2(x) + \cos(2x) = 1$. Thus, just because two integrals $F(x)$ and $G(x)$ for $f(x)$ *must* differ by a constant doesn't mean that they are *easily recognizable* as differing by a constant (and are both correct answers to the same integration problem). This has obvious bad implications for anyone who has to grade homework and exams of students doing problems on integration (early insanity, rapid aging, etc.).

Linearity Of The Integral

The most basic property of integrals is "linearity," which is stated in Theorem 4.2.

4.2 THEOREM   Let $f(x)$ and $g(x)$ be functions and α and β numbers. Then

$$\int (\alpha f(x) + \beta g(x))dx = \alpha \int f(x)dx + \beta \int g(x)dx$$

Proof: This follows directly from the definition of the integral together with linearity of $\dfrac{d}{dx}$. Let $F(x)$ and $G(x)$ be the antiderivatives of $f(x)$ and $g(x)$. Then $\dfrac{d}{dx}(\alpha F(x) + \beta G(x)) = \alpha f(x) + \beta g(x)$, which means, by definition of the integral, that

$$\alpha F(x) + \beta G(x) = \int (\alpha f(x) + \beta g(x))dx.$$

Substituting $F(x) = \int f(x)dx$ and $G(x) = \int g(x)dx$ gives the result.

Our first task in learning to find antiderivatives or integrals will be to develop some systematic ways to improve our ability to reduce new problems to ones we have already solved. THEOREM 4.2 is a start in this direction. We have already noticed that $\int 2x\cos(x^2)dx = \sin(x^2)$ and $\int \sin(2x) = \sin^2(x)$.

Thus we can, using THEOREM 4.2, evaluate an integral such as $\int(2\pi x\cos(x^2)$ $+ 25\sin(2x))dx$ getting

$$\pi\int 2x\cos(x^2)dx + 25\int \sin(2x)dx = \pi\sin(x^2) + 25\sin^2(x).$$

You Already Know Many Integrals

As we begin to compute more integrals, you will discover that you will use the rule of THEOREM 4.2 automatically. Now is the time to take note of the fact that you already know many integrals! Every differentiation formula you have memorized gives rise to a corresponding integration formula:

$$\frac{d}{dx}\sin(x) = \cos(x) \text{ becomes } \sin(x) = \int \cos(x)dx$$

$$\frac{d}{dx}\ln(x) = 1/x \text{ becomes } \ln(x) = \int \frac{1}{x}dx$$

$$\frac{d}{dx}x^n = nx^{n-1} \text{ becomes } x^n = \int nx^{n-1}dx$$

This latter integral has been memorized by millions of calculus students as

$$\int x^n dx = \frac{x^{n+1}}{n+1} \text{ for } n \neq -1.$$

The reason for not allowing n to be -1 is that the right-hand side of this equation is not defined for $n = -1$. We know, however, that for $n = -1$, $\int x^{-1}dx = \ln(x)$

The Chain Rule In Reverse

By applying THEOREM 4.2, you can now compute an impressive class of integrals such as

$$\int(34\sin(x) + 23x^3 + 45\sec^2(x))dx = -34\cos(x) + 23\frac{x^4}{4} + 45\tan(x).$$

But this is not nearly good enough! To really get started on the problem of computing integrals, we must learn how to do the CHAIN RULE in reverse. In particular,

$$\frac{d}{dx}f(g(x)) = f'(g(x))g'(x) \text{ becomes } f(g(x)) = \int f'(g(x))g'(x)dx.$$

This innocent-looking statement seems to be the source of much difficulty for beginning calculus students. For this reason, we shall look at a number of examples of the type that cause headaches for beginning students.

Let's start with an easy example. Consider $\int\cos(x^2)(2x)dx$. It appears that $f'(x) = \cos(x)$ and $g(x) = x^2$ here. This is determined by guessing or by inspection with the aid of past experience. Thus, $\int\cos(x^2)(2x)dx = \int f'(g(x))g'(x)dx = f(g(x)) = \sin(x^2)$. We had to guess that $g(x) = x^2$ was the proper choice since the function g is not explicitly mentioned. The way we make such a guess is by knowing our derivative formulas well. We know that $\dfrac{d}{dx} x^2 = 2x$. As our eyes scan the expression $\cos(x^2)(2x)$, we spot the pair x^2 and its derivative $2x$. This is the clue that prompts us to try $g(x) = x^2$.

In addition to guessing correctly that $g(x) = x^2$ in the previous example, we had to know how to integrate $f'(x) = \cos(x)$ to get $f(x) = \sin(x)$. As another example, let's try to calculate $\int\ln(x^2)(2x)dx$.

Again, we see the pair $g(x) = x^2$ and $g'(x) = 2x$. Thus $f'(x) = \ln(x)$ so that $f'(g(x))g'(x) = \ln(x^2)(2x)$. To compute $f(g(x))$, we must compute the integral $f(x)$ of $f'(x)$. You may not know how to do this, in which case you are stuck! At this point, some students think "If $f'(x) = \ln(x)$ then $f(x) = 1/x$ . . .," but this is going the wrong way as $f''(x) = 1/x$. We shall learn later how to find the integral $f(x)$ of $f'(x) = \ln(x)$. The answer is $f(x) = x\ln(x) - x$ (check that $f'(x) = \ln(x)$ for this function). From this fact, we conclude that $\int\ln(x^2)(2x)dx = (x^2)\ln(x^2) - x^2$. Check this statement by computing the derivative of the expression on the right.

Another complication that occurs is seen in the following two integrals:

$$\int x\cos(x^2)dx = ? \quad \text{and} \quad \int 5x\ln(x^2)dx = ?$$

In these integrals, we see the $g(x) = x^2$ all right, but the $g'(x) = 2x$ is not there. Instead, we see x in the first integral and 5x in the second integral. This may be confusing, but it's not fatal. We know that for any number α and any function h(x), $\int\alpha h(x)dx = \alpha\int h(x)dx$. Thus,

$$\int x\cos(x^2)dx = \int (1/2)(2x)\cos(x^2)dx = (1/2)\int \cos(x^2)(2x)dx$$

We have already discovered that $\int\cos(x^2)(2x)dx = \sin(x^2)$, so we have $\int x\cos(x^2)dx = (1/2)\sin(x^2)$.

Some beginning calculus students like this trick so much that they try the following type of calculation:

$$\int 2x^2\cos(x^2)dx \;=\; x\int 2x\cos(x^2)dx \;=\; x(\sin(x^2)).$$

What they do is bring the variable outside the integral sign. This doesn't work as you can check that $\dfrac{d}{dx}\,x\sin(x^2) \neq 2x^2\cos(x^2)$. Only constants can be brought outside the integral sign in the rule $\int\alpha h(x)dx \;=\; \alpha\int h(x)dx$.

You should now be in a position to understand the following calculation:

$$\int 5x\ln(x^2)dx \;=\; \int (5/2)2x\ln(x^2)dx \;=\; (5/2)\int \ln(x^2)(2x)dx.$$

This gives, by our previous discussion,

$$\int 5x\ln(x^2)dx \;=\; (5/2)(x^2\ln(x^2) - x^2).$$

Differential Notation

Certain notations of calculus are designed to help in the task of applying the CHAIN RULE in reverse. We know that $\int\cos(x)dx = \sin(x)$. There is, of course, nothing special about the x here:

$$\int \cos(t)dt \;=\; \sin(t) \qquad \int \cos(\tau)d\tau \;=\; \sin(\tau) \qquad \int \cos(A)dA \;=\; \sin(A)$$

and

$$\int \cos(CALCULUS)d(CALCULUS) \;=\; \sin(CALCULUS)$$

$$\int \cos(JUNK)d(JUNK) \;=\; \sin(JUNK).$$

The general rule is that if F(x) is an integral or an antiderivative of f(x) (i.e., $F'(x) = f(x)$) then

$$\int f(JUNK)d(JUNK) \;=\; F(JUNK)$$

where JUNK stands for anything for which these formulas make sense. JUNK can be quite complicated. For example

$$\int \cos\left(\frac{\ln(x)e^x\tan(\sin(x))}{x^5 + 2x^3 + 5x + 1}\right)d\left(\frac{\ln(x)e^x\tan(\sin(x))}{x^5 + 2x^3 + 5x + 1}\right) =$$

$$\sin\left(\frac{\ln(x)e^x\tan(\sin(x))}{x^5 + 2x^3 + 5x + 1}\right)$$

In this formula

$$\text{JUNK} = \left(\frac{\ln(x)e^x\tan(\sin(x))}{x^5 + 2x^3 + 5x + 1} \right).$$

Do you get the idea? As long as we know $F(x)$ and that $F'(x) = f(x)$, then $F(\text{JUNK}) = \int f(\text{JUNK})d(\text{JUNK})$. What does $d(\text{JUNK})$ mean? Consider, for example, the function $g(x) = x^3 + x^2$. We know that $g'(x) = 3x^2 + 2x$. In our other notation

$$\frac{dg}{dx} = 3x^2 + 2x.$$

By multiplying by dx on both sides (what this means can be made precise, but for now we think of this as a notational device only) gives

$$d(x^3 + x^2) = (3x^2 + 2x)dx.$$

In other words,

d(some function of x) = (the derivative of that function)dx.

4.3   DEFINITION (differential notation)   Let $g(x)$ be a function of x with $\dfrac{dg}{dx} = g'(x)$. Then define $dg = g'(x)dx$. This is the *differential notation for the derivative of g*. The term dg is called the *differential of g* and the term dx is called the *differential of x*.

We can state our above discussion as a (trivial) theorem:

4.4   THEOREM   Let $F(x)$ be such that $F'(x) = f(x)$ and let $g(x)$ be a function with differential $dg = g'(x)dx$. Then,

$$\int f(g)dg = F(g).$$

This "theorem" is, of course just a restatement of the CHAIN RULE as $(F(g(x)))' = F'(g(x))g'(x)$. For any function $h(x)$ we have, by the definition of the integral, $h(x) = \int h'(x)dx = \int dh$. Applying this to the CHAIN RULE (where $F'(x) = f(x)$) gives $F(g(x)) = \int d(F(g(x))) = \int F'(g(x))g'(x)dx = \int f(g(x))dg$. Thus $\int f(g)dg = F(g)$. We are just playing with notation here, but such "fooling around" is important. If you find this confusing, skip it for now and reread this section after you have worked a number of examples in the exercises that follow. In the solutions of Appendix 3, we emphasize differential notation.

How We Make Up Hard Problems

Before beginning to work exercises based on THEOREM 4.4, let's take a look at the sinister process by which calculus instructors the world over torment their students with this type of problem. Using the notation of THEOREM 4.4, we first pick some function that we know how to integrate. Let's take our old friend, $\cos(x)$ to be $f(x)$. Thus, $F(x) = \sin(x)$. Next we pick any function $g(x)$ that we can differentiate. Let's take $g(x) = \ln(x^2 + 1)$. Then, $g'(x) = \dfrac{2x}{x^2 + 1}$ or $dg = \dfrac{2x}{x^2 + 1}dx$. Substituting into the identity $\int f(g)dg = F(g)$ we obtain

$$\int \cos(\ln(x^2 + 1))d(\ln(x^2 + 1)) = \sin(\ln(x^2 + 1)).$$

But this is the same as

$$\int \cos(\ln(x^2 + 1))\,\frac{2x}{x^2 + 1}\,dx = \sin(\ln(x^2 + 1)).$$

Now, the mean old calculus instructor gives the poor innocent student the following problem:

$$\int \frac{x\cos(\ln(x^2 + 1))}{x^2 + 1}\,dx = ?$$

What does the poor student have to know to solve this awful-looking mess? What a good student will do is notice that $\ln(x^2 + 1)$ has been substituted into $\cos(x)$ for x. This will suggest that $g(x) = \ln(x^2 + 1)$. Next, the student will quickly compute $g'(x) = \dfrac{2x}{x^2 + 1}$. This will be good news to the student because this is almost, *except for a constant multiple of 2*, what appears in the expression to be integrated. Using $g(x) = \ln(x^2 + 1)$ the student will write the mean old calculus instructor's problem as $(1/2)\int\cos(g)dg = ?$ But this is easy; the answer is $(1/2)\sin(g)$. Substituting for g, the student then obtains the correct answer, $(1/2)\sin(\ln(x^2 + 1))$. Differentiate this expression to make sure it's right! This method of solving integrals is called the "method of substitution" and will be the subject of our first set of exercises.

The following exercises are designed to give you practice with the method "integration by substitution" discussed in the previous paragraphs. In EXERCISE 4.5, you are to find functions $f(x)$ and $g(x)$ such that the given problem is of the form $\int f(g)dg$. The phrase "of the form" means that the given problem may differ from $\int f(g)dg$ by a constant. In problem 1 of

EXERCISE 4.5, for example, take $g(x) = x^2 + 2$ and $f(x) = x^{1/2}$. Then $dg = 2xdx$ and the given integral $\int x(x^2 + 2)^{1/2}dx = (1/2)\int f(g)dg$. In the problems of EXERCISE 4.5, we give hints. After working these problems, try the VARIATIONS on these exercises. There, you are on your own! Now is the time to start getting acquainted with the TABLE OF INTEGRALS in Appendix 1, in particular FUNDAMENTAL FORMS.

4.5   EXERCISES   Evaluate the following integrals:

1. $\displaystyle\int x\sqrt{x^2 + 2}\ dx =$ $(f(x) = x^{1/2}$ and $g(x) = x^2 + 2)$

2. $\displaystyle\int \frac{x^{1/4}}{5 + x^{5/4}}\ dx =$ $(f(x) = 1/x$ and $g(x) = 5 + x^{5/4})$

3. $\displaystyle\int \frac{\sqrt{x} - 1}{\sqrt{x}}\ dx =$

(Beware of this type of problem! This is $\int dx - \int x^{-1/2}dx$ and no tricky substitutions are required.)

4. $\displaystyle\int \frac{8x}{1 + e^{2}x}\ dx =$

(Don't be fooled by e^2.)

5. $\displaystyle\int \frac{x\cos\sqrt{5x^2 + 1}}{\sqrt{5x^2 + 1}}\ dx =$ $(f(x) = \cos(x)$ and $g(x) = (5x^2 + 1)^{1/2})$

6. $\displaystyle\int \sin^{3/4} 2x \cos 2x\ dx =$ $(f(x) = x^{3/4}$ and $g(x) = \sin(2x))$

7. $\displaystyle\int \tan^2 3t \sec^2 3t\ dt =$ $(f(t) = t^2$ and $g(t) = \tan(3t))$

8. $\displaystyle\int e^x\sec^2(e^x)\ dx =$ $(f(x) = \sec^2(x)$ and $g(x) = e^x)$

9. $\displaystyle\int \frac{2\ln^3(x)}{x}\ dx =$ $(g(x) = \ln(x))$

10. $\displaystyle\int \frac{6x}{(x^2 + 9)^3}\ dx =$ $(g(x) = x^2 + 9)$

11. $\displaystyle\int \frac{e^{\log_2(x)}}{x}\ dx =$ $(g(x) = \log_2(x)$ or try $\log_2(x) = \log_2(e)\log_e(x))$

12. $\displaystyle\int x\sec(x^2)\tan(x^2)dx =$ $(g(x) = x^2)$

13. $\displaystyle \int \frac{u}{(u^2 + 29)^{40}}\, du\ =$ $(g(u) = u^2 + 29)$

14. $\displaystyle \int \frac{\csc^2 x}{\sqrt{1 + 5\cot x}}\, dx\ =$ $(g(x) = 1 + 5\cot(x))$

15. $\displaystyle \int \frac{(\arctan(x))^{3/2}}{1 + x^2}\, dx\ =$ $g(x) = \arctan(x))$

Now check the solutions to EXERCISE 4.5 in Appendix 3. Note, in particular, the "notational style" used in these solutions. This is a matter of personal taste. Strive to develop your own style.

4.6   VARIATIONS ON EXERCISE 4.5   Either g(x) or f(x) is missing —you find it. Evaluate the integrals.

1. $\displaystyle \int \frac{x^7}{(x^8 + 1)^6}\, dx$ $(f(x) = x^{-6},\ g(x) =$ $)$

2. $\displaystyle \int \sin^m x \cos x\, dx$ $(f(x) =$ $,\ g(x) = \sin x)$

3. $\displaystyle \int \frac{(x + 1)^2}{x}\, dx \left(\text{reduce } \frac{(x + 1)^2}{x} = x + 2 + 1/x \right)$

4. $\displaystyle \int \sin^3 x\, dx$ $(g(x) = \cos(x),\ f(x) = 1 - \cos^2 x)$

5. $\displaystyle \int (x + 1/x)^2 dx$ (square the integrand)

6. $\displaystyle \int \frac{\csc^2 x}{\cot x + 1}\, dx$ $(f(x) = 1/x,\ g(x) =$ $)$

7. $\displaystyle \int \sec^3 x \tan x\, dx$ $(f(x) = x^2,\ g(x) =$ $)$

8. $\displaystyle \int \frac{d\theta}{\cos \theta} \left(\text{let } g(\theta) = \sin\theta, \text{ then } \int \frac{d\theta}{\cos \theta} = \int \frac{d\sin\theta}{1 - \sin^2\theta} \right)$

9. $\displaystyle \int x \sin^2 x^2 \cos x^2 dx$ $(f(x) = x^2,\ g(x) =$ $)$

10. $\displaystyle \int e^{\cos x^2} x \sin x^2 dx$ $(f(x) = e^x,\ g(x) =$ $)$

11. $\displaystyle \int \frac{(\arctan x)^\alpha}{1 + x^2}\, dx$ $(f(x) = x^\alpha,\ g(x) =$ $)$

12. $\displaystyle \int (\alpha \sin rt)^\beta \cos rt\, dt$ $(f(t) =$ $,\ g(t) = \alpha \sin rt)$

13. $\int (x + 1)e^{x^2 + 2x} dx$ $(f(x) = \quad , g(x) = (x + 1)^2)$

14. $\int \dfrac{\ln^m x}{x} dx$ $(m > 0)$ $(f(x) = \quad , g(x) = \ln x)$

15. $\int (x \ln x \cdot \ln\ln x)^{-1} dx$ (use $g(x) = \ln\ln x$)

4.7   VARIATIONS ON EXERCISE 4.5   Here are some additional tricks
to study. Evaluate the integrals.

1. $\int \dfrac{x^\alpha}{(x^{\alpha+1} + \beta)} dx$ $(g(x) = x^{\alpha+1} + \beta, f(x) = 1/x)$

2. $\int \sin 2x \cos^2 x \, dx$ (use $\sin 2x = 2 \sin x \cos x$, $g(x) = \cos x$)

3. $\int \dfrac{x^2}{x + 1} dx$ (use $x^2 = (x + 1 - 1)^2$, and $g(x) = x + 1$)

4. $\int \cos^2 x \, dx$ $\left(\text{use } \cos^2 x = \dfrac{1 + \cos 2x}{2}\right)$

5. $\int (x + 1)(x^2 + 2x + 1)^9 dx$ $(g(x) = (x + 1)^2)$

6. $\int \dfrac{\cos x}{\sin x} dx$ $(g(x) = \sin x)$

7. $\int \csc^m x \cot x \, dx$ $(g(x) = \csc x)$

8. $\int \dfrac{1}{\sin \theta} d\theta$ $\left(\text{use } \dfrac{1}{\sin \theta} = \dfrac{\sin \theta}{1 - \cos^2 \theta}, g(\theta) = \cos \theta\right)$

9. $\int x \tan^2 x^2 \sec^2 x^2 dx$ $(g(x) = \tan x^2)$

10. $\int \ln(\sin x^2) \cdot x \cos x^2 \, dx$ $(g(x) = \sin x^2)$

11. $\int \dfrac{(\text{arccot } x^2)^\alpha x}{1 + x^4} dx$ $(g(x) = \text{arccot } x^2)$

12. $\int t^{\alpha-1} \cos t^\alpha \cdot \sin t^\alpha \, dt$ $(g(t) = \cos t^\alpha)$

13. $\int e^{\log_{10} x} dx$ (use $e^{\log_{10} x} = x^{\log_{10} e}$)

14. $\int \dfrac{1}{x \ln x} dx$ (use $g(x) = \ln x$)

15. $\int \dfrac{(1 + 1/x^2)^m}{x^3} dx$ (use $g(x) = x^{-2}$)

4.8   VARIATIONS ON EXERCISE 4.5

1. $\displaystyle \int \frac{x}{(x^2 + 1)}\, dx =$

2. $\displaystyle \int \frac{\cos\sqrt{x}\,\sin\sqrt{x}}{\sqrt{x}}\, dx =$

3. $\displaystyle \int \frac{x^2 + 1}{x^3}\, dx =$

4. $\displaystyle \int \frac{\sin x}{\cos x}\, dx =$

5. $\displaystyle \int \frac{6x^3 + 5x}{3x^4 + 5x^2}\, dx =$

6. $\displaystyle \int \frac{\sec^2 x}{\tan x + 1}\, dx =$

7. $\displaystyle \int \pi\csc^2\pi^2 x \cot\pi^2 x\, dx =$

8. $\displaystyle \int \frac{(3x + 4)^{1/5}}{(x + 4)^{4/5}}\, dx =$

9. $\displaystyle \int (4x + 8)(x^2 + 4)\, dx =$

10. $\displaystyle \int x(x^2 + 1)^{3/2}(x^2 + 1)^{1/2}\, dx =$

11. $\displaystyle \int \frac{(1 + \sec x \tan x)}{(x + \sec x)}\, dx =$

12. $\displaystyle \int \sin 4x \cos 4x\, dx =$

13. $\displaystyle \int xe^{x^2} \cdot e^{x^2}\, dx =$

14. $\displaystyle \int \frac{\log_5 4x^3}{x}\, dx =$

15. $\displaystyle \int 2^{\cot 3x^2}(x\csc^2 3x^2) 2^{\cot 3x^2} dx =$

4.9   VARIATIONS ON EXERCISE 4.5

1. $\displaystyle \int x\cos x^2 (e^{\sin x^2})^2\, dx =$

2. $\displaystyle \int \frac{x\ln(3x^2 + 2)}{6x^2 + 4}\, dx =$

3. $\displaystyle \int [\log_3(\sin x)] \cot(x)\, dx =$

4. $\displaystyle \int ((\sin x)e^{x\sin x} + x(\cos x)e^{x\sin x})\,dx$

5. $\displaystyle \int xe^{5x^2}\sec^2(e^{5x^2})\tan(e^{5x^2})\, dx =$

6. $\displaystyle \int (x^4 + 1)^{49}x^3\, dx =$

7. $\displaystyle \int x\sin^5(x^2)\cos(x^2) =$

8. $\displaystyle \int \frac{\cos \sqrt{x}}{2\sqrt{x}}\, dx =$

9. $\displaystyle \int 5x^2(4x^3 + 1)^{4/5}\, dx =$

10. $\displaystyle \int \frac{\ln^6 (\sqrt{2x + 1})}{(2x + 1)}\, dx =$

11. $\displaystyle \int \sqrt{\sin(e\theta)}\, \cos(e\theta)\, d\theta =$

12. $\displaystyle \int (10\sin t)^7\cos t\, dt =$

13. $\displaystyle \int (\sin x + x\cos x)e^{x\sin x}\, dx =$

(Compare with 4)

14. $\displaystyle \int \sin(\sin\theta) \cos\theta\, d\theta =$

15. $\displaystyle \int \frac{\sin\theta}{(5 - \cos\theta)^{10}}\, d\theta =$

Will Computers Save Us From Integral Calculus?

The rest of this chapter will be devoted to two tasks. One task is to learn some applications of integral calculus. The other is to learn some additional techniques for evaluating integrals. We have just learned a few techniques, most notably the method of "substitution" or recognizing the CHAIN RULE in reverse. In order to make things more interesting, we shall first give some applications of integral calculus. Then we'll learn some more techniques of integration, followed by a few more applications. It is now the case, and it will be even more the case in the future, that sophisticated computer programs will perform the routine computation of integrals. Strange as it may seem, this does not diminish the importance of learning the basic techniques of integration. The reason for this is that the user of algebraic symbol manipulation programs must still intercede to put the input and output into "canonical forms." In other words, one must still be able to transform problems and answers into equivalent forms that may appear quite different at first glance. The techniques for making these various transformations are exactly the techniques we shall study in this chapter.

The Fundamental Theorem Of Calculus

   As our first task, we shall study a simple but very important idea called "the fundamental theorem of calculus." Look at FIGURE 4.10. There you see a solid graph of f and another dashed graph of a function $F(x) = S_a^x(f)$. The function f was just drawn with a smooth curve and is not the result of computing the values of some expression involving the variable x. In other words, the function f is given graphically. There is nothing wrong with that! This is the way we began our study of derivatives in FIGURE 1.2. The next thing we did was choose a point $a = -3$ which we use as a "base point" for computing $S_a^x(f)$. The function $F(x) = S_a^x(f)$ is called the "signed area" function for f with base point $a = -3$ (we could have picked any other value for a in the domain of f). For $x > a$, the signed area $S_a^x(f)$ is, by definition, the area between the graph of f and the portion of the horizontal axis between a and x.

   In computing this area, areas of regions below the horizontal axis are given negative sign and areas of regions above are given positive sign. For example, the region between the graph of f and the interval $[-3, -2]$ on the horizontal axis has area -1.93. The region between the graph of f and the interval $[+2, +3]$ has area $+1.52$. The dashed curve in FIGURE 4.10 is the graph of $S_a^x(f)$ as a function of x. Note that at $x = +6$ the value of the signed area function with base point -3 is $+5$. In other words, $S_a^6(f) = +5$. This means that the total accumulated signed area between the graph of f and the interval $[-3, +6]$ on the horizontal axis is $+5$.

   The values for each unit interval are computed in FIGURE 4.10 (by tedious inspection!) to get -1.93 for the area between f and $[-3, -2]$, -1.29 for

FIGURE 4.10   The Signed Area Function

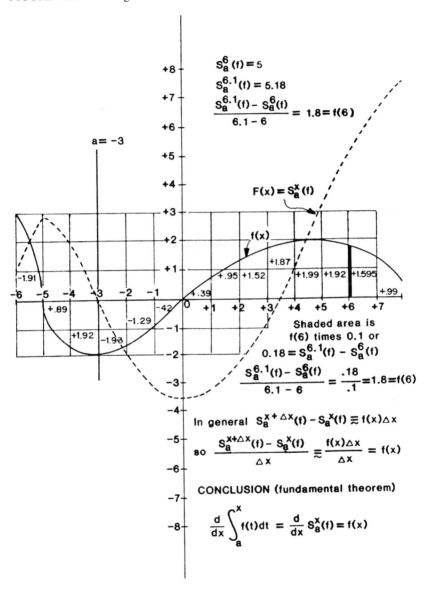

the area between f and $[-2, -1]$, etc. Adding these values, we get approximately $+5$ for the accumulated signed area $S_a^6(f)$. If $x < a$, we compute the signed area in the same way except that regions below the horizontal axis get positive weight and regions above get negative weight.

Again, look carefully at FIGURE 4.10. In particular, look at $+6$ on the horizontal axis. At 6, $f(6) = 1.8$ and $S_a^6(f) = 5$. A careful check reveals that at 6.1, $S_a^{6.1}(f) = 5.18$ and the difference, $S_a^{6.1}(f) - S_a^6(f) = .18$. The number .18 represents the difference between the accumulated area between the graph of f and the interval $[-3, 6.1]$ and the accumulated area between the graph of f and the interval $[-3, 6]$. This area, .18, is represented by the shaded region shown in FIGURE 4.10 between the graph of f and the interval $[6, 6.1]$. This shaded region looks like a tiny rectangle (almost) and has area about $f(6)$ times .1, or 1.8 times .1, which is .18. If we replace 6 by a generic point x and .1 by Δx this fact becomes $S_a^{x+\Delta x}(f) - S_a^x(f) = f(x)\Delta x$ or

$$\frac{S_a^{x+\Delta x}(f) - S_a^x(f)}{\Delta x} \approx f(x).$$

The equality in the above expression is approximate but better and better as Δx gets smaller (for graphs of continuous functions such as defined in DEFINITION 3.12). We can write this as

$$\lim_{\Delta x \to 0} \frac{S_a^{x+\Delta x}(f) - S_a^x(f)}{\Delta x} = \frac{d}{dx}S_a^x(f) = f(x).$$

We have used here ERUDITE OBSERVATION 1.14 applied to the function $S_a^x(f)$ of x (instead of to $f(x)$ as in the definition). We summarize these ideas in THEOREM 4.11.

4.11 THEOREM (FUNDAMENTAL THEOREM OF CALCULUS)
Let f be defined and continuous for all x in the interval $[c, d] = \{t : c \leq t \leq d\}$ and let a be a number in $[c, d]$. Let $S_a^x(f)$ be the signed area function with base point a. For every point x in the interval $(c, d) = (t : c < t < d)$, the function $F(x) = S_a^x(f)$ has a derivative and $\frac{d}{dx}F(x) = f(x)$. Thus the signed area function is an antiderivative or integral of $f(x)$.

The Envelope Game

It's time to play the ENVELOPE GAME again. Imagine that you have an envelope and inside is the signed area function of $f(x) = x^2$ based at 0. What will you see when you open the envelope? One possibility is that you will see a graph such as that of $S_a^x(f)$ shown in FIGURE 4.10. Someone could have drawn a careful graph of the function x^2 and tediously measured the area, making a graph of the resulting signed area function $F(x) = S_0^x(f)$ with base point $a = 0$. If this is what you find in the envelope you

will know that whoever put it there wasn't thinking very hard! We learned
at the very beginning of this chapter that if F(x) and G(x) are two antiderivatives for f(x) then they differ by the constant function, or F(x) = G(x) +
C. We know that G(x) = $x^3/3$ is an antiderivative function for x^2. This means
that all this measuring of signed area was a total waste of time in this case.
The signed area function must be $x^3/3$ + C. In this case we easily see that
C must be zero since F(0) is zero by definition of signed area with base point
0 and, obviously, G(0) = 0 so F(0) = G(0) + C implies that C = 0.

   You should think very carefully about what we have just done. This basic
idea has many amazing variations and accounts in large part for the usefulness
of calculus. A tedious task of computing areas under graphs of functions f(x)
has been replaced by a seemingly very different task of guessing antiderivatives F(x) for such functions. We can become very clever at guessing antiderivatives and thus very good at computing areas. The area between the
graph of f(x) = x^2 and the interval [0, 1] is $S_0^1(f)$ = $(1)^3/3$ = 1/3. Try
computing the same area without calculus! Probably, it sounds sort of interesting but not really all that useful to you to spend your time computing areas
under curves. What happens is that many problems in physics, chemistry,
engineering, astronomy, etc., are just this sort of problem in disguise. Before
going on, we need to develop some notation.

Some Very Important Ideas

4.12   GENERAL REMARKS AND NOTATION   We have learned that,
given a function f(x), any function G(x) such that $\frac{d}{dx}G(x)$ = f(x) is called
an antiderivative or integral of f(x). We know that all antiderivatives differ
by constant functions. Thus we can construct all other antiderivatives by
forming G(x) + C where C is some constant. We learned in THEOREM
4.11 that antiderivative functions can be constructed graphically. These antiderivative functions are called the "signed area" functions and are denoted
by $S_a^x(f)$. If F(x) = $S_a^x(f)$ is the signed area function of f based at a, then
clearly F(a) = 0.

   You can think of the graphs of all antiderivatives of f as being exactly the
same shape but shifted up or down with respect to the vertical axis. Two such
functions are either equal nowhere or equal everywhere. Said in another way,
if F(x) and G(x) are antiderivatives of f and F(a) = G(a) for some number
a, then F(x) = G(x) for all values of x (for which the functions are defined).
In particular, any antiderivative G(x) which has G(a) = 0 must be exactly

the signed area function $F(x) = S_a^x(f)$. This means that the set of all signed area functions for f is just the set of all antiderivatives $G(x)$ that are zero for some value a (satisfy $G(a) = 0$ for some a). If $G(a) = 0$ then $G(x) = S_a^x(f)$ for all values of x. It might happen that $G(b) = 0$ also for some b different from a. In this case $G(x) = S_a^x(f) = S_b^x(f)$. It is easy to see that there are antiderivative functions which are not signed area functions. For example, $x^2 + 2$ is an antiderivative for x, but not a signed area function for x^2 since it does not vanish for any value of x.

The common notation for the signed area function $S_a^x(f)$ in calculus is $\int_a^x f$, or $\int_a^x f(t)dt$ if we wish to be explicit about the dependent variable of f. It does not make any difference what we call this dependent variable: $\int_a^x f(t)dt = \int_a^x f(y)dy = \int_a^x f(z)dz = \ldots$. Even $\int_a^x f(x)dx$ is o.k. but best avoided! They are all equal to $S_a^x(f)$. The signed area functions $\int_a^x f(t)dt$ are called the "definite integrals" of f. Specifically, we say $\int_a^x f(t)dt$ is the "definite integral of f from a to x." The signed area function $\int_a^x f(t)dt$ viewed as a function of x is, as we have seen, nothing else than the antiderivative of f that vanishes at $x = a$. If c is a number, then the value of the definite integral at c is $\int_a^c f(t)dt$. This value is just a number. Thus $\int_0^1 t^2 dt = 1/3$.

The numbers such as a and x or 0 and 1 that appear at the top and bottom of the integral sign ($\int_a^x$ or $\int_0^1$) are called the "limits of integration." If the integral sign has no limits, such as $\int f(x)dx$, then this stands for just any old antiderivative or integral of f. Sometimes the phrase "indefinite integral $\int f$" is used to mean "integral of f" or "antiderivative of f." As we remarked before, the notation of calculus ranges from bad to horrible, but by working many examples, you will learn to tolerate it if not to love it!

If you want to compute a signed area function or definite integral $\int_a^x f(t)dt$, you can do so by finding any antiderivative H for f. Then we must have that $G(x) = H(x) - H(a)$ is also an antiderivative. But $G(a) = 0$, so, by what we have said above, $G(x) = \int_a^x f(t)dt$.

The Riemann Sum

As we have already remarked, in computing the signed area function of the function f of FIGURE 4.10 we used tedious numerical approximations to the areas bounded by the graph of f and small intervals on the horizontal axis. In the case of FIGURE 4.10, each subinterval was chosen to be of length one. The various areas that were computed for each subinterval are shown in FIGURE 4.10. It is very important to describe this process a little more carefully. Even in cases where we can find an antiderivative for f, it is a useful intuitive guide to think about how the computation would be done geometrically.

Consider FIGURE 4.13. There we see the graph of a function f. We are interested in computing the definite integral (signed area) from a to b. In other words, we want to compute $\int_a^b f(t)dt$. To approximate this integral, we have divided the interval [a, b] into subintervals $[x_1, x_2]$, $[x_2, x_3]$, . . . , $[x_{11}, x_{12}]$. In each subinterval $[x_i, x_{i+1}]$, we have chosen a number t_i. We then approximate the definite integral by

$$\int_a^b f(t)dt = \sum_{i=1}^{11} f(t_i)(x_{i+1} - x_i).$$

If we let $\Delta t_i = x_{i+1} - x_i$ this sum becomes

$$\int_a^b f(t)dt = \sum_{i=1}^{11} f(t_i)\Delta t_i.$$

A sum such as that used above to approximate a definite integral is called a "Riemann sum."

FIGURE 4.13   Riemann Sum

$$\sum_{i=1}^{n} f(t_i)\Delta t_i \text{ where } \Delta t_i = x_{i+1} - x_i \text{ and } n = 11.$$

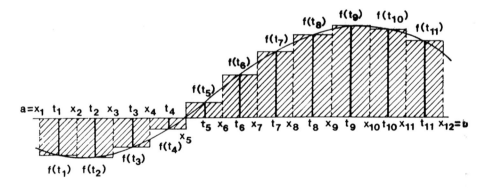

4.14 NOTATION (RIEMANN SUM)   Let f be a continuous function defined on the interval [a, b]. Let $a = x_1 < x_2 < . . . < x_{n+1} = b$ be points in the interval [a, b]. Define $\Delta t_i = x_{i+1} - x_i$. Choose t_i in $[x_i, x_{i+1}]$, $i = 1, . . . , n$. The following sum is called the Riemann sum based on the points x_i and t_i:

$$\sum_{i=1}^{n} f(t_i)\Delta t_i.$$

As we let n get larger and larger in such a way that the Δt_i go to zero for all i, the Reimann sum becomes a better and better approximation to the signed area $\int_a^b f(t)dt$. We write this as

$$\text{limit} \sum_{i=1}^{n} f(t_i)\Delta t_i = \int_a^b f(t)dt.$$

In using the Riemann sum to approximate an integral, a little common sense must be applied. In 4.14, for example, we have

$$\sum_{i=1}^{n} f(t_i)\Delta t_i = \int_a^b f(t)dt.$$

But from the definition of the signed area function

$$\int_a^b f(t)dt = -\int_b^a f(t)dt.$$

But for the finite sum, $\sum_{i=1}^{n} f(t_i)\Delta t_i = \sum_{i=n}^{1} f(t_i)\Delta t_i.$

The above statement simply says that the order of summation doesn't make any difference. The order in which the limits on the integral are written does make a sign difference. If $a < b$ as in 4.14, then the Riemann sum is a correct approximation to the definite integral $\int_a^b f(t)dt$. To approximate $\int_b^a f(t)dt$, approximate first $\int_a^b f(t)dt$ as in 4.14, and change the sign. This is equivalent to changing Δt_i to $-\Delta t_i$, $i = 1, 2, \ldots, n$, in the Riemann sum. We shall use Riemann sums quite a bit in Chapter 5.

We now attempt some exercises using the relationship between integrals and signed areas. Try each of the exercises in EXERCISE 4.15. If you have trouble, look at the solution of that exercise in SOLUTIONS TO EXERCISE 4.15. You should then make some change, however minor, in the problem that caused you trouble and rework it. After this, try the VARIATIONS ON EXERCISE 4.15.

4.15 EXERCISES In these exercises "area" means actual area, not signed area.

(1) Find the area bounded by the graph of $y = x^3$, the lines $x = -1$, $x = +2$, and the horizontal axis.

(2) Find the area bounded by the graph of $y = x^3$, the lines $y = +8$, $y = -1$, and the vertical axis.

(3)   Find the area of the bounded region between the curves $y = -x^2 + 2$ and $y = 2^x$.

(4)   Find the area enclosed by the curve $\{(1 + 2\cos(t), 2 + 3\sin(t)): 0 \le t < 2\pi\}$.

(5)   Using polar coordinates, find the area enclosed by the curve $r(\phi) = 1 + \cos(\phi), 0 \le \phi < 2\pi$.

4.16   SOLUTIONS TO EXERCISE 4.15

(1)   A sketch of the situation looks as follows:

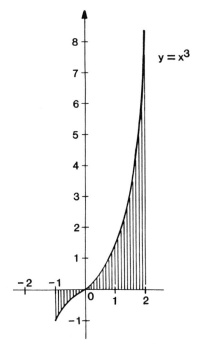

We are trying to find the area of the shaded region. The expression $\int_{-1}^{+2} x^3 dx$ would give the *signed area* between the graph of $f(x) = x^3$ and the interval $[-1, +2]$. The area between the curve and the interval $[-1, 0]$ would have negative weight. The problem as stated should be interpreted as finding the actual area. If we choose the base point for the signed area function to be $a = 0$, then the actual area is

$$\int_0^{-1} x^3 dx + \int_0^{+2} x^3 dx.$$

An antiderivative of x^3 is $F(x) = x^4/4$. Thus the required area is $F(-1) - F(0)$ plus $F(+2) - F(0)$ or $1/4 + 4 = 17/4$. One standard way of writing $F(b) - F(a)$ is

$$\left. \frac{x^4}{4} \right]_a^b .$$

Thus, we may write $(F(-1) - F(0)) + (F(2) - F(0))$ as

$$\left. \frac{x^4}{4} \right]_0^{-1} + \left. \frac{x^4}{4} \right]_0^2 = 1/4 + 4 = 17/4.$$

(2) We are trying to find the actual area of the dotted region in the following sketch:

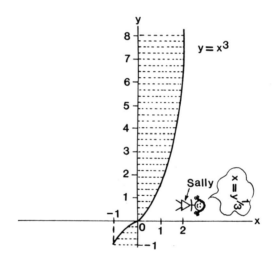

The easiest way to work this problem is to view it as Sally is in the sketch above. She thinks of y as the dependent variable and is looking at the curve $x = y^{1/3}$. She computes the signed area function based at $y = 0$ to get

$$\int_0^{-1} y^{1/3} dy + \int_0^8 y^{1/3} dy = \left. \frac{y^{4/3}}{4/3} \right]_0^{-1} + \left. \frac{y^{4/3}}{4/3} \right]_0^8 = 51/4.$$

Another way to evaluate this same area is by computing the following definite integrals (why?):

$$\int_{-1}^0 (x^3 + 1) dx + \int_0^8 (8 - x^3) dx.$$

(3) This is a typical calculus problem with an atypical twist. The graph of $-x^2 + 2$ is a parabola with the line $x = 0$ as axis of symmetry. If you've forgotten about parabolas, etc., it doesn't make much difference, since by graphing a few points you should discover the following picture:

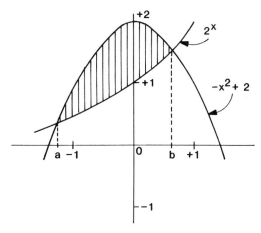

Thus the required integral is $\int_a^b (-x^2 + 2 - 2^x)dx$. To evaluate this integral, we must know a and b. In other words, we must know when $-x^2 + 2 - 2^x = 0$. Most calculus books set up these problems so that it is easy to find a and b (using, say, the quadratic formula). This problem is more realistic in that no such simple method works. Still, with a computer the problem is easy to solve. It is clear from the sketch that $-2^{1/2} < a < b < 2^{1/2}$ and $2^{1/2}$ is about 1.4. The following BASIC program gives us a closer look at the function $-x^2 + 2 - 2^x$ in the interval of interest.

```
10 FOR X = -1.4 TO +1.4 STEP .1
20 PRINT X, -X*X+2-2^X
30 NEXT X
```

X	$-x^2+2-2^x$
-1.4	$-.3389292$
-1.3 $\longleftrightarrow$	$-9.612608E\text{-}02$
-1.2	$.1247248$
-1.1	$.3234836$
$-.9999999$	$.5000002$
$-.8999999$	$.6541135$
$-.7999998$	$.785651$
$-.6999998$	$.8944279$
$-.5999998$	$.9802461$

− .4999998	1.042893
− .3999998	1.082142
− .2999998	1.097748
− .1999998	1.089449
− 9.999981E-02	1.056967
1.937151E-07	.9999999
.1000002	.9182264
.2000002	.8113015
.3000002	.6788553
.4000002	.5204919
.5000002	.3357861
.6000002	.1242831
.7000003 ⟷	− .1145051
.8000003	− .3811016
.9000002	− .6760665
1	− 1.000001
1.1	− 1.353548
1.2	− 1.737397
1.3	− 2.15229

By inspecting the output of this program, we see that $-1.3 < a < -1.2$ and $.6 < b < .7$. The next two BASIC programs give us some more accuracy.

(a)

```
10 FOR X = − 1.3 TO − 1.2 STEP .01
20 PRINT X, − X*X + 2 − 2^x
30 NEXT X
```

x	$-x^2 + 2 - 2^x$
− 1.3	− 9.612608E-02
− 1.29	− 7.305104E-02
− 1.28	− 5.019555E-02
− 1.27	− 2.755976E-02
− 1.26 ⟷	− .005144
− 1.25	1.705173E-02
− 1.24	3.902725E-02
− 1.23	6.078249E-02
− 1.22	8.231711E-02
− 1.21	.1036312
− 1.2	.1247246

(b)

```
10 FOR X = .6 TO .7 STEP .01
20 PRINT X, − X*X + 2 − 2^x
30 NEXT X
```

X	$-x^2 + 2 - 2^x$
.6	.1242836
.61	.1016408
.62	7.872498E-02
.63	.0555352
.64	.032071
.65 ⟵——————⟶	8.332015E-03
.66	− 1.568234E-02
.67	− 3.997278E-02
.68	− 6.453955E-02
.69	− 8.938336E-02
.7	− .1145045

It seems that $a = -1.26$ and $b = .65$ are reasonably good approximations. Thus we compute

$$\int_{-1.26}^{.65} (-x^2 + 2x - 2^x)dx = -x^3/3 + 2x - 2^x/\ln(2) \Big]_{-1.26}^{.65}$$

$$= -1.06 - (-5.12) = 4.06$$

(4) This is the parametric equation of an ellipse with center at $(1, 2)$. The same ellipse with center at $(0, 0)$ has parametric equation

$$\{(2\cos(t), 3\sin(t)): 0 < t < 2\pi\}.$$

This ellipse, which obviously has the same area as the original one, looks as shown at the top of page 174.

We compute the shaded area and multiply by 4 to get the total area. Thus we compute

$$\int_0^2 ydx = \int_{\pi/2}^0 (3\sin(t))\, d(2\cos(t)) = -6\int_{\pi/2}^0 \sin^2(t)dt = 6\int_0^{\pi/2} \sin^2(t)dt.$$

Using the trigonometric identity $2\sin^2(t) = 1 - \cos(2t)$ we compute

$$6\int_0^{\pi/2} \sin^2(t)dt = 3\int_0^{\pi/2} (1 - \cos(2t))dt$$

$$= 3(t - (\sin(2t))/2) \Big]_0^{\pi/2} = 3\pi/2.$$

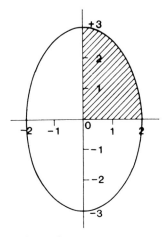

The total area is 4 times this or 6π.

(5)   This curve is called a "limacon" and looks as follows:

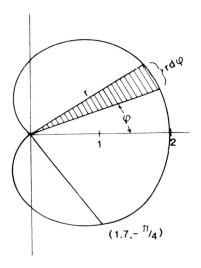

$(1.7,-{}^{\pi}/_4)$

The small shaded triangle has area $dA = r(\phi)(r(\phi)d\phi/2)$ where $A(\phi)$ is the area as a function of ϕ. Thus we compute

$$A(\phi) = (1/2)\int_0^{2\pi}(1 + \cos(\phi))^2d\phi = \int_0^{\pi}(1 + \cos(\phi))^2d\phi.$$

But $(1 + \cos(\phi))^2 = 1 + 2\cos(\phi) + \cos^2(\phi)$ and $\cos^2(\phi) = (\cos(2\phi) + 1)/2$. This gives

$$\int_0^\pi (1 + \cos(\phi))^2 d\phi = \phi + 2\sin(\phi) + (1/4)\sin(2\phi) + \phi/2 \Big]_0^\pi = 3\pi/2.$$

4.17   VARIATIONS ON EXERCISE 4.15

(1)   Find the area bounded by the graph $y = x^2$, the x-axis, and the lines $x = 0$ and $x = 2$. (Answer: 8/3)

(2)   Find the area bounded by the graph of $y = x^2$, the lines $y = 4$ and $y = 9$, and the y-axis. (Answer: 38/3)

(3)   Find the area between the curves $y = x^2$ and $y = x^4$. (Answer: 4/15)

(4)   Find the area enclosed by the curve $\{(4\cos(2\pi t), 4\sin(2\pi t)):0 < t < 1\}$. (Answer: 16π)

(5)   Using polar coordinates, find the area enclosed by the curve $r(\phi) = \phi$ and the ray $\phi = 0$, $0 \le \phi \le \pi$. (Answer: $\pi^3/6$).

4.18   VARIATIONS ON EXERCISE 4.15

(1)   Find the area bounded by the graph $y = e^x$, the x-axis, and the lines $x = 0$ and $x = 1$. (Answer: $e - 1$)

(2)   Find the area bounded by the graph of $y = e^x$, the lines $y = e$ and $y = e^2$, and the y-axis. (Answer: e^2)

(3)   Find the area between the curves $y = e^{2x}$ and $y = e^x$, $-2 < x < +2$.

(4)   Find the area enclosed by the curve

$$\{(1 + \cos(t), 3 + 4\sin(t)):0 < t < 2\pi/3\}$$

and the straight line segments that join the endpoints of this curve to the point $(1, 3)$.

(5)   Using polar coordinates, find the area enclosed by the curve $r(\phi) = 1 + \sin(\phi)$, $0 < \phi < \pi$, and the ray $\phi = 0$.

4.19   VARIATIONS ON EXERCISE 4.15

(1)   Find the area bounded by the graph, $y = x^{-1/2}$, the x-axis, the y-axis, and the line $x = 1$. (Answer: 2)

(2)   Find the area bounded by the graph of $y = x^{-1/2}$, the line $y = 1$, and the y-axis. (Answer: 1)

(3) Find the area between the curves $y = \sin(x)$ and $y = 2x/\pi$. (Answer: $2(1 - \pi/4)$)

(4) Find the area under the curve $\{(t^2, te^t):0<t<1\}$. (Answer: $(2/3)(e - 1)$)

(5) Using polar coordinates, find the area enclosed by the curve $r(\phi) = \phi + 1$ and the rays $\phi = 0$ and $\phi = \pi/2$. (Answer: $(1/6)(((\pi/2) + 1)^3 - 1)$)

4.20 VARIATIONS ON EXERCISE 4.15

(1) Find the area bounded by the graph $y = x^{-3}$, the x-axis, $x \geq 1$, and the line $x = 1$. Hint: Find the area under the curve $y = x^{-3}$ for $1 < x < \alpha$. Then, take the limit as α tends to infinity. (Answer: $1/2$)

(2) Find the area bounded by the graph of $y = x^{-3}$, the line $y = 1$, the x-axis, and the y-axis. Hint: Use the same trick used in problem (1). (Answer: $3/2$)

(3) Find the area between the curves $y = e^{-x}$ and $y = e^{-2x}$, $x > 0$. Hint: Once again, find the area between the curves for $0 < x < \alpha$ and let α tend to infinity. (Answer: $1/2$)

(4) Find the area between the curve $\{(t^{1/2}\sin(t), t^{1/2});0 < t < \pi/2\}$ and the y-axis. (Answer: $1/2$)

(5) Using polar coordinates, find the area enclosed by the curve $r(\phi) = 2\sin(\phi)$, $0 < \phi < \pi/2$, and the ray $\phi = \pi/2$. (Answer: $\pi/2$)

4.21 VARIATIONS ON EXERCISE 4.15

(1) Find the area bounded by the graph of $y = 3^x$, the lines $x = 0$, $x = +1$, and the horizontal axis.

(2) Find the area bounded by the graph of $y = 3^x$, the lines $y = 3/2$, $y = +3$, and the vertical axis.

(3) Find the area of the bounded region between the curves $y = -x^2 + 3$ and $y = 3^x$.

(4) Find the area enclosed by the curve $\{(a + b\cos(t), c + d\sin(t)):0 \leq t < 2\pi\}$ where $a, b > 0$, c and $d > 0$ are real numbers.

(5) Using polar coordinates, find the area enclosed by the curve $r(\phi) = 1 + 2\cos(\phi)$, $0 \leq \phi < \pi/2$.

4.22 VARIATIONS ON EXERCISE 4.15

(1) Find the area bounded by the graph of $y = x^3 + 1$, the line $x = +2$, and the horizontal axis. Explain the relationship between this problem and EXERCISE 4.15(1). In particular, how can you use the calculations of 4.15(1) to simplify the calculations of this problem.

(2) Find the area bounded by the graph of $y = x^3 + 1$, the lines $y = +9$, and the line $x = -1$.

(3) Find the area of the bounded region between the curves $y = -x^4 + 16$ and $y = (1.5)^x$.

(4) Find the area enclosed by the curve $\{(t - \sin(t), \ 1 - \cos(t)):0 \le t < 2\pi\}$ and the horizontal axis.

(5) Using polar coordinates, find the area enclosed by the curve $r(\phi) = \cos(2\phi)$, $0 \le \phi < \pi/2$.

4.23 VARIATIONS ON EXERCISE 4.15

(1) Find the area bounded by the graph of $y = x^4/4 - 2x^3/3$, the line $x = -1$, the line $x = 8/3$, and the horizontal axis.

(2) Find the area bounded by the graph of $y = x^4/4 - 2x^3/3$, the line $y = -4/3$, and the vertical axis.

(3) Find the area of the bounded region between the curve $y = -x^4/4 - 2x^3 + 4x^2 - 2$ and the horizontal axis. (Hint: y has two real roots, one near -10, the other near -0.6. Use your computer!)

(4) Find the area enclosed by the curve $\{(1 + \sin(t), \ \cos(t)(1 + \sin(t))): -\pi/2 \le t \le 3\pi/2\}$ and the horizontal axis.

(5) Using polar coordinates, find the area bounded by the line segment $\{(r, \phi): -1 < r < +1, \ \phi = 0\}$, the curve $r_1(\phi) = e^{\phi/\pi}$, $0 \le \phi \le \pi/2$, and the curve $r_2(\phi) = e^{1-\phi/\pi}$, $\pi/2 < \phi \le \pi$.

4.24 COMPUTING INTEGRALS Having had a glimpse of some of the applications of integration, it is now time to improve our technical ability to compute integrals. We first concentrate on three very important techniques: integration by parts, trigonometric identities, and trigonometric substitution. Mastering these techniques is almost entirely a matter of practice. The practice centers around EXERCISES 4.26, 4.27, and 4.28, and the VARIATIONS that follow. After learning these basic techniques of integration, we shall look at a fourth method, integration by partial fractions. Most students find this method extremely boring. Fortunately, this is a class of integrals that algebraic symbol manipulation packages do well with. The ideas we shall learn are still necessary for communicating with such software.

Integration By Parts—The Product Rule In Reverse

The method of substitution, the theme of EXERCISE 4.5 and its VARI-ATIONS, is concerned with what we called the "chain rule in reverse." The method of "integration by parts" is the "product rule in reverse." The idea is simple. We know that $(fg)' = f' g + fg'$. Hence $\int (fg)' = \int f' g + \int fg'$. We write this as

$$fg = \int f'g \, dx + \int fg' \, dx.$$

If, by hook or by crook, we know any two of the terms in the above expression, we can then find the third. What usually happens is that we have some function $h(x)$, for example, $h(x) = x\cos(x)$, and we are asked to evaluate $\int x\cos(x)dx$. We are stuck. In our mind we split $h(x)$ into a product of two functions $h(x) = p(x)q(x)$ such that we know how to differentiate $p(x)$ and we know how to integrate $q(x)$. To dramatize this fact, we call $p(x) = f(x)$ and $q(x) = g'(x)$. Just how to split $h(x)$ in this way may not be clear at first glance. You may have to fool around a bit. In the case of $h(x) = x\cos(x)$, we try $f(x) = x$ and $g'(x) = \cos(x)$. Then $f'(x) = 1$ and $g(x) = \sin(x)$. Thus the identity

$$fg = \int f'g \, dx + \int fg' \, dx$$

becomes

$$x\sin(x) = \int (1)\sin(x)dx + \int x\cos(x)dx.$$

Solving for $\int x\cos(x)dx$ gives

$$\int x\cos(x)dx = x\sin(x) + \cos(x)$$

where we have used the simple fact that $\int \sin(x)dx = -\cos(x)$. This sequence of events represents the method of integration by parts working perfectly. We state the method as follows:

4.25   INTEGRATION BY PARTS

$$\int f(x)g'(x)dx = f(x)g(x) - \int f'(x)g(x)dx.$$

For the method to be useful, we must know how to split up a given function $h(x)$ into a product of two functions $f(x)g'(x)$ such that $f(x)$ is reasonable to differentiate to obtain $f'(x)$, $g'(x)$ is reasonable to integrate to obtain $g(x)$, and (miracle of miracles) we can integrate $f'(x)g(x)$. To see how we can foul up in this process, suppose we had thought of $x\cos(x)$ as $f(x)g'(x)$ with $f(x) = \cos(x)$ and $g'(x) = x$. We find easily that $f'(x) = -\sin(x)$ and $g(x) = x^2/2$. In this case, $f'(x)g(x) = (-1/2)x^2\sin(x)$ is not easy to integrate. You should note that the identity of 4.25 is still valid:

$$\int x\cos(x)dx = \cos(x)x^2/2 + \int (1/2)x^2\sin(x)dx.$$

The fact that we can't evaluate $\int(1/2)x^2\sin(x)$ as an explicit expression just means that we can't use 4.13 (with this particular choice of $f(x)$ and $g'(x)$) to evaluate $\int x\cos(x)dx$.

It helps some students to remember the integration by parts rule in conjunction with the following table:

$f(x)$	$g(x)$
$f'(x)$	$g'(x)$

$$\int f(x)g'(x)dx = f(x)g(x) - \int f'(x)g(x)dx.$$

In the above table, the integral of the product of the two functions along any diagonal is the product of the two functions in the top row minus the integral of the product of the functions along the other diagonal. Now try to work the problems in EXERCISE 4.26. EXERCISE 4.26(3) contains a useful trick. You apply integration by parts twice. After the second application, the original integral appears a second time to give an equation in which the original integral occurs in two places. Solve this equation for the original integral.

4.26 EXERCISES Work the following problems using integration by parts. In each case a HINT is given showing how to factor the integrand as a product fg'. Other factorizations may work as well.

(1) $\displaystyle\int \arctan(x)dx =$ HINT:

$\arctan(x)$	x
$\dfrac{1}{1 + x^2}$	1

RELATED INTEGRALS: $\int dx$ and $\int \dfrac{x}{1 + x^2} dx$

(2) $\int x \arctan(x) dx =$ HINT:

$\arctan(x)$	$\dfrac{x^2}{2}$
$\dfrac{1}{1 + x^2}$	x

RELATED INTEGRALS: $\int \dfrac{x^2}{(1 + x^2)} dx$

(3) $\int \sin(\ln(x)) dx =$ HINT:

$\sin(\ln(x))$	x
$\dfrac{\cos(\ln(x))}{x}$	1

HINT:

$\cos(\ln(x))$	x
$\dfrac{-\sin(\ln(x))}{x}$	1

(4) $\int \dfrac{x}{(2 + 3x)^{1/2}} dx$ HINT:

x	$\dfrac{2}{3}(2 + 3x)^{1/2}$
1	$(2 + 3x)^{-1/2}$

RELATED INTEGRALS: $\int (2 + 3x)^{-1/2} dx$ and $\int (2 + 3x)^{1/2} dx$

(5) $\int \sec^3(x) dx =$ HINT:

$\sec(x)$	$\tan(x)$
$\sec(x)\tan(x)$	$\sec^2(x)$

RELATED INTEGRALS: $\int \sec(x) dx$ and $\int \sec(x)\tan^2(x) dx$ (use $1 + \sec^2(x)$ $= \tan^2(x)$)

Trigonometric Integrals

In the next set of exercises, 4.27, we work with trigonometric integrals. The method used to solve these integrals is based on the use of trigonometric identities. You should review BASIC TRIGONOMETRIC FACTS 2.22. Learning this technique of integration is done by practice as there is nothing much to the sage advice "use trig identities." As an example, consider EXERCISE 4.23(3). using $\sin(x)\cos(x) = \sin(2x)$, replace $\sin^2(x)\cos^2(x)$ by $\sin^2(2x)$ to obtain

$$\int \sin^2(x)\sin^2(2x)dx.$$

Now use the identity $\sin^2(x) = (1 - \cos(2x))/2$ to obtain

$$(1/2) \int (1 - \cos(2x))\sin^2(2x)dx =$$

$$(1/2) \int \sin^2(2x)dx - (1/2) \int \sin^2(2x)\cos(2x)dx.$$

Using substitution on the second integral (as in EXERCISE 4.5) gives

$$(1/2) \int \sin^2(2x)\cos(2x) = (1/12)\sin^3(2x).$$

In the first integral, replace $\sin^2(2x)$ by $(1 - \cos(4x))/2$ to give

$$(1/4) \int (1 - \cos(4x))dx = x/4 - (1/16)\sin(4x).$$

This gives as a final answer

$$\int \sin^2(x) \sin^2(2x)dx = x/4 - (1/16)\sin(4x) - (1/12)\sin^3(2x) + C.$$

Sooner or later, better sooner, you should become acquainted with the Table of Integrals in Appendix 1. Now try the trigonometric integrals of EXERCISE 4.27 with and without the Table of Integrals. Check the solutions in Appendix 3.

4.27 EXERCISES Work the following trigonometric integrals:

(1) $\int \sec(x)dx =$ HINT: $\sec(x)$

$$= \frac{\sec(x)(\sec(x) + \tan(x))}{\sec(x) + \tan(x)}$$

(2) $\int \cos^3(x)\sin^2(x)dx =$ HINT: $\cos^2(x) = 1 - \sin^2(x)$

(3) $\int \sin^4(x)\cos^2(x)dx =$

(4) $\int\cos(5x)\cos(x)dx$ =   HINT: $\cos(6x) = \cos(5x + x)$
$\cos(4x) = \cos(5x - x)$

(5) $\int\sin(3x)\cos(2x)dx$ HINT: $\sin(5x) = \sin(3x + 2x)$
$\sin(x) = \sin(3x - 2x)$

Trigonometric Substitution

The next class of integrals that we consider is related to the technique called "trigonometric substitution." Take a look at the five problems of EXERCISE 4.28. They all involve a number plus or minus the square of the variable. Look at the HINT provided for each case. The given triangle specifies a relation between the function to be integrated and a trigonometric function of the variable τ. In 4.28(1) we see that $x = 4\tan(\tau)$ and thus $dx = 4\sec^2(\tau)d\tau$. The integrand in 4.28(1) is evidently $((1/4)\cos(\tau))^4$. Making these substitutions transforms the integral of 4.28(1) into $4^{-3}\int\cos^2(\tau)d\tau$. Evaluating this integral gives $4^{-3}((1/4)\sin(2\tau) + \tau/2)$. We now express this in terms of the variable x. Note first that $\sin(2\tau) = 2\sin(\tau)\cos(\tau)$. Then, by looking at the triangle in the HINT, we see that $\tau = \arcsin(x/(16 + x^2)^{1/2})$ and also $\tau = \arccos(4/(16 + x^2)^{1/2})$. Substituting the first expression for τ into $\sin(\tau)$ and the second into $\cos(\tau)$ gives

$$\sin(2\tau) = 8x/(16 + x^2).$$

Using $\tau = \arctan(x/4)$ gives

$$\int(16 + x^2)^{-2}dx = 4^{-3}(2x/(16 + x^2) + (1/2)\arctan(x/4)).$$

You should check this answer by differentiation with respect to x. Now try EXERCISE 4.28.

4.28 EXERCISES Work each of the following exercises by trigonometric substitution:

(1) $\displaystyle\int\frac{1}{(16 + x^2)^2}dx$ =   HINT: x

RELATED INTEGRALS: $\int\sin^2(\tau)d\tau$ and $\int\cos^2(\tau)d\tau$

(2) $\displaystyle\int \frac{x^2}{(4 - x^2)^{5/2}}\, dx \; =$ HINT: x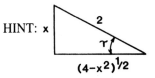

RELATED INTEGRALS: $\int\tan^2(\tau)\sec^2(\tau)d\tau$ and $\int\cot^2(\tau)\csc^2(\tau)d\tau$

(3) $\displaystyle\int \frac{1}{(9 - x^2)^{3/2}}\, dx \; =$ HINT: x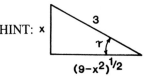

RELATED INTEGRALS: $\int\sec^2(\tau)d\tau$ and $\int\csc^2(\tau)d\tau$

(4) $\displaystyle\int (x^2 + 25)^{1/2}dx \; =$ HINT: x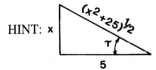

RELATED INTEGRALS: $\int\sec^3(\tau)d\tau$ and $\int\csc^3(\tau)d\tau$

(5) $\displaystyle\int \frac{1}{(9 + x^2)^2}\, dx \; =$ HINT: x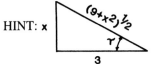

RELATED INTEGRALS: $\int\cos^4(\tau)d\tau$ and $\int\sin^4(\tau)d\tau$

We now give a series of VARIATIONS on EXERCISES 4.26, 4.27, and 4.28. Each VARIATION contains problems on integration by parts, trigonometric integrals, and trigonometric substitution. EXERCISES 4.29–4.31 and

EXERCISES 4.32–4.34 follow the pattern of EXERCISES 4.26, 4.27, and 4.28 but with fewer hints. In EXERCISES 4.35–4.37, we still classify the problems by techniques but no hints are given. In EXERCISE 4.38 we don't tell you the technique to use, but the answers are given. Finally, in EXERCISE 4.39 you are entirely on your own. You'll soon be an expert on techniques of integration. Good luck!

4.29   VARIATIONS ON EXERCISES 4.26, 4.27, 4.28

Integration by parts:

(1) $\int \arcsin(x)dx =$ HINT:

arcsinx	
	1

(2) $\int x^2 \arctan(x)dx =$ HINT:

arctan(x)	
	x^2

(3) $\int \cos(\ln x)dx =$ HINT:

cos(lnx)	
	1

HINT:

sin(lnx)	
	1

(4) $\int x\sqrt{2 + 3x}\, dx =$ HINT:

x	
	$(2 + 3x)^{1/2}$

(5) $\int \csc^3(x)dx$ HINT:

csc(x)	
	$\csc^2(x)$

(HINT: Use $\cot^2 x + 1 = \csc^2 x$. $\int \csc(x)dx$ is to be worked in Trigonometric Integrals 4.30(1). Check also TABLE OF INTEGRALS in Appendix 1.)

4.30 VARIATIONS ON EXERCISES 4.26, 4.27, 4.28

Trigonometric integrals:

(1) $\displaystyle\int \csc(x)dx =$ **(4)** $\displaystyle\int \sin(5x)\sin x\,dx =$

(2) $\displaystyle\int \cos^3(x)\sin^3(x)dx =$ **(5)** $\displaystyle\int \sin(5x)\cos(2x) =$

(3) $\displaystyle\int \sin^2(x)\cos^2(x)dx =$

4.31 VARIATIONS ON EXERCISES 4.26, 4.27, 4.28

Trigonometric substitutions:

(1) $\displaystyle\int \frac{dx}{(25 + x^2)^2}$

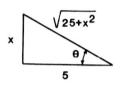

(2) $\displaystyle\int \frac{x^2 dx}{(2 - x^2)^{3/2}}$

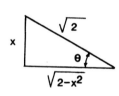

(3) $\displaystyle\int \frac{dx}{(16 - x^2)^{5/2}}$

(4) $\displaystyle\int (x^2 - 4)^{1/2}dx$

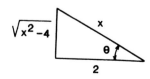

(5) $\displaystyle\int \frac{dx}{(x^2 - 9)^2}$

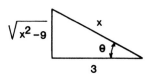

4.32   VARIATIONS ON EXERCISES 4.26, 4.27, 4.28   Integration by parts:

(1) $\displaystyle\int \arccos(x)dx =$ HINT:

arccos(x)	
	1

(2) $\displaystyle\int x\arcsin(x)dx =$ HINT:

arcsin(x)	
	x

(3) $\displaystyle\int \frac{x^2 dx}{\sqrt{1 - x^2}} =$ HINT:

x	
	$\dfrac{x}{\sqrt{1 - x^2}}$

(Use an inverse trig substitution)

(4) $\displaystyle\int \frac{x^2 dx}{(5 + 2x)^{3/2}} =$ HINT:

x^2	
	$(5 + 2x)^{-3/2}$

HINT:

x	
	$-(5 + 2x)^{-1/2}$

(5) $\displaystyle\int \cos^2(x)dx =$ HINT:

cosx	
	cosx

RELATED INTEGRALS: $\int \sec^2(x)dx$ and $\int \sec(x)\tan^2(x)dx$ (Use $\sec^2 x = \tan^2 x + 1$.

4.33   VARIATIONS ON EXERCISES 4.26, 4.27, 4.28

Trigonometric integrals:

(1) $\displaystyle\int \tan(x)dx =$ **(4)** $\displaystyle\int \cos(5x)\cos(3x)dx =$

(2) $\displaystyle\int \cos^2(x)\sin^3(x)dx =$ **(5)** $\displaystyle\int \sin x\cos 2x dx =$

(3) $\displaystyle\int \sin^2(x)\cos^4(x)dx =$

4.34   VARIATIONS ON EXERCISES 4.26, 4.27, 4.28

Trigonometric substitutions:

(1) $\displaystyle\int \frac{dx}{36 + x^2} =$ **(4)** $\displaystyle\int (5 - x^2)^{1/2}dx =$

(2) $\displaystyle\int \frac{xdx}{(5 - x^2)^{5/2}} =$ **(5)** $\displaystyle\int \frac{dx}{(x^2 - 16)^3} =$

(3) $\displaystyle\int \frac{dx}{(16 - x^2)^{3/2}} =$

4.35   VARIATIONS ON EXERCISES 4.26, 4.27, 4.28

Integration by parts:

(1) $\displaystyle\int \text{arcsec}(x)dx =$ **(4)** $\displaystyle\int xe^{(2x + 3)}dx =$

(2) $\displaystyle\int x^2\arccos(x)dx =$ **(5)** $\displaystyle\int \sin^2(x)dx =$

(3) $\displaystyle\int e^x\sin x dx =$

4.36   VARIATIONS ON EXERCISES 4.26, 4.27, 4.28

Trigonometric integrals:

(1) $\displaystyle\int \cot(x)dx =$ **(4)** $\displaystyle\int \sin(3x)\sin(6x)dx =$

(2) $\int \cos(x)\sin^2(x)dx \ =$ **(5)** $\int x\sin x^2\cos 2x^2 \ =$

(3) $\int \sin^4(x)\cos^4(x)dx \ =$

4.37   VARIATIONS ON EXERCISES 4.26, 4.27, 4.28

Trigonometric substitutions:

(1) $\int \dfrac{dx}{(36 + x^2)^{5/2}}$ **(4)** $\int (x^2 + 9)^{1/2}dx \ =$

(2) $\int \dfrac{x^2 \ dx}{(36 - x^2)^{3/2}}$ **(5)** $\int \dfrac{dx}{(x^2 - 5)^2}$

(3) $\int \dfrac{dx}{(36 - x^2)^{5/2}}$

4.38   VARIATIONS ON EXERCISES 4.26, 4.27, 4.28   We give you
the answers but the problems are scrambled so you have to guess the method.
Before working the problems, go through and mark them IP (for integration
by parts), TI (for trigonometric identities), or TS (for trigonometric substitution). It will be fun to see how accurate your guesses as to method were.
(Check them against those of Appendix 3.)

(1) $\int \sqrt{x^2 + 16} \ dx \ =$

$$\frac{1}{2}\left(\frac{\sqrt{x^2 + 16}}{4}\right)\frac{x}{4} + \frac{1}{2}\ln\left|\frac{\sqrt{x^2 + 16} + x}{4}\right| + c$$

(2) $\int \cos^6(x)dx \ = \ \dfrac{1}{4}\sin(2x) \ - \ \dfrac{1}{48}\sin^3(2x) \ + \ \dfrac{3x}{16} \ + \ \dfrac{3}{64}\sin(4x)$

$+ \ \dfrac{x}{8} + c$

(3) $\int x\sin x \ dx \ = \ -x\cos x \ + \ \sin x \ + \ c$

(4) $\int \sin^4 x\cos^5 x dx \ = \ \dfrac{\sin^5 x}{5} \ - \ \dfrac{2\sin^7 x}{7} \ + \ \dfrac{\sin^9 x}{9} \ + \ c$

(5) $\displaystyle\int \sqrt{x^2 - 4}\, dx = \frac{1}{2}\left(\frac{\sqrt{x^2 - 4}}{2}\right)\frac{x}{2} - \ln\left|\frac{x + \sqrt{x^2 + 4}}{2}\right| + c$

(6) $\displaystyle\int \sin x \cos 2x\, dx = \cos x - \frac{2}{3}\cos^3 x + c$

(7) $\displaystyle\int \sin(\ln 2x)\, dx = \frac{x}{2}(\sin(\ln 2x) - \cos(\ln 2x)) + c$

(8) $\displaystyle\int \arcsin(2x)\, dx = x\arcsin(2x) + \frac{1}{2}(1 - 4x^2)^{1/2} + c$

(9) $\displaystyle\int \frac{dx}{x^2\sqrt{a - x^2}} = \frac{-1}{a}\frac{\sqrt{a - x^2}}{x} + c$

(10) $\displaystyle\int x\arctan(2x)\, dx = \frac{x^2}{2}\arctan(2x) - \frac{x}{4} + \frac{1}{8}\arctan(2x) + c$

(11) $\displaystyle\int \frac{\sqrt{36 - x^2}}{x}\, dx =$
$$- 6\ln\left|\frac{6}{x} + \frac{\sqrt{36 - x^2}}{x}\right| + 6\left(\frac{\sqrt{36 - x^2}}{x}\right) + c$$

(12) $\displaystyle\int \sin^2 x\cos^4 x\, dx = \frac{x}{16} - \frac{\sin(4x)}{64} + \frac{\sin^3(2x)}{48} + c$

(13) $\displaystyle\int \sin^5 x\cos^2 x\, dx = \frac{2\cos^5 x}{5} - \frac{3\cos^3 x}{3} - \frac{\cos^7 x}{7} + c$

(14) $\displaystyle\int \arctan(6x)\, dx = x\arctan(6x) - \frac{1}{12}\ln(1 + 36x^2) + c$

(15) $\displaystyle\int \frac{dx}{(9 - x^2)^{3/2}} = \frac{1}{3}\frac{x}{\sqrt{9 - x^2}} + c$

(16) $\displaystyle\int \frac{x^2 dx}{(a^4 - x^2)^{3/2}} = \frac{x}{\sqrt{a^4 - x^2}} - \arcsin\left(\frac{x}{a^2}\right) + c$

(17) $\displaystyle\int \sin^3 x \cos^5 x dx = \frac{\cos^8 x}{8} - \frac{\cos^6 x}{6} + c$

(18) $\displaystyle\int x^3 \sqrt{1 - x}\, dx = -\frac{2x^3}{3}(1 - x)^{3/2} - \frac{4x^2}{5}(1 - x)^{5/2} - \frac{16x}{35}$

$(1 - x)^{7/2} - \dfrac{16}{35}\left(\dfrac{2}{9}\right)(1 - x)^{9/2} + c$

(19) $\displaystyle\int x^3 e^{-x} dx = -x^3 e^{-x} - 3x^2 e^{-x} - 6xe^{-x} - 6e^{-x} + c$

4.39 VARIATIONS ON EXERCISES 4.26, 4.27, 4.28 Now you must find both the method and the answers!

(1) $\displaystyle\int \cos^4(x) dx =$

(2) $\displaystyle\int \frac{dx}{(16 - x^2)^2} =$

(3) $\displaystyle\int \sec x \tan^2 x dx =$

(4) $\displaystyle\int \frac{x^2}{(6 + 2x)^{5/2}}\, dx$

(5) $\displaystyle\int \sec(x) dx =$

(6) $\displaystyle\int \cos(2x)\cos(3x) dx =$

(7) $\displaystyle\int \frac{x^3 dx}{(16 - x^2)^{3/2}} =$

(8) $\displaystyle\int \text{arccsc}(x) dx =$

(9) $\displaystyle\int \cos^2(x)\sin^5(x) dx =$

(10) $\displaystyle\int (x^2 - 16)^{1/2} dx =$

(11) $\displaystyle\int x\ln x dx =$

(12) $\displaystyle\int e^w \cos(3e^w)\sin(5e^w) dw =$

(13) $\displaystyle\int \frac{dx}{(4 - x^2)^{5/2}} =$

(14) $\int \text{xarcsec(x)dx} =$ **(15)** $\int \dfrac{dx}{(25 + x^2)^3} =$

Partial Fractions

Before returning to the applications of integration, we consider one more basic technique of integration, called "partial fractions." This technique is concerned with computing integrals of rational functions. A rational function is a ratio of two polynomials. Thus, a rational function is an expression of the form $f(x) = p(x)/q(x)$ where $p(x)$ and $q(x)$ are polynomials. In your precalulus course, you learned how to divide one polynomial by another. If the degree of $p(x)$ is larger than the degree of $q(x)$, then, by polynomial division, you can always write $p(x)/q(x) = a(x) + r(x)/q(x)$ where the "remainder" $r(x)$ has degree less than $q(x)$. For example, if $p(x) = 2x^4 - 3x^3 + x^2 + x - 2$ and $q(x) = x^2 - 3x + 2$, then $a(x) = 2x^2 + 3x + 6$ and $r(x) = 13x - 14$. You should carry out the polynomial division for this example to make sure you haven't forgotten how to do it. Polynomial division has obvious implications for the integration of rational functions. In the example just given we can easily integrate the polynomial $a(x)$ to compute

$$\int \frac{2x^4 - 3x^3 + x^2 + x - 2}{x^2 - 3x + 2}dx = \frac{2x^3}{3} + \frac{3x^2}{2} + 6x + \int \frac{13x - 14}{x^2 - 3x + 2}dx.$$

You may not think that this is great progress, but really the job is now almost done. First of all, we could write the second of the above integrals as follows:

$$\int \frac{13x - 14}{x^2 - 3x + 2}dx = \int \frac{13x}{x^2 - 3x + 2}dx + \int \frac{-14}{x^2 - 3x + 2}dx.$$

The two integrals on the right are in any standard integral table. In the integral table at the end of the book, look under "Expressions Involving $a + bx + cx^2$."

Another approach is to write

$$\frac{13x - 14}{x^2 - 3x + 2} = \frac{1}{x - 1} + \frac{12}{x - 2}.$$

This gives easily that

$$\int \frac{2x^4 - 3x^3 + x^2 + x - 2}{x^2 - 3x + 2}dx$$

$$= \frac{2x^3}{3} + \frac{3x^2}{2} + 6x + \ln(x - 1) + 12\ln(x - 2) + C.$$

An obvious question that you should ask at this point is "How in the world did anyone think to write the following:"

$$\frac{13x - 14}{x^2 - 3x + 2} = \frac{1}{x - 1} + \frac{12}{x - 2}$$

Well, the first thing we did is to notice that $x^2 - 3x + 2 = (x - 1)(x - 2)$. We then used a theorem, a special case of a theorem on "partial fraction expansions" that we shall discuss below, which says that given any polynomial $q(x)$ which factors into distinct linear factors $(x - a_1)(x - a_2) \ldots (x - a_n)$ and any polynomial $r(x)$ of degree less than $q(x)$, then

$$\frac{r(x)}{q(x)} = \frac{A_1}{x - a_1} + \frac{A_2}{x - a_2} + \cdots + \frac{A_n}{x - a_n}$$

where $A_1, \ldots, A_n$ are numbers. This result sounds good at first glance, but at second glance it has some very annoying aspects. First of all, it can be very hard to factor polynomials. Everybody (almost!) knows how to factor polynomials of degree two using the quadratic formula or just guessing. But what about some polynomial of higher degree? How do you factor this:

$$43x^{23} + 12x^{13} + 2x^{10} - 31x^5 + 21x^2 - 31$$

We don't know how either. If you know some numerical analysis, you might be able to approximate the roots, but it might be extremely difficult to settle questions of multiplicity of roots. For example, the polynomial $x^2 - 2x + 1$ has one root (the integer $+1$) of multiplicity 2 as $x^2 - 2x + 1 = (x - 1)^2$. The polynomial $x^2 - (2.0000001)x + (1.0000001)$, on the other hand, has two distinct roots, $x = 1.0000001$ and $x = 1$.

It is easy to see that arbitrarily close (in terms of coefficients) to any polynomial with multiple roots is one with distinct roots. This means that any result that depends in a general way on finding multiplicities of roots of arbitrary polynomials has to be looked at critically from a computational point of view. The main result on partial fraction expansions of rational functions is just this type of result. Still, it is a very worthwhile result to know. The reason for this is that frequently polynomials are *defined* by specifying their roots. In these cases, the result we are about to study gives a tedious but useful method for computing certain integrals.

Factoring A Polynomial Can Be Very Hard

A general theorem in algebra states that any polynomial with real coefficients can be factored into a product of linear terms $px + q$ and irreducible quadratic terms $a + bx + cx^2$. These factors also have real coefficients. As we noted above, it may be very hard to actually do this factorization on a given polynomial. A quadratic term $a + bx + cx^2$ is irreducible if it has no real roots (e.g., $b^2 - 4ac < 0$). The polynomial $(2x + 3)^3(1 + x + x^2)^4$ is factored in this way. Suppose the polynomial is written as

$$(L_1)^{n_1} \ldots (L_s)^{n_s}(Q_1)^{m_1} \ldots (Q_t)^{m_t}$$

where the L_i are the linear factors and the Q_i are the irreducible quadratic factors. For each linear factor L_i in this expression, let

$$\phi(L_i) = \frac{C_i(1)}{(L_i)^1} + \frac{C_i(2)}{(L_i)^2} + \cdots + \frac{C_i(n_i)}{(L_i)^{n_i}}$$

and for each quadratic factor Q_i in this expression, let

$$\phi(Q_i) = \frac{A_i(1) + B_i(1)x}{(Q_i)^1} + \frac{A_i(2) + B_i(2)x}{(Q_i)^2} + \cdots + \frac{A_i(m_i) + B_i(m_i)x}{(Q_i)^{m_i}}$$

where the $A_i(j)$, $B_i(j)$, and $C_i(j)$, $j = 1, 2, \ldots$, are constants to be determined.

Note that $\phi(Q_i)$ is a sum of rational functions where each numerator is linear and each denominator is a power of the irreducible quadratic Q_i. Correspondingly, $\phi(L_i)$ is a sum of rational functions where each numerator is a constant and each denominator is a power of the linear function L_i. We now state the important THEOREM 4.40. Following this theorem there are a number of examples. You might want to look at these examples before reading the proof of the theorem.

4.40   THEOREM (Partial Fractions Expansion)   Let

$$q(x) = (L_1)^{n_1} \ldots (L_s)^{n_s}(Q_1)^{m_1} \ldots (Q_t)^{m_t}$$

be a polynomial where the L_i are linear factors and the Q_i are irreducible quadratic factors. Let $r(x)$ be a polynomial with degree less than $q(x)$. Then it is possible to choose the constants $A_i(j)$, $B_i(j)$, $C_i(j)$ $j = 1, 2, \ldots$, such that the rational function $r(x)/q(x)$ equals

$$\phi(L_1) + \cdots + \phi(L_s) + \phi(Q_1) + \cdots + \phi(Q_t).$$

Proof: Rather than go through all of the details, we shall just be content to point out the main idea of the proof. Assume $q(x) = (L_1)^{n_1} \ldots (L_s)^{n_s}$ contains just linear factors. Pick any of the factors $(L_i)^{n_i}$ and call it L^n for simplicity, with $L = px + q$. Write $q(x) = L^n \hat{q}(x)$. Let $\tau = -q/p$ and let $C = r(\tau)/\hat{q}(\tau)$. Note that $L(\tau) = 0$ and hence $q(\tau) = 0$ but $\hat{q}(\tau) \neq 0$. The polynomial $r(x) - C\hat{q}(x)$ has τ as a root and is thus divisible by L. Write $r(x) - C\hat{q}(x) = L\hat{r}(x)$ and note that

$$\frac{r(x)}{q(x)} = \frac{r(x) - C\hat{q}(x)}{L^n \hat{q}(x)} + \frac{C}{L^n} = \frac{\hat{r}(x)}{L^{n-1}\hat{q}(x)} + \frac{C}{L^n} .$$

The polynomial $L^{n-1}\hat{q}(x)$ divides $q(x)$ and the polynomial $\hat{r}(x)$ has degree less than $L^{n-1}\hat{q}(x)$. Thus we may repeat the process (or apply induction) until, eventually, we produce $r(x)/q(x)$ as a sum $\phi(L_1) + \phi(L_2) + \ldots + \phi(L_s)$, as required by the theorem. To take care of the case where the irreducible quadratic factors Q_i appear, a little complex variable theory plus this same basic idea will suffice. We skip that part of the proof.

4.41 EXAMPLES OF PARTIAL FRACTIONS

(1) What is the form of $\dfrac{2x^2 + 3}{(2x + 1)^3(x^2 + x + 1)^3}$ when expanded by partial fractions? To solve this, let $L_1 = 2x + 1$ and $Q_1 = 1 + x + x^2$. The numerator is $r(x) = 2x^2 + 3$ and this has degree less than the denominator (this is important!). We find, from THEOREM 4.40, that

$$\phi(L_1) = \frac{C_1(1)}{(2x + 1)} + \frac{C_1(2)}{(2x + 1)^2} + \frac{C_1(3)}{(2x + 1)^3} .$$

Similarly, we find that

$$\phi(Q_1) = \frac{A_1(1) + B_1(1)x}{1 + x + x^2} + \frac{A_1(2) + B_1(2)x}{(1 + x + x^2)^2} + \frac{A_1(3) + B_1(3)x}{(1 + x + x^2)^3} .$$

In this example, we asked for "the form of" the rational function when expanded by partial fractions. By this we meant that the actual calculation of the constants $C_1(i)$, $B_1(i)$, and $A_1(i)$ was not required. For EXAMPLE 4.41(1), these constants could have been computed by multiplying both the original rational function and its partial fraction expansion by $(2x + 1)^3(x^2 + x + 1)^3$, equating coefficients, and solving for the constants, which would be quite tedious in this case.

(2) Find the partial fractions expansion of $\dfrac{5x^2 - 3x - 4}{(3x - 2)(x^2 - 2x - 1)}$ · Note

that $5x^2 - 3x - 4$ is not irreducible and in fact has roots $3 \pm 89^{1/2}$. Thus, we could write

$$\frac{7x^2 - 25x + 1}{(3x - 2)(x^2 - 2x - 1)} = \frac{A}{x - 3 - 89^{1/2}} + \frac{B}{x - 3 + 89^{1/2}} + \frac{C}{3x - 2}.$$

This is a good time to point out that, although $x^2 - 2x - 1$ is reducible, it can still be left as a quadratic in the partial fractions expansion. Thus, we could write

$$\frac{5x^2 - 3x - 4}{(3x - 2)(x^2 - 2x - 1)} = \frac{A + Bx}{x^2 - 2x - 1} + \frac{C}{3x - 2}.$$

Multiplying both sides by the expression $(3x - 2)(x^2 - 2x - 1)$ gives the identity

$$5x^2 - 3x - 4 = (A + Bx)(3x - 2) + C(x^2 - 2x - 1)$$

$$= (3B + C)x^2 + (3A - 2B - 2C)x + (-2A - C).$$

Solving the three equations $3B + C = 5$, $3A - 2B - 2C = -3$, $-2A - C = -4$, gives $A = 1$, $B = 1$, and $C = 2$. The lesson to be learned in this example is that, even though a quadratic factor in the denominator of a rational function may be reducible, it may be preferable to leave it unfactored.

(3) Find the partial fraction expansion of $\dfrac{x^4 + x^2 + 1}{(x^2 + 1)^3}$. This type of problem can be very embarrassing. The first thought is to write the form of the partial fraction expansion.

$$\frac{x^4 + x^2 + 1}{(x^2 + 1)^3} = \frac{A + Bx}{(x^2 + 1)^1} + \frac{C + Dx}{(x^2 + 1)^2} + \frac{E + Fx}{(x^2 + 1)^3}$$

and solve for the constants A, B, ... In this case, however, there is a useful trick for avoiding such a messy process. Let $y = x^2 + 1$ so that $x^2 = y - 1$. The rational function then becomes

$$\frac{(y - 1)^2 + (y - 1) + 1}{y^3} = \frac{y^2 - y + 1}{y^3} = \frac{1}{y} + \frac{1}{y^2} + \frac{1}{y^3}.$$

Thus we obtain the partial fraction expansion

$$\frac{x^4 + x^2 + 1}{(x^2 + 1)^3} = \frac{1}{x^2 + 1} - \frac{1}{(x^2 + 1)^2} + \frac{1}{(x^2 + 1)^3}.$$

(4)   Evaluate the integral $\int \dfrac{1 + x}{(1 + x + x^2)^2} dx$. This is one of the basic types of integrals that occur in partial fraction expansions. In these examples, and in the exercises and variations that follow, we keep separate the tedium of finding partial fraction expansions from the task of computing the resulting integrals. To evaluate these integrals, we shall make use of the integral tables in Appendix 1. Look there under the section "Expressions involving $(a + bx + cx^2)$." The denominator of this particular integral is $X = 1 + x + x^2$, so $a = b = c = 1$ and $q = 4ac - b^2 = 3$. Our integral breaks up into two integrals, both of which we find in the table. The first is

$$\int \frac{dx}{X^2} = \frac{2x + 1}{3X} + (2/3) \int \frac{dx}{X} \text{ where}$$

$$\int \frac{dx}{X} = \frac{2}{3^{1/2}} \arctan\left(\frac{2x + 1}{3^{1/2}}\right).$$

The second integral to be evaluated is

$$\int \frac{xdx}{X^2} = -\frac{x + 2}{3X} - (1/3) \int \frac{dx}{X} = -\frac{x + 2}{3X} - \frac{2}{3^{3/2}} \arctan\left(\frac{2x + 1}{3^{1/2}}\right).$$

Putting these facts together, we obtain

$$\int \frac{1 + x}{(1 + x + x^2)^2} dx = \frac{x - 1}{3(1 + x + x^2)} + \frac{2}{3^{3/2}} \arctan\left(\frac{2x + 1}{3^{1/2}}\right).$$

4.42   EXERCISES

(1)   What is the form of the partial fraction expansion of the following?

$$\frac{2x^2 + 3}{(2x + 1)^5(x^2 + x + 1)^2}$$

(2)   Find the partial fraction expansion of

$$\frac{3x^3 - 8x^2 + 2x + 3}{(3x - 2)(x^2 - 2x - 1)}$$

(HINT: First multiply factors in denominator and divide into numerator.)

(3)   Find the partial fraction expansion of

$$\frac{x^3 + 1}{(x^3 - 1)^2}.$$

(4)   Evaluate the following integrals:

(a) $\int \dfrac{dx}{(1 + x + x^2)^3} =$

(b) $\int \dfrac{x^4\,dx}{(x^2 + 1)^2} =$

(c) $\int \dfrac{x\,dx}{(x - 1)^2(x + 1)^2} =$

(d) $\int \dfrac{dx}{(x - 1)^2(x + 1)^2} =$

4.43   VARIATIONS ON EXERCISE 4.42

(1)   What is the form of the partial fraction expansion of the following?

$$\frac{2x^2 + 3}{(2x + 1)^5(x^2 + x - 1)^2}$$

(2)   Find the partial fraction expansion of

$$\frac{3x^3 - 8x^2 + 2x + 3}{(3x - 2)^2(x^2 - 2x - 1)}$$

(3)   Find the partial fraction expansion of

$$\frac{x^2 + 3}{(x^4 + 2x^2 + 1)^2}$$

(4)   Evaluate the following integrals:

(a) $\int \dfrac{dx}{(1 + 2x + x^2)^3} =$

(b) $\int \dfrac{x^3\,dx}{(x^2 + 1)^2} =$

(c) $\int \dfrac{dx}{(x - 1)^3(x + 1)} =$

(d) $\displaystyle\int \frac{x^2 dx}{(x - 1)^2(x + 1)^2} =$

4.44 VARIATIONS ON EXERCISE 4.42

(1) What is the form of the partial fraction expansion of the following?

$$\frac{2x^9 + 3x^5 + 2x^2 + 1}{(2x + 1)^5(x^2 + x - 1)^2}$$

(2) Find the partial fraction expansion of

$$\frac{3x^3 - 8x^2 + 2x + 3}{(x^2 + 1)(x^2 - 2x + 2)}$$

(3) Find the partial fraction expansion of

$$\frac{(x + 1)^2}{(x^2 + 2x + 2)^2}$$

(4) Evaluate the following integrals:

(a) $\displaystyle\int \frac{x^4 dx}{(1 + x + x^2)^2} =$

(b) $\displaystyle\int \frac{x^4 dx}{(x^2 + 1)^2} =$

(c) $\displaystyle\int \frac{x^4 dx}{(x - 1)^3(x + 1)} =$

(d) $\displaystyle\int \frac{x^4 dx}{(x - 1)^2(x + 1)^2} =$

4.45 VARIATIONS ON EXERCISE 4.42

(1) What is the form of the partial fraction expansion of the following?

$$\frac{3x^8 + 5x^5 + 10x}{(x^2 + 1)^2 (x^2 - 1)^2}$$

(2) Find the partial fraction expansion of

$$\frac{9}{(x^2 - 1)^2 (x^2 + 1)}$$

(3) Find the partial fraction expansion of

$$\frac{1}{(x^2 - 1)(x^2 + 1)^2}$$

(4) Evaluate the following integrals:

(a) $\displaystyle\int \frac{dx}{(x^2 - 1)^2 (x^2 + 1)} =$

(b) $\displaystyle\int \frac{dx}{(x^2 - 1)(x^2 + 1)^2} =$

(c) $\displaystyle\int \frac{x^4}{(x^2 + 1)^3}\, dx =$

(d) $\displaystyle\int \frac{x^5}{(x - 1)^3 (x + 1)^2}\, dx =$

4.46 VARIATIONS ON EXERCISE 4.42

(1) What is the form of the partial fraction expansion of the following?

$$\frac{x^9 + 2x^5 + x^3 + 2x^2 + 1}{(x + 1)^2 (x - 1)^3 (x^6 + 4)}$$

(2) Find the partial fraction expansion of

$$\frac{x^3}{(x^2 - 1)^2 (x^2 + 1)}$$

(3) Find the partial fraction expansion of

$$\frac{x^4}{(x^2 - 1)(x^2 + 1)^2}$$

(4) Evaluate the following integrals:

(a) $\displaystyle\int \frac{x^3}{(x^2 - 1)^2 (x^2 + 1)}\, dx =$

(b) $\displaystyle\int \frac{x^4}{(x^2 - 1)(x^3 + 1)^2} =$

(c) $\displaystyle \int \frac{x^6}{(x^2 + 1)^3}\, dx =$

(d) $\displaystyle \int \frac{x^4 - 2x^2 + 1}{(x - 1)^3\, (x + 1)^2}\, dx =$

4.47 GEOMETRIC APPLICATIONS OF INTEGRATION

By now, you should be pretty adept at calculating integrals. Thus, we return to the applications of integration, confident that you can handle any integral that may arise! We shall restrict our attention to the geometric rather than the scientific or engineering applications. The most basic scientific and engineering applications follow easily from the techniques we shall learn. More specialized applications to science and engineering are best done in courses devoted to these topics.

Arclengths, Surface Areas, Volumes

The applications we consider are those of computing arclengths of curves and finding surface areas and volumes of solid objects. We follow the usual procedure of a set of exercises, solutions to these exercises, and a series of variations on these exercises. You should quickly read over EXERCISE 4.48 and SOLUTIONS 4.49. After doing this, return to EXERCISE 4.48 and make a serious attempt to work each exercise. When you get stuck, study again the solution to that exercise and immediately make up and work your own variation. Your variation can be quite minor, but it should be enough to at least require some new calculations to obtain the solution. When you have thoroughly mastered EXERCISE 4.48, go on to the VARIATIONS.

4.48 EXERCISES

(1) A solid object lies in three-dimensional space,

$$R^3 = \{(x, y, z): x, y, z \text{ real numbers}\}.$$

For each $x > 0$, a cross-sectional slice through the object and perpendicular to the x-axis has the shape of an ellipse

$$\frac{y^2}{a^2(1 - x)^2} + \frac{z^2}{b^2(1 + x)^2} = 1$$

where a and b are positive numbers. Find the volume of the object that lies between the planes $x = 0$ and $x = 1$.

(2)   Consider the graph of the parabola $y = x^2$, $x \geq 0$. Imagine this graph revolved about the x-axis to create a surface in R^3. This surface is called the "surface of revolution of the curve $y = x^2$, $x \geq 0$, about the x-axis." By using the method of discs, compute the volume, $V(x)$, of the region bounded by this surface and the plane perpendicular to the x-axis at the point x.

(3)   Compute the volume of revolution of the parabola, $V(x)$ of problem (2), by the method of cylinders.

(4)   Compute the arclength of the curve $y = x^2$, $x > 0$, as a function of x.

(5)   Compute the area of the surface of revolution of the curve $y = x^2$, $x \geq 0$, about the x-axis that lies between the planes perpendicular to the x-axis at 0 and x.

(6)   Compute the arclength of the curve $r(\phi) = e^{\phi/2}$, $0 \leq \phi \leq \pi$.

(7)   Compute the surface area obtained by revolving the curve $r(\phi) = e^{\phi/2}$, $0 \leq \phi \leq \pi$, about the ray $\phi = 0$.

(8)   Compute the volume bounded by the surface of revolution of $r(\phi) = e^{\phi/2}$, $0 \leq \phi \leq \pi$, about the ray $\phi = 0$.

4.49   SOLUTIONS TO EXERCISE 4.48

(1)   The object whose volume is to be found is shown in FIGURE 4.50 (with $a = 1/2$ and $b = 1/4$). In 4.15(4), we computed the area of an ellipse with

FIGURE 4.50   Solid with Elliptical Cross-Sections

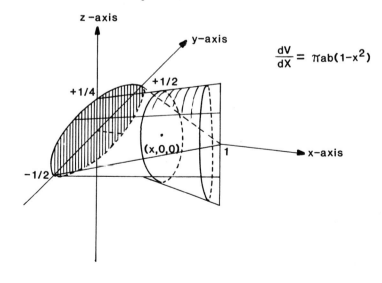

semi-major axis 3 and semi-minor axis 2. If we do the same computation with the constant A replacing 3 and the constant B replacing 2, we get πAB for the area of the ellipse. This means that the cross-sectional area of the slice through the object at the point x has area $\pi ab(1 - x)(1 + x) = \pi ab(1 - x^2)$. If the slice has thickness Δx then the volume of the slice is $\Delta V \approx \pi ab(1 - x^2)\Delta x$. This equality is approximate but gets better and better as Δx gets small. In other words, if we let $V(x)$ denote the volume between the planes perpendicular to the x-axis at 0 and x, then $V'(x) = \pi ab(1 - x^2)$. Usually, when thinking about this type of problem intuitively, one imagines the cross-sectional slice at x to have thickness dx and volume $dV =$ (area of the slice)dx. The area of the slice in this case is $\pi ab(1 - x^2)$. In other words, $\dfrac{dV}{dx} =$ (area of slice at x). We easily integrate the expression $\pi ab(1 - x^2)$ to obtain $V(x) = \pi ab(x - x^3/3) + c$, where the constant c must be set equal to 0 as $V(0) = 0$. Thus, $V(1) = 2\pi ab/3$.

(2) The solid object that has volume $V(x)$ is shown in FIGURE 4.51. Imagine a very thin slice of thickness dx through this object perpendicular to the x-axis at the point $(x, 0, 0)$. This slice is a disc of radius $y = x^2$. The

FIGURE 4.51 The Method of Discs

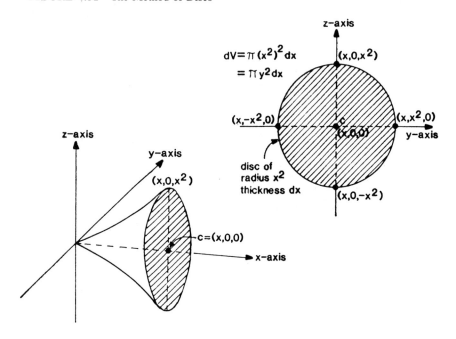

disc has slanted sides, but this fact can be ignored in what follows if dx is imagined to be very small. The volume of this disc is $dV = \pi y^2 dx = \pi x^4 dx$. In other words, as a function of x, $V'(x) = \pi x^4$. Hence, $V(x) = \pi x^5/5$. If you're a serious student and worried about the "slanted sides" of our thin disc, you might want to compute the volume taking this slant into consideration. A disc with slanted sides is more properly called a "frustrum of a cone." The volume of a frustrum of a cone is $\pi(h/3)(s^2 + st + t^2)$ where h is the height, s is the radius of one of the circular faces, and t is the radius of the other circular face. In our case, take $h = \Delta x$, $s = x^2$, and $t = (x + \Delta x)^2$. Think about what happens as Δx goes to zero and you'll see why we ignore the slanted sides.

(3)   Take a look at FIGURE 4.52. There we see the solid object with volume $V(x)$ where $x = a$ is fixed. Let $V_a(y)$ be the volume of the portion of this object that lies inside the cylinder of radius y and axis coincident with the x-axis. Obviously, $V_a(a^2) = V(a)$. Think of the surface of the cylinder as a thin cylindrical knife or "apple corer" cutting through the solid object. If the thickness of this knife is dy, then the amount of material cut out is $2\pi y(a - y^{1/2})dy$. Using the same reasoning as in problems (1) and (2) above, we have $dV_a(y) = 2\pi y(a - y^{1/2})dy$ or $\dfrac{d}{dy}V_a(y) = 2\pi y(a - y^{1/2})$. Thus, $V_a(y) = \pi ay^2 - (4/5)\pi y^{5/2}$. Now, set $y = a^2$ to get

$$V(a) = \pi a^5 - (4/5)\pi a^5 = \pi a^5/5.$$

FIGURE 4.52   The Method of Cylinders

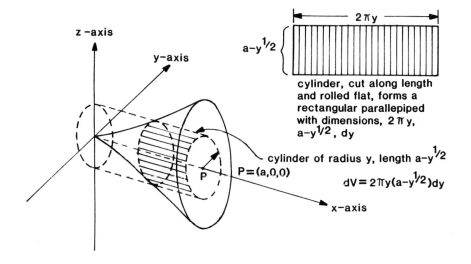

$2\pi y$

$a-y^{1/2}$

cylinder, cut along length and rolled flat, forms a rectangular parallelepiped with dimensions, $2\pi y$, $a-y^{1/2}$, dy

cylinder of radius y, length $a-y^{1/2}$

$P = (a,0,0)$

$dV = 2\pi y(a-y^{1/2})dy$

z –axis

y–axis

x–axis

Thus, $V(x) = \pi x^5/5$ in agreement with (2). This method seems harder in this example, but there are times when the method of cylinders is the easier of the two methods.

(4) The general idea is shown in FIGURE 4.53. As before, think of dx as a small number. In FIGURE 4.53, the portion of the curve over the interval from x to x + dx is shown. It's almost a straight line. If it were a straight line, it would be the hypotenuse of the small right triangle of sides dx and dy. Calling the length of this hypotenuse dL, we obtain $(dL)^2 = (dx)^2 + (dy)^2$. Thus, $\dfrac{dL}{dx} = (1 + (\dfrac{dy}{dx})^2)^{1/2}$. With $y = x^2$, we obtain $\dfrac{dL}{dx} = (1 + 4x^2)^{1/2}$. By now, you should recognize this integral as one that can be solved by trigonometric substitution followed by integration by parts. But, we've suffered enough! Let's use the integral tables in the appendix. Let $z = 2x$ so that

$$\int (1 + 4x^2)^{1/2}dx = (1/2) \int (1 + z^2)dz$$
$$= (z/4)(z^2 + 1) + (1/4)\ln(z + (z^2 + 1)^{1/2}).$$

Thus we obtain

$$L(x) = (x/2)(4x^2 + 1) + (1/4)\ln(2x + (4x^2 + 1)^{1/2}).$$

(5) The geometric situation is shown in FIGURE 4.54. Imagine a knife of thickness dx cutting the surface perpendicular to the x-axis at the point x. The shaded area in FIGURE 4.54 is what is removed. If this shaded area is cut and rolled out, it forms a rectangle of width dL and length $2\pi y$, as shown in FIGURE 4.54. Common sense tells you that if you actually made the cut

FIGURE 4.53 Arclength

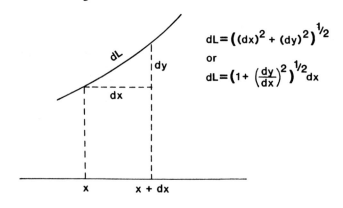

$$dL = \left((dx)^2 + (dy)^2\right)^{1/2}$$
or
$$dL = \left(1 + \left(\frac{dy}{dx}\right)^2\right)^{1/2}dx$$

FIGURE 4.54 Surface Area

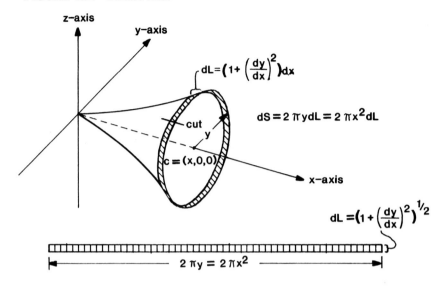

as shown and tried to roll the strip out, funny things would happen at the ends. If dx is very small, these funny things can be ignored, just like the slanted sides of problem (2) can be ignored. If $S(x)$ is the surface area as a function of x, then $dS = 2\pi y dL$ or $S'(x) = 2\pi y L'(x) = 2\pi x^2 L'(x)$. We computed $L'(x) = (1 + 4x^2)^{1/2}$ in Exercise (4). Thus,

$$S(x) = 2\pi \int x^2 (1 + 4x^2)^{1/2} dx.$$

Setting $z = 2x$, $S(x) = (\pi/4)\int z^2 (1 + z^2)^{1/2} dz$. We look up this latter integral in the appendix to obtain

$$\int z^2 (1 + z^2)^{1/2} dz =$$

$$(z/4)(z^2 + 1)^{3/2} - (z/8)(z^2 + 1)^{1/2} - (1/8)\ln(z + (z^2 + 1)^{1/2}).$$

This gives $S(x) =$

$$(\pi/4)\left((x/2)(4x^2 + 1)^{3/2} - (x/4)(4x^2 + 1)^{1/2} - \ln(2x + (4x^2 + 1)^{1/2})\right).$$

(6) The basic idea here is the same as in problem (4) except that we are using polar coordinates. A sketch of the curve $r(\phi) = e^{\phi/2}$, $0 \le \phi \le \pi$, is shown in FIGURE 4.55. Imagine that the angle ϕ changes by a small amount $d\phi$ and the distance $r(\phi)$ changes by an amount dr. The arclength changes

FIGURE 4.55   Arclength in Polar Coordinates

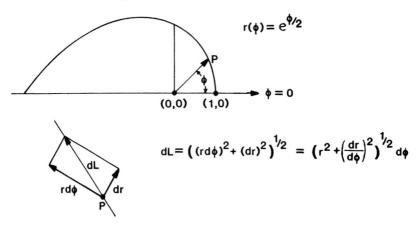

by an amount dL represented as the hypotenuse of the small right triangle of sides dr and rdϕ shown in FIGURE 4.55. We thus obtain

$$L(\pi) = \int_0^\pi \left(r^2 + \left(\frac{dr}{d\phi}\right)^2\right)^{1/2} d\phi = \int_0^\pi (5/4)^{1/2} e^{\phi/2} d\phi = 5^{1/2}(e^{\pi/2} - 1).$$

(7)   This problem involves the same idea as in problem (5), except we use polar coordinates. FIGURE 4.56 shows half of the surface of revolution. Note

FIGURE 4.56   Surface Area in Polar Coordinates

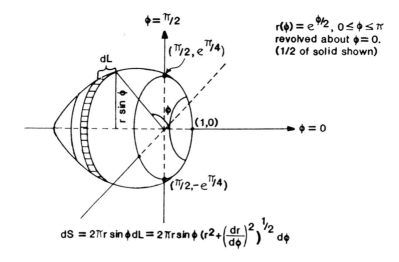

the small increment of arclength denoted by dL. As this increment of arclength revolves about the ray $\phi = 0$, it traces out the shaded area. Cutting out this shaded area (for the full revolution) and laying it flat, as in FIGURE 4.54, gives a rectangle of width dL and length $2\pi r\sin(\phi)$. If dS denotes the total shaded area, then $dS = 2\pi r\sin(\phi)dL$ where dL is as in problem (6). If S denotes the surface area, then

$$S = \int_0^\pi 2\pi e^{\phi/2}\sin(\phi)(5/4)^{1/2}e^{\phi/2}d\phi = 5^{1/2}\pi\int_0^\pi e^\phi\sin(\phi)d\phi.$$

The integral $\int e^\phi\sin(\phi)d\phi$ can be evaluated by two applications of integration by parts or by using the table in the appendix to get $e^\phi(\sin(\phi) - \cos(\phi))/2$. Thus, $S = 5^{1/2}(e^\pi + 1)\pi/2$ square units.

(8)   Look at FIGURE 4.57. Imagine the small shaded area revolved about the ray $\phi = 0$. It traces out a thin circular wire with an almost rectangular cross-section. The circumference of this wire is $2\pi r\sin(\phi)$. If this thin wire is cut and unfolded, it forms a long, thin rectangular parallelepiped with dimensions dr, $rd\phi$, and $2\pi r\sin(\phi)$. Let $dV = 2\pi r\sin(\phi)rd\phi dr$ be the volume of this thin parallelepiped. If we were to fill the region bounded by the curve $r(\phi) = e^{\phi/2}$, $0 < \phi < \pi$, with millions of non-overlapping such little shaded regions and add up the volumes generated by all of them, we would obtain the volume bounded by the surface of revolution of this curve. Making this idea precise is the topic of "multiple integrals" studied in more advanced

FIGURE 4.57   Volume in Polar Coordinates

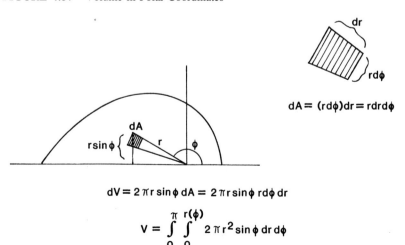

$$dA = (rd\phi)dr = rdrd\phi$$

$$dV = 2\pi r\sin\phi\, dA = 2\pi r\sin\phi\, rd\phi\, dr$$

$$V = \int_0^\pi \int_0^{r(\phi)} 2\pi r^2\sin\phi\, dr\, d\phi$$

courses. The basic ideas required to understand multiple integrals are the final topic of this chapter. In our case, we express this idea as the following "double integral":

$$V = \int_0^\pi \int_0^{r(\phi)} 2\pi r^2 \sin(\phi) dr d\phi.$$

The limits on the outer integral are from 0 to π and refer to the variable ϕ. The limits on the inner integral are from 0 to $r(\phi) = e^{\phi/2}$ and refer to the variable r. To evaluate this integral, we first evaluate the inner integral with respect to r, treating ϕ as a constant:

$$\int_0^{r(\phi)} 2\pi r^2 \sin(\phi) dr = 2\pi \sin(\phi)(r(\phi))^3/3 = (2\pi/3)e^{3\phi/2} \sin(\phi).$$

We now evaluate the outer integral

$$\int_0^\pi (2\pi/3)e^{3\phi/2} \sin(\phi) d\phi = (2\pi/3)e^{3\phi/2}(3/2)(\sin(\phi) - \cos(\phi))(4/13) \Big]_0^\pi.$$

Here, we used the integral $\int e^{ax} \sin(px) dx$ given in the integral tables in the appendix to evaluate $\int e^{3\phi/2} \sin(\phi) d\phi$. Thus the volume bounded by this surface of revolution is $(8\pi/39)(e^{3\pi/2} + 1)$ cubic units.

After having worked EXERCISE 4.48 and studied carefully the solutions to these exercises, you should now be ready to try the VARIATIONS that follow. In all cases involving arclengths, surface areas, and volumes, draw sketches of the figures corresponding to your calculations.

4.58   VARIATIONS ON EXERCISE 4.48

(1)   A solid object has the property that each cross-sectional slice perpendicular to the x-axis is an isosceles triangle with base 3^x and height 2^x. The x-axis passes through the midpoint of the base and the vertex not on the base has coordinates $(x, 0, 2^x)$. Draw a sketch of the solid. Compute the volume of the solid bounded by the planes $x = 0$ and $x = 2$. What is the total volume of the solid in the region $x < 0$?

(2)   The curve $y = x^{3/2}$, $x \geq 0$, is revolved about the x-axis. Using the method of discs, compute the volume, $V(x)$, of the region bounded by this surface of revolution and the plane perpendicular to the x-axis at the point x.

(3)   Compute the volume $V(x)$ of problem (2) using the method of cylinders.

(4)   Compute the arclength of the curve $y = x^{3/2}$, $x \geq 0$, as a function of x.

(5)   Compute the area of the surface of revolution of the curve $y = x^{3/2}$, $x \geq 0$, about the x-axis, as a function of x. (Hint: See Appendix 3.)

(6)   Compute the arclength of the curve $r(\phi) = 2^{\phi/2}$, $0 \leq \phi \leq \pi$.

(7)   Compute the surface area obtained by revolving the curve $r(\phi) = 2^{\phi/2}$, $0 \leq \phi \leq \pi$, about the ray $\phi = 0$.

(8)   Compute the volume bounded by the surface of revolution of $r(\phi) = 2^{\phi/2}$, $0 \leq \phi \leq \pi$, about the ray $\phi = 0$.

4.59   VARIATIONS ON EXERCISE 4.48

(1)   A solid object has the property that each cross-sectional slice perpendicular to the x-axis is a rectangle with base $\log_2(x)$ and height 2^x where $x \geq 1$. The x-axis passes through the midpoint of the base and the midpoint of the side opposite the base has coordinates $(x, 0, 2^x)$. Draw a sketch of the solid. Compute the volume of the solid bounded by the planes $x = 1$ and $x = 2$.

(2)   The curve $y = \sin(x)$, $0 \leq x \leq \pi$, is revolved about the x-axis. Using the method of discs, compute the volume, $V(x)$, of the region bounded by this surface of revolution.

(3)   The curve $y = \sin(x)$, $0 \leq x \leq \pi/2$, is revolved about the y-axis. Using the method of cylinders, compute the volume bounded by this surface and the plane $y = 1$.

(4)   Compute the arclength of the curve $y = \sin(x)$, $0 \leq x \leq \pi$. If you can't evaluate the integral, try a Riemann sum approximation.

(5)   Compute the area of the surface of revolution of the curve $y = \sin(x)$, $0 \leq x \leq \pi$, about the x-axis, as a function of x.

(6)   Compute the arclength of the curve $r(\phi) = \phi$, $0 < \phi < \pi$.

(7)   Compute the surface area obtained by revolving the curve $r(\phi) = \phi$, $0 \leq \phi \leq \pi$, about the ray $\phi = 0$. If you can't evaluate the integral, try a Riemann sum approximation.

(8)   Compute the volume bounded by the surface of revolution of $r(\phi) = \phi$, $0 \leq \phi \leq \pi$, about the ray $\phi = 0$.

4.60   VARIATIONS ON EXERCISE 4.48

(1)   A solid object has the property that each cross-sectional slice perpendicular to the x-axis is a rectangle with base x^2 and height 2^x. The x-axis passes through the midpoint of the base and the midpoint of the side opposite the base has coordinates $(x, 0, 2^x)$. Draw a sketch of the solid. Compute the volume of the solid bounded by the planes $x = 0$ and $x = 2$.

(2) The curve $y = 2^x$, $x \geq 0$, is revolved about the x-axis. Using the method of discs, compute the volume, $V(x)$, of the region bounded by this surface of revolution, the plane perpendicular to the x-axis at the point 0, and the plane perpendicular to the x-axis at the point x.

(3) Compute the volume $V(x)$ of problem (2) using the method of cylinders.

(4) Compute the arclength of the curve $y = 2^x$, $x \geq 0$, as a function of x.

(5) Compute the area of the surface of revolution of the curve $y = 2^x$, $x \geq 0$, about the x-axis, as a function of x.

(6) Compute the arclength of the curve $r(\phi) = \sin(\phi)$, $0 \leq \phi \leq \pi$.

(7) Compute the surface area obtained by revolving the curve $r(\phi) = \sin(\phi)$, $0 \leq \phi \leq \pi$, about the ray $\phi = 0$.

(8) Compute the volume bounded by the surface of revolution of $r(\phi) = \sin(\phi)$, $0 \leq \phi \leq \pi$, about the ray $\phi = 0$.

4.61 VARIATIONS ON EXERCISE 4.48

(1) Compute the volume generated by revolving the indicated planar region about the indicated axis.

(a) $y^2 = x^2 - 4$, $x = 4$, $y = 0$ about the x-axis

(b) $y = 2x^2$, $y = 0$, $x = 3$ about the x-axis

(c) $x = \theta - \sin \theta$, $y = 1 - \cos \theta$, $0 \leq \theta \leq 2\pi$, $y = 0$, about the x-axis

(d) Area between $- x^2 - 3x + 6$ and $y = - x/2 + 3$ about the x-axis.

(2) Compute the arclength of the following curves:

(a) $y = \dfrac{1}{2} x^2 - \dfrac{1}{4} \ln x$, $1 \leq x \leq 2$

(b) $y = \dfrac{x^3}{6} + \dfrac{1}{2x}$, $1 \leq x \leq 2$

(c) $r(\phi) = \cos^2(\phi/2)$, $0 \leq \phi \leq \dfrac{\pi}{2}$

(d) $r(\phi) = \sin^3(\phi/3)$, $0 \leq \phi \leq \dfrac{\pi}{2}$

(3) Compute the surface area obtained by revolving the indicated curve about the indicated axis.

(a) $y = x^3$, $0 \le x \le 2$ about the x-axis.

(b) $y = x^3$, $0 \le x \le 2$ about the y-axis.

4.62 VARIATIONS ON EXERCISE 4.48

(1) Compute the volume generated by revolving the indicated planar region about the indicated axis.

(a) $y^2 = x^3$, $y = 0$, $x = 2$ about $y = -1$

(b) $y = x^3$, $y = 0$, $x = 2$ about $x = -1$

(c) $x = \theta - \sin \theta$, $y = 1 - \cos \theta$, $0 \le \theta \le \pi$, $x = \pi$, $y = 0$ about $x = 0$

(d) Area between $-x^2 - 3x + 6$ and $y = -\dfrac{x}{2} + 3$ about $y = 8.25$

(2) Compute the arclengths of the following curves:

(a) $y = \ln(x)$, $1 \le x \le 2$.

(b) $y = e^t \sin t$, $x = e^t \cos t$, $0 \le t \le 2$.

(c) $r(\phi) = \cos^2(\phi)$, $0 \le \phi \le \dfrac{\pi}{2}$.

(d) $r(\phi) = \sin^3(\phi)$, $0 \le \phi \le \dfrac{\pi}{2}$.

(3) Compute the surface area obtained by revolving the indicated curve about the indicated axis.

(a) $y = \cosh(x)$, $-1 \le x \le +1$, about the x-axis

(b) $y^2 = 4x$, $0 \le x \le 2$, about the x-axis

(c) $y = e^{-x}$, $0 \le x < \infty$, about the x-axis

(d) $y = (24 - 4x)^{1/2}$, $3 \le x \le 6$, about the x-axis

4.63 DOUBLE INTEGRALS
As our next application of integration, we consider the problem of finding the volume under a surface and above a planar region. We use these ideas to introduce the notion of a double integral.

The Volume Under A Surface

The general geometric idea is shown in FIGURE 4.64. A surface with equation $z = f(x, y)$ is shown in three-dimensional space with rectangular coordinates. For example, $f(x, y) = xy^2$ might be one possibility. The "surface defined by $z = f(x, y)$" is the set of all points with coordinates $(x, y, f(x, y))$ where the pairs (x, y) range over all values where $f(x, y)$ is defined

FIGURE 4.64   The Volume Under a Surface

$$V(a) = \int_0^a \left(\int_0^{t(x)} f(x,y)dy \right) dx$$

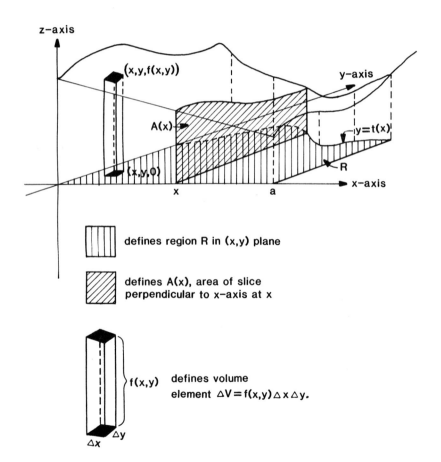

defines region R in (x,y) plane

defines A(x), area of slice perpendicular to x-axis at x

$f(x,y)$ defines volume element $\Delta V = f(x,y)\Delta x \Delta y.$

(i.e., the domain of f). In FIGURE 4.64, we are interested in computing the volume of the solid above the region R and below the surface defined by z = f(x, y). Imagine a plane perpendicular to the x-axis cutting through this solid at x. The points of the solid on this plane are shown in FIGURE 4.64. The area of this region of intersection is called A(x). Let V(x) denote the volume of the portion of the solid that lies to the left of this region of intersection. If you worked EXERCISE 4.48(1), you will see immediately that $\dfrac{dV}{dx}$ = A(x). Thus

$$\int_0^a A(x)dx = V(a)$$

In EXERCISE 4.48 and all of its corresponding VARIATIONS, we made it easy for you to compute A(x). Note that in FIGURE 4.64, the region R, which lies in the (x, y) plane, is bounded by the y-axis, the x-axis, the line x = a, and the curve y = t(x). Thus,

$$A(x) = \int_0^{t(x)} f(x, y)dy.$$

In this integral, we treat x as a constant because it does not change in computing the integral. We thus obtain

$$V(a) = \int_0^a \left(\int_0^{t(x)} f(x, y)dy \right) dx.$$

As an example, let z = f(x, y) = 4 − x be a plane. Let t(x) = $(4 − x^2)^{1/2}$. We shall find the area under the surface z = 4 − x and above the region in the x, y-plane bounded by the semicircle y = $(4 − x^2)^{1/2}$ and the interval [−2, +2] on the x-axis. You should sketch the picture. We compute

$$\int_{-2}^{+2} \int_0^{(4-x^2)^{1/2}} (4 - x)dydx = \int_{-2}^{+2} \left[(4 - x)y \right]_0^{(4 - x^2)^{1/2}}$$

$$= \int_{-2}^{+2} (4 - x)(4 - x^2)^{1/2}dx.$$

This integral is the sum of two integrals

$$4 \int_{-2}^{+2} (4 - x^2)^{1/2}dx + (1/2) \int_{-2}^{+2} (4 - x^2)d(4 - x^2).$$

From the integral tables in the appendix we find that the first integral is

$$\left[2x(4 - x^2)^{1/2} + 8 \arcsin(x/2)\right]_{-2}^{+2} = 8\pi.$$

The second integral is zero. Thus, the total volume under the plane and above the semicircular region is 8π cubic units.

The Two-Dimensional Riemann Sum

There is another and equivalent way to think about computing the volume over a region and under a surface R. Look at FIGURE 4.65. We are looking at the same region R of FIGURE 4.64, but now we are looking straight down onto the region from above (in a direction parallel to the z-axis). The region has been divided into many small squares or "area elements." If we imagine (x_i, y_i) to be the coordinates of the center of the ith little square, then $\Delta V = f(x_i, y_i)\Delta A$ is the volume between the little square of area $\Delta A = \Delta x \Delta y$ and the surface $z = f(x, y)$. This volume element is approximately a tall, thin parallelepiped as shown in FIGURE 4.64. If we write VOLUME to mean the volume under the surface and above the region R, then we have the approximate equality

$$\text{VOLUME} = \sum_{\text{all squares}} f(x_i, y_i)\Delta x \Delta y.$$

FIGURE 4.65   Partitioning the Region R into Lots of Little Squares

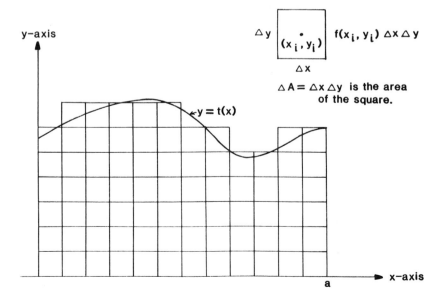

In using the squares of FIGURE 4.65 to approximate VOLUME we are going to be off from the true VOLUME by a certain amount. if we take smaller squares, our intuition tells us correctly that the approximation gets better and better. We want to make sure that the entire region is covered by little squares and that no little square lies entirely outside of the region. This process of better and better approximating sums is what is used to define the double integral. Thus, we write

$$\iint\limits_{R} f(x, y)dA = \lim_{\text{squares}\to 0} \sum_{\text{all squares}} f(x_i, y_i)\Delta A \text{ where } \Delta A = \Delta x \Delta y.$$

There is an obvious similarity, at least in spirit, between approximating the double integral of the function f(x, y) with finer and finer meshes of little squares, and the idea of a RIEMANN SUM (4.14). These approximating sums could be used to compute the double integral, just as Riemann sums can be used to compute integrals. The way calculus is used to compute double integrals is through the idea of the "iterated" multiple integral of FIGURE 4.64. The fact that double integrals can be computed in this way is really a theorem in more advanced courses. The key idea involved in proving this theorem is contained in the transition from FIGURE 4.64 to FIGURE 4.65.

The idea of thinking of double integration in terms of the volume under a surface z = f(x, y) was just a way to give us geometric feeling for the process. The function f(x, y) can be any continuous ("smoothly changing") function and the process

$$\iint\limits_{\text{REGION}} f(x, y)dA = \lim_{\text{squares}\to 0} \sum_{\text{all squares}} f(x_i, y_i)\Delta A \text{ where } \Delta A = \Delta x \Delta y$$

can be used to define the double integral of f(x, y) over a REGION. The process of making all of these ideas "precise" can be quite technical and is beyond the scope of this book. Anyone with the right intuition and a little mathematical sophistication can "make things precise." If you have the wrong intuition, no amount of precision and attention to technical details will save you!

Iterated Integrals

One thing we need to do now is discuss a little more carefully the process of going from double integrals, thinking of them as a limit of approximating sums, and the iterated integrals of FIGURE 4.64. In symbols, we are concerned about the following sort of identities:

$$\iint_{\text{REGION}} f(x, y)dA = \int_c^d \left(\int_{a(x)}^{b(x)} f(x, y)dy \right) dx$$

or perhaps

$$\iint_{\text{REGION}} f(x, y)dA = \int_c^d \left(\int_{a(y)}^{b(y)} f(x, y)dx \right) dy$$

In each of the above two equalities, the left-hand side is conceptually a limit of approximating sums and the right-hand side is a technical process of evaluating two integrals. Taking the first of the above equalities as an example, the right-hand side involves computing the integral

$$g(x) = \int_{a(x)}^{b(x)} f(x, y)dy$$

as if x were a constant. Next we treat x as a variable and compute

$$\int_c^d g(x)dx.$$

For an example of this process, see the computation of the integral

$$\int_{-2}^{+2} \left(\int_0^{(4 - x^2)^{1/2}} (4 - x)dy \right) dx$$

done in connection with FIGURE 4.64.

Determining The Limits Of Integration

One difficulty for the beginner in all of this is in determining the limits of integration, such as $a(x)$, $b(x)$, c, and d from the description of REGION. Take a look at FIGURE 4.66. To help describe what we see there, let's introduce some notation. Given a region R in the x,y-plane, let $(\underline{x}, -)_R$ denote the set $\{x:(x, y) \text{ is in R for some } y\}$ and let $(x, \underline{y})_R$ denote the set $\{y:(x, y) \text{ is in R for that fixed } x\}$. Similarly, we let $(-, \underline{y})_R$ denote the set $\{y:(x, y) \text{ is in R for some } x\}$ and $(\underline{x}, y)_R = \{x:(x, y) \text{ is in R for that fixed } y\}$. In FIGURE 4.66,

$$(\underline{x}, -)_{R_1} = \{x:a \le x \le b\} = [a, b] \quad \text{and} \quad (\underline{x}, -)_{R_2}$$

$$= \{x:q \le x \le r\} = [q, r].$$

For the fixed values of x shown in FIGURE 4.66

FIGURE 4.66   Determining Limits of Integration

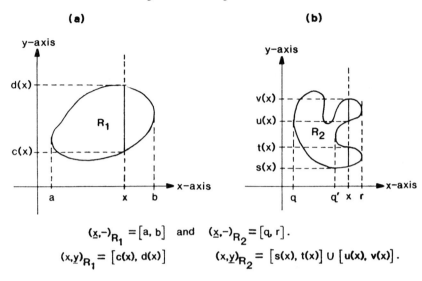

$$(\underline{x},-)_{R_1} = [a, b] \quad \text{and} \quad (\underline{x},-)_{R_2} = [q, r].$$
$$(x,\underline{y})_{R_1} = [c(x), d(x)] \qquad (x,\underline{y})_{R_2} = [s(x), t(x)] \cup [u(x), v(x)].$$

$$(x, \underline{y})_{R_1} = \{y : c(x) \le y \le d(x)\} = [c(x), d(x)]$$

and

$$(x, \underline{y})_{R_2} = \{y : s(x) \le y \le t(x) \text{ or } u(x) \le y \le v(x)\}$$

$$= [s(x), t(x)] \cup [u(x), v(x)].$$

In terms of this notation, we have

$$\iint\limits_{R} f(x, y)dA = \int\limits_{(\underline{x},-)_R} \left(\int\limits_{(x,\underline{y})_R} f(x, y)dy \right) dx = \int\limits_{(-,\underline{y})_R} \left(\int\limits_{(\underline{x},y)_R} f(x, y)dx \right) dy.$$

In particular, for R_1 of FIGURE 4.66, we have

$$\iint\limits_{R_1} f(x, y)dA = \int_a^b \int_{c(x)}^{d(x)} f(x, y)dy \, dx.$$

For R_2 of FIGURE 4.66, the situation is more complicated. For $q \le x \le q'$, the set $(x, \underline{y})_{R_2}$ is a single interval and for $q' < x \le r$ a union of two intervals. The integral looks something like the following:

$$\iint\limits_{R_2} f(x, y)dA = \int_q^{q'} \left(\int_{s(x)}^{v(x)} f(x, y)dy \right) dx + \int_{q'}^r \left(\int_{s(x)}^{t(x)} f(x, y)dy \right.$$

$$+ \int_{u(x)}^{v(x)} f(x, y)dy \bigg) dx.$$

You should identify exactly which portions of the boundaries of the region R_2 of Figure 4.66 represent the graphs of the functions s(x), t(x), u(x), and v(x)

Iterated Integration In Polar Coordinates

The iterated integration used to evaluate the double integral over a region R can also be done in polar coordinates. In FIGURE 4.67, we are using polar coordinates to describe a region $R = \{(r, \theta): 0 \le r \le e^{\theta/2}, 0 \le \theta \le \pi/2\}$. As in the case of rectangular coordinates, we define

$(\mathbf{r}, -)_R = \{r: (r, \theta)$ is in R for some $\theta\}$,
$(-, \boldsymbol{\theta})_R = \{\theta: (r, \theta)$ is in R for some $r\}$,
$(r, \underline{\theta})_R = \{\theta: (r, \theta)$ is in R for that fixed $r\}$,
and $(\underline{r}, \theta)_R = \{r: (r, \theta)$ is in R for that fixed $\theta\}$.
Expressed in terms of this notation, we have

$$\iint\limits_R f(r, \theta)dA = \int_{(r,-)_R} \left(\int_{(r,\underline{\theta})_R} f(r, \theta)d\theta \right) dr = \int_{(-,\underline{\theta})_R} \left(\int_{(\underline{r},\theta)_R} f(r, \theta)dr \right) d\theta.$$

FIGURE 4.67   A Region in Polar Coordinates

$$\left\{ \theta: 2\ln(2) \le \theta \le \frac{\pi}{2} \right\} = \left[2\ln(2), \frac{\pi}{2} \right]$$

$2.2 = e^{\pi/4}$

$(2, 2\ln(2)) = (2, 1.386)$

$r(\theta) = e^{\theta/2}$

$r = 2$

$\theta = 2\ln(2)$

$1 = e^0$

$$R = \left\{ (r, \theta): 0 \le r \le e^{\theta/2}, 0 \le \theta \le \frac{\pi}{2} \right\}$$

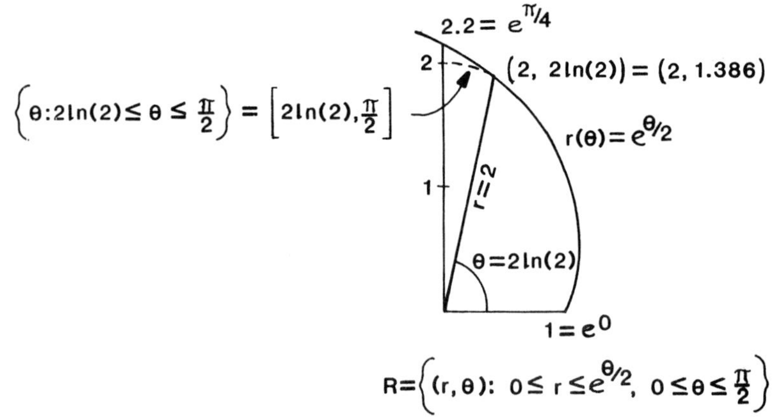

In the case of FIGURE 4.67, $(\underline{r}, -)_R = [0, e^{\pi/4}]$, $(-, \underline{\theta})_R = [0, \pi/2]$, $(\underline{r}, \theta)_R = [0, e^{\theta/2}]$, and $(r, \underline{\theta})_R = [0, \pi/2]$ for $0 \le r \le 1$ and $(r, \underline{\theta})_R = [2\ln(r), \pi/2]$ for $1 \le r \le e^{\pi/4}$]. The last set is illustrated in FIGURE 4.67 for $r = 2$. These sets lead to the following iterated integrals:

$$\iint\limits_R f(r, \theta)dA = \int_0^{\pi/2} \left(\int_0^{e^{\theta/2}} f(r, \theta)\, dr \right) d\theta$$

and

$$\iint\limits_R f(r, \theta)dA = \int_0^1 \left(\int_0^{\pi/2} f(r, \theta)d\theta \right) dr + \int_1^{e^{\pi/4}} \int_{2\ln(r)}^{\pi/2} f(r, \theta)d\theta\, dr.$$

Integrals In Three Dimensions

Integration over regions in three and higher dimensions is treated in a manner analogous to that of two dimensions. We now give a brief discussion of integration over regions in three-dimensional space. Our purpose is to give you some intuitive feeling for the situations that can arise. The idea of volume under a surface, used to give us an intuitive feeling for integrals over planar regions, will be replaced by the model of computing the mass or weight of a three-dimensional object from its mass density function.

Have you ever been to a county fair hobby show where some enthusiastic hobbyist has made a model of the cathedral Notre-Dame de Paris, or some other famous building, out of sugar cubes? If so, you've seen the basic idea of defining integrals in three-dimensional space. Imagine a simpler object than Notre-Dame de Paris, say a watermelon, made out of sugar cubes. Call the volume of each sugar cube ΔV cubic centimeters. We have the obvious fact that

$$\text{VOLUME OF WATERMELON} = \sum_{\text{all cubes}} \Delta V.$$

Actually, it is the volume of the *model* of the watermelon that we get by the above expression. That's not quite the same as the real volume of the watermelon, but it's close. Suppose we have function $f(x, y, z)$, which gives the density of the watermelon in grams per cubic centimeter at each point (x, y, z) in the watermelon. We have put some coordinates in the watermelon. The weight of the ith sugar cube sized piece of the watermelon is $f(x_i, y_i, z_i)\Delta V$ where (x_i, y_i, z_i) are the coordinates of a point in that particular cube. Then the weight of the whole melon is given by

$$\text{WEIGHT OF WATERMELON} = \sum_{\text{all cubes}} f(x_i, y_i, z_i)\Delta V.$$

Again, this is only an approximation as the sugar cube watermelon, having a rather bumpy surface, is only an approximation to the real thing. But, just as with the Riemann sum or the integral over two-dimensional regions, we can imagine using smaller and smaller sugar cubes and thus getting a better and better approximation to the watermelon. We say,

$$\iiint\limits_{\text{REGION}} f(x, y, z)dV = \lim_{\text{cubes}\to 0} \sum_{\text{all cubes}} f(x_i, y_i, z_i)\Delta V.$$

This "triple integral" is thus defined by approximating sums. Intuitively, when one thinks about a triple integral over some region R in three-dimensional space (corresponding to the watermelon above), one writes

$$\iiint\limits_{R} f(x, y, z)dV$$

and speaks of dV as the "volume element" or "element of volume." As we shall see below, there are several common ways to describe the volume element, but in the standard x, y, z-rectangular coordinates, one usually writes dV = dxdydz and imagines a very small cube with sides dx, dy, and dz. The intuitition, of course, comes from the sum of small cubes of volume ΔV = $\Delta x\Delta y\Delta z$. Describing the region of the integration and evaluating the integral over this region can be a bit trickier than in the case of regions in the plane. The basic idea, however, is the same. Although integrals over regions in three-dimensional space are conceptualized as limits of partitions of regions into smaller and smaller cubes, these integrals are computed by iterated integration. Our notation for describing the limits of integration in the two-dimensional case extends directly to the three-dimensional case.

Limits Of Integration In Three Dimensions

Let R be a region in three-dimensional space. We use the notation $(\underline{x}, -, -)_R$ to denote the set $\{x:(x, y, z) \text{ is in R for some y and z}\}$. Similarly define $(-, \underline{y}, -)_R$ and $(-, -, \underline{z})_R$. Define $(x, \underline{y}, -)$ to be $\{y:(x, y, z) \text{ is in R for that x and some z}\}$. There are a total of six sets of this type, the others being $(\underline{x}, y, -)_R$, $(-, \underline{y}, z)_R$, $(-, y, \underline{z})_R$, $(x, -, \underline{z})_R$, and $(\underline{x}, -, z)_R$. Finally, let $(x, y, \underline{z})_R$ be the set $\{z:(x, y, z) \text{ is in R for that x and y}\}$. There are two other sets of this type, $(x, \underline{y}, z)_R$ and $(\underline{x}, y, z)_R$. If f(x, y, z) is a continuous function defined on the region R in three-dimensional space, then the integral

$$\iiint_R f(x,\ y,\ z)dV\ =\ \int_{(\underline{x},-,-)_R} \left(\int_{(x,\underline{y},-)_R} \left(\int_{(x,y,\underline{z})_R} f(x,\ y,\ z)dz \right) dy \right) dx.$$

There are five other ways to express this integral based on the orders of integration in the iterated integral of dzdxdy, dxdydz, dxdzdy, dydxdz, and dydzdx.

As an example of setting up the limits of integration for a region R in three-dimensional space, consider the region bounded by the planes $x = 0$, $y = 0$, $z = 0$, and the plane passing through the three points $(3, 0, 0)$, $(0, 2, 0)$, and $(0, 0, 1)$. This latter plane has equation $(x/3) + (y/2) + z = 1$. These planes and the bounded region are shown in FIGURE 4.68. We shall set up the iterated integral according to the order of integration dzdydx. The set $(\underline{x}, -, -)_R = [0, 3]$. The set $(x, \underline{y}, -)_R = [0, 2(1 - x/3)]$ because, for a fixed x, the largest value of y still contained in the region R is $2(1 - x/3)$, as shown in FIGURE 4.68. Finally, $(x, y, \underline{z})_R = [0, 1 - (x/3) - (y/2)]$. These intervals are shown in FIGURE 4.68. Thus, we have

$$\iiint_R f(x,\ y,\ z)dzdydx\ =\ \int_0^3 \left(\int_0^{2(1-x/3)} \left(\int_0^{(1-x/3-y/2)} f(x,\ y,\ z)dz \right) dy \right) dx.$$

FIGURE 4.68 The Limits of Integration on a Three-Dimensional Region

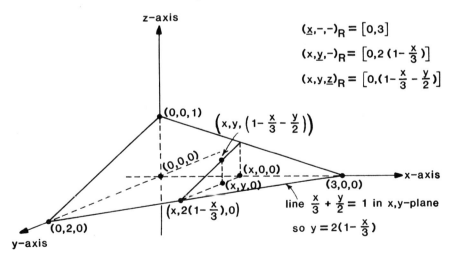

Cylindrical And Spherical Coordinates

Depending on the shape of the region R in three-dimensional space, it may be better to use cylindrical or spherical coordinates rather than the standard rectangular coordinates to describe the boundaries of the region. In FIGURE 4.69(a), we see cylindrical coordinates. The plane perpendicular to the z-axis and passing through z = 0 has standard polar coordinates (r, θ). Every point in three-dimensional space can be assigned a triple (r, θ, z) where (r, θ) are the polar coordinates of the projection of the point onto the plane perpendicular to the z-axis. Because polar coordinates are not unique, neither are cylindrical coordinates. If it is desirable to have unique cylindrical coordinates, restrictions must be put on the values of r and θ. For example, r ≥ 0, 0 ≤ θ < 2π, results in unique cylindrical coordinates. The basic volume element, $\Delta V = r\Delta r\Delta\theta\Delta z$, for cylindrical coordinates is shown in FIGURE 4.69(a).

In FIGURE 4.69(b), we see spherical coordinates. The lines θ = 0, θ = π/2, and φ = 0 correspond to the lines θ = 0, θ = π/2, and the z-axis in FIGURE 4.69(a). These lines intersect at right angles. A point in three-dimensional space may be assigned a triple (ρ, θ, φ) where ρ is the distance to the point, θ is the same as in cylindrical coordinates, and φ is the angle between the line joining the point to the origin and the line φ = 0. To make spherical coordinates unique, we can make the restrictions ρ ≥ 0, 0 ≤ θ < 2π, and 0 ≤ φ < π. As shown in FIGURE 4.69(b), the volume element in spherical coordinates is $\Delta V = \rho^2\sin(\phi)\Delta\rho\Delta\theta\Delta\phi$.

Limits Of Integration In Cylindrical And Spherical Coordinates

In integrating a function f(r, θ, z) expressed as function of cylindrical coordinates, or a function g(ρ, θ, φ) expressed in terms of spherical coordinates, over a region R, we again must perform an iterated integration. We can adopt the same notational conventions for describing the limits of integration in the iterated integral. Thus, in spherical coordinates, $(-, \underline{\theta}, -)_R = \{\theta:(\rho, \theta, \phi)$ is in R for some ρ and some φ}, $(-, \theta, \underline{\phi})_R = \{\phi:(\rho, \theta, \phi)$ is in R for that θ and some ρ}, etc. To fix these ideas in your mind, we'll set up the limits of integration for the same three-dimensional region in both cylindrical and spherical coordinates. Take a look at FIGURE 4.70. The region R will be the region above the shaded region and below the cone. In spherical coordinates, the shaded region is {(ρ, θ, φ): 0 ≤ ρ ≤ 2, 0 ≤ θ ≤ π/2, φ = π/2}. The cone is the set {(ρ, θ, φ): φ = π/4}. In cylindrical coordinates, the shaded region is the set {(r, θ, z): 0 ≤ r ≤ 2, 0 ≤ θ ≤ π/2, z = 0}. The cone, in cylindrical coordinates, is the set {(r, θ, z): r = z}.

FIGURE 4.69 Cylindrical and Spherical Cooordinates

(a) Cylindrical Coordinates

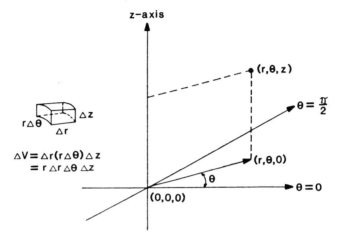

(b) Spherical Coordinates

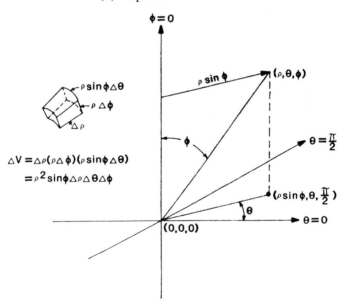

First, let's set up the limits of integration for the iterated integrals that evaluate the integral of $f(r, \theta, z)$ over the region R of FIGURE 4.70 using cylindrical coordinates. We have $(r, -, -)_R =$

FIGURE 4.70   The Region R Above the Shaded Area and Below the Cone

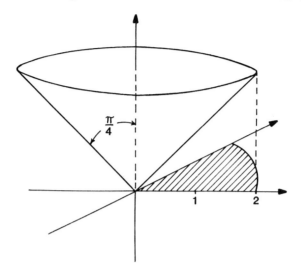

[0, 2], $(r, \underline{\theta}, -)_R = [0, \pi/2]$, and $(r, \theta, \underline{z})_R = [0, r]$. Thus the integral is

$$\iiint\limits_R f(r, \theta, z)dV = \int_0^2 \left(\int_0^{\pi/2} \left(\int_0^r f(r, \theta, z)r\,dz \right) d\theta \right) dr.$$

There are five other ways to set up this integral depending on the order of interated integration, corresponding to rdzdrdθ, rdθdzdr, rdθdrdz, rdrdzdθ, and rdrdθdz. You should set up the limits of integration for at least two of these ways.

Finally, let's set up the limits of integration for the region R of FIGURE 4.70 in spherical coordinates. First, $(-, \underline{\theta}, -)_R = [0, \pi/2]$. Next, $(-, \theta, \underline{\phi}) = [\pi/4, \pi/2]$. Finally, $(\underline{\rho}, \theta, \phi) = [0, 2\csc(\phi)]$. This gives

$$\iiint\limits_R g(\rho, \theta, \phi)dV = \int_0^{\pi/2} \int_{\pi/4}^{\pi/2} \int_0^{2\csc(\phi)} g\rho^2\sin(\phi)d\rho\,d\phi\,d\theta.$$

Again, there are five other ways to set up this integral. You should describe these five ways and set up the limits of integration for at least two of them.

Changing Coordinates

Of course, any time you integrate the function $f(x, y) = 1$ over a region in the plane, you obtain the area of the region, and any time you integrate the function $f(x, y, z) = 1$ over a region in three-dimensional space, you obtain the volume of the region. Using FIGURE 4.69, it is easy to transform the coordinates of a point from rectangular to cylindrical or spherical coordinates and vice versa. For example, from rectangular to cylindrical we have the relations

$$x = r\cos(\theta) \qquad y = r\sin(\theta) \qquad z = z.$$

From rectangular to spherical coordinates, we have

$$x = \rho\sin(\phi)\cos(\theta) \qquad y = \rho\sin(\phi)\sin(\theta) \qquad z = \rho\cos(\phi).$$

If you are faced with a task of integrating a function over some region in three-dimensional space, you can choose the coordinate system that makes the work of computing the iterated integral the easiest. Once the coordinate system is chosen, you must express the limits of integration and the function to be integrated in terms of that coordinate system. If the function to be integrated is $f(x, y, z) = x^2 + y^2 + z^2$, for example, then in cylindrical coordinates the function becomes $g(r, \theta, z) = r^2 + z^2$, and in spherical coordinates, the function becomes $h(\rho, \theta, \phi) = \rho^2$.

4.71 EXERCISES

(1) In each of the following problems, change the order of integration. Choose one of the two iterated double integrals and evaluate it. Sketch the region of integration.

(a) $\displaystyle\int_0^2 \int_{x^2}^x dy\,dx$ (Answer: $-2/3$)

(b) $\displaystyle\int_0^1 \int_y^{3y} (1 + y)dx\,dy$ (Answer: 2)

(c) $\displaystyle\int_0^{\pi/2} \int_0^{\cos(\theta)} r\sin(\theta)dr\,d\theta$ (Answer: 1/6)

(d) $\displaystyle\int_0^{\pi/2} \int_0^2 r^2\cos(\theta)dr\,d\theta$ (Answer: 8/3)

(e) $\displaystyle\int_{-1}^2 \int_{x^2}^{x+2} 2x^2dy\,dx$ (Answer: 6.3)

(f) $\displaystyle\int_0^{\pi/2} \int_2^{4\cos(\theta)} r^3\cos^2 dr\,d\theta$ (Answer: $49\pi/2$)

(HINT: $\int\cos^k d\theta$ is π for $k = 2$, 0 for $k = 3$, $3\pi/4$ for $k = 4$, 0 for $k = 5$, and $5\pi/8$ for $k = 6$.)

(2) Try to evaluate each of the following iterated double integrals. If you have trouble, change the order of integration and evaluate that iterated integral.

(a) $\displaystyle\int_0^1 \int_0^y e^{-x^2} dxdy$

(b) $\displaystyle\int_0^1 \int_{3y}^3 e^{x^2} dxdy$

(3) Each of the following iterated integrals results in a sum of two integrals when the order of integration is changed. Find the indicated constants a and b, the functions c and d, and sketch the region of integration for each of the following problems:

(a) $\displaystyle\int_{-1}^2 \int_{x^2}^{x+2} f(x,\,y)dydx\ =$

$$\int_0^1 \int_{-\sqrt{y}}^{\sqrt{y}} f(x,\,y)dxdy\ +\ \int_a^b \int_{c(y)}^{d(y)} f(x,\,y)dxdy$$

(b) $\displaystyle\int_0^1 \int_y^{2y} f(x,\,y)dxdy\ =\ \int_0^1 \int_{x/2}^x f(x,\,y)dydx\ +\ \int_a^b \int_{c(x)}^{d(x)} f(x,\,y)dydx$

(c) $\displaystyle\int_{-1}^{+1} \int_0^{x^2} f(x,\,y)dydx\ =$

$$\int_0^1 \int_1^{-\sqrt{y}} f(x,\,y)dxdy\ +\ \int_a^b \int_{c(y)}^{d(y)} f(x,\,y)dxdy$$

(4) Evaluate each of the following iterated triple integrals:

(a) $\displaystyle\int_{-1}^1 \int_0^2 \int_0^1 (x^2y\ +\ xy^2)dxdydz$ (Answer: 4/3)

(b) $\displaystyle\int_0^1 \int_0^1 \int_0^1 y^2dxdydz$ (Answer: 1/3)

(c) $\displaystyle\int_0^1 \int_0^1 \int_0^{1-x} xyz\ dxdydz$ (Answer: 1/48)

(d) $\displaystyle\int_0^1 \int_0^2 \int_0^3 (x^2\ +\ yz)dxdydz$ (Answer: 11)

(e) $\displaystyle\int_0^2 \int_0^{\left(\frac{4-x^2}{2}\right)^{1/2}} \int_{x^2+3y^2}^{8-x^2-y^2} dzdydx$ (Answer: $2^{3/2}\pi$)

(f) $\displaystyle\int_0^\pi \int_0^3 \int_0^{a^2} r^3dzdrd\theta$ (Answer: $\pi r^4(a^2/4\ -\ r^2/6)$)

(g) $\int_0^{2\pi} \int_0^{\pi/4} \int_0^{a\cos(\phi)} \rho^2\sin(\phi)d\rho d\phi d\theta$ (Answer: $\pi a^3/4$)

(5) In each of the following, sketch the region of integration and verify that the first change of order of integration is correct. Supply the limits on the second change of order of integration.

(a) $\int_0^1 \int_0^x \int_0^{x-y} f(x, y, z)dxdydz =$

$\int_0^1 \int_y^1 \int_0^{x-y} f(x, y, z)dydxdz = \int\int\int f(x, y, z)dxdzdy$

(b) $\int_{-2}^2 \int_0^{4-y^2} \int_0^{y+2} f(x, y, z)dxdydz =$

$\int_0^4 \int_{-\sqrt{4-x}}^{\sqrt{4-x}} \int_0^{y+2} f(x, y, z)dydxdz = \int\int\int f(x, y, z)dxdzdy$

(c) $\int_{-1}^1 \int_0^{1-y^2} \int_{-\sqrt{x}}^{\sqrt{x}} f(x, y, z)dxdydz =$

$\int_0^1 \int_{-\sqrt{1-x}}^{\sqrt{1-x}} \int_{-\sqrt{x}}^{\sqrt{x}} f(x, y, z)dydxdz = \int\int\int f(x, y, z)dxdzdy$

(6) For each of the following regions R, compute the integral of the function f over R using iterated integration. Set up the problem in rectangular, cylindrical, and polar coordinates and evaluate whichever one you find to be the easiest.

(a) $R = \{(x, y, z): x \geq 0, y \geq 0, z \geq 0, x + y + z \leq 1\}$,
$f(x, y, z) = 12xy$ (Answer: 1/10)

(b) $R = \{(x, y, z): 0 \leq x \leq 1, 0 \leq y \leq 1, 0 \leq z \leq -x^2\}$,
$f(x, y, z) = 6yz$ (Answer: 2/3)

(c) $R = \{(x, y, z): x^2 + y^2 \leq 1, 0 \leq z \leq y\}$, $f(x, y, z) = 1$ (Answer: 2/3)

(d) $\{(x, y, z): x \geq 0, y \geq 0, z \geq 0, x^2 + y^2 + z^2 \leq 1\}$, $f(x, y, z) = 12xyz$ (Answer: 1/4)

(e) $\{(x, y, z): x \geq 0, y \geq 0, z \geq 0, x^2 + y^2 + z^2 \leq 16\}$, $f(x, y, z) = z(16 - x^2 - y^2)^{1/2}$ (Answer: $2^8\pi/5$)

(7) A cylindrical hole of radius $r < s$ is drilled through the center of a sphere $S = \{(x, y, z): x^2 + y^2 + z^2 \leq s^2\}$. The axis of symmetry of the cylinder is the z-axis. Express the volume of material removed from the sphere as a triple integral in rectangular, cylindrical, and spherical coordinates. Evaluate the integral in the coordinate system of your choice. Express both the volume of material removed from the sphere and the volume of material left in the sphere as a function of the length of the hole, $h = 2(s^2 - r^2)^{1/2}$.

Chapter 5

INFINITE SERIES

Let's Get Acquainted With The Math Tables In The Appendix

By now, you have had much practice with differential and integral calculus. You have surely noticed that the techniques of integration were harder to master than the techniques of differentiation. Don't feel bad about this situation. It's the same for everyone who studies calculus. If you don't believe that this is the case, look at the math tables in the appendix of this book. Notice how much more space is devoted to tables of integrals than to tables of derivatives! In fact, now is a good time to take a few minutes and browse through the math tables in the appendix. In particular, look at the section labled SERIES and the TABLE OF INTEGRALS. Look in the TABLE OF INTEGRALS under the subsection TRANSCENDENTAL FUNCTIONS and find $\int \dfrac{\sin(x)}{x} dx$.

What you should have found is the statement

$$\int \frac{\sin(x)}{x} = x - \frac{x^3}{3 \cdot 3!} + \frac{x^5}{5 \cdot 5!} - \frac{x^7}{7 \cdot 7!} + \frac{x^9}{9 \cdot 9!} \cdots$$

This strange expression for $\int \dfrac{\sin(x)}{x}$ is called an "infinite series expansion of the integral of $\dfrac{\sin(x)}{x}$." For a moment, let's ignore the question of how this infinite series expansion was derived and just try to understand what it means. Notice first of all that the series is a sum of powers of x. In this case,

228

all powers are odd. The terms alternate in sign. For any odd number n, the power x^n is divided by n · n!. The number n · n! is obtained by multiplying the integer n times the integer n!. The integer n! is the product $(n)(n - 1)(n - 2)$ · · · $(2)(1)$. These numbers n! grow very rapidly as n increases. The series ends with three little dots ". . ." which is meant to imply that anyone should be able to figure out the general term from the terms already given. You should be aware, however, that sometimes series are terminated with . . . without anybody really knowing the general rule! Students, in particular, love to do this. Sometimes, the general rule for forming the series is quite complicated. Look in the math tables in the section labeled SERIES at the infinite series for tan(x) and ctn(x) where the general term is expressed as a function of the Bernoulli numbers.

An Infinite Series Specifies Better And Better Polynomial Approximations

We still haven't discussed what the infinite series for $\int \dfrac{\sin(x)}{x}$ really means. Recall that in the beginning of CHAPTER 4, INTEGRATION, we discussed the idea of the signed area function. FIGURE 4.6 is particularly relevant here. There we saw that given a function f(x), we could always produce an integral F(x), the "signed area function," by graphical methods. In particular, we could apply this technique to compute $F(x) = \int \dfrac{\sin(x)}{x}$. Imagine that we have done this with great accuracy and we have graphed F(x), as was done in FIGURE 4.6. Assume that we have taken the base point for the signed area function to be x = 0, so F(0) = 0. Now look once again at the infinite series expansion for $\int \dfrac{\sin(x)}{x}$. Notice, if we throw away all terms in the infinite series involving powers of x larger than k, we obtain a polynomial. For k = 1, 3, 5, and 7, for example, we would obtain the polynomials x, x −
$\dfrac{x^3}{3 \cdot 3!}$, $x - \dfrac{x^3}{3 \cdot 3!} + \dfrac{x^5}{5 \cdot 5!}$, and $x - \dfrac{x^3}{3 \cdot 3!} + \dfrac{x^5}{5 \cdot 5!} - \dfrac{x^7}{7 \cdot 7!}$. Let's call
these polynomials $p_1(x)$, $p_3(x)$, $p_5(x)$, and $p_7(x)$. If we were to compare the graphs of these polynomials with the graph of F(x), starting at x = 0 and moving away, we would see that $p_1(x)$ looks almost identically like the graph of F(x) for very small values of x. We would see that the graph of $p_3(x)$ looks almost identically like the graph of F(x) for an even wider interval of values about x = 0.

Similarly, $p_5(x)$ and $p_7(x)$ are even better approximations to $F(x)$ over a wider range of values of $F(x)$ centered at $x = 0$. This is the fundamental idea of what the statement of equality between any function $F(x)$ and its infinite series (sometimes called its "power series") is to mean. The infinite series is simply a concise way of specifying a sequence of polynomials that are better and better approximations to the function about some value of x ($x = 0$ in our example).

Some Simple-Looking Functions Are Very Nasty To Integrate

But wait a minute! Why did we bother to express the integral $\int \dfrac{\sin(x)}{x}$ as an infinite series? We didn't do that sort of thing in CHAPTER 4. We wrote $\int \sin(x)dx = -\cos(x)$, for example. We didn't express $\int \sin(x)dx$ as an infinite series. Why don't we write $\int \dfrac{\sin(x)}{x} = -\dfrac{\cos(x)}{x}$ or, if that's not quite right, something like that? The answer is that, try as they may, no one has been able to find a reasonable-looking expression for $\int \dfrac{\sin(x)}{x}dx$ in terms of the familiar functions used to express integrals in CHAPTER 4. Mathematicians express this fact by saying "the integral $\int \dfrac{\sin(x)}{x}dx$ cannot be expressed in terms of elementary functions." This may sound a little condescending to you, as the functions we have been using thus far in this book may not seem so "elementary," but this is what is said. Thus, the way we deal with "nasty to integrate functions" like $\dfrac{\sin(x)}{x}$ is to approximate them with polynomials. Polynomials are nice functions. The bad news about this is that any given polynomial is only a good approximation for certain restricted values of the variable x. Usually, these values are specified as lying in some interval about a fixed value of x. We speak of these polynomials as "local approximations" to our nasty function. When we do our calculus tricks on these local approximations, such as find areas, volumes, line integrals, etc., we have to be careful. We must always remember that we are doing "local calculus." This is not to say that you should be paralyzed with fear at the thought of making a mistake with local calculus. By far the best strategy is to plow ahead and be daring. But always be a little suspicious. The techniques of power series were developed largely in the nineteenth century when no computers were available. For you, it's a different ball game! If you wonder if some calculations you have done are correct it may be very simple for you to write a little program to test things out. You may learn much more in this way than trying to rely on some nineteenth-century theorem.

Infinite Series Help Us With Nasty Functions And Local Calculus

We have much to learn about the mathematics of infinite series. Once you get into the spirit of it, it won't be too bad. Before we get into the more pedantic aspects of the subject, we are going to spend some time fooling around with power series approximations to functions. There are many possibilities here, but since we are fresh from studying integral calculus, we shall look at some applications of integration in the local calculus setting. In doing this, we shall work with nasty functions since nice functions can be treated by the methods of CHAPTER 4. One problem is to recognize the nasty as opposed to the nice functions. By "nasty" we mean "its integral is not expressible in terms of elementary functions." As you browse through the TABLE OF INTEGRALS in the appendix, you will see many horrible-looking functions that clever people have been able to integrate in terms of elementary functions. Still, you will notice that many reasonable-looking functions are missing from the table. Indefinite integrals involving $(a + bx)^{1/2}$ and $(a + bx^2)^{1/2}$ are given, but not of $(a + bx^3)^{1/2}$ or $(a + bx^4)^{1/2}$. Probably these latter two functions are nasty. Integrals of $\sin(\ln(x))$ and $\cos(\ln(x))$ are given but not of $\ln(|\sin(x)|)$ and $\ln(|\cos(x)|)$. Probably these are nasty. And where in the world are the integrals of $\sin(x^2)$ and $\cos(x^2)$? Also, no anti-derivatives are given for e^{x^2} or e^{-x^2}. These are very important functions in statistics. Just as with humans, functions can be important even though they are nasty!

Following our usual pattern, we now give a series of exercises involving some nasty functions and local calculus. After the exercises, we give the solutions. Study them carefully. Then we give variations on these exercises for you to try.

Now We Try Some Local Calculus

5.1 EXERCISES

(1) Find the area under the graph of $f(x) = \dfrac{\sin(x)}{x}$ for $0 \leq x \leq \pi$.

(2) Find the volume of the solid of revolution obtained by revolving the curve $f(x) = \dfrac{\sin(x)}{x}$, $0 \leqslant x \leqslant \pi$, about the x-axis.

(3) Find the arclength of the curve $y(x) = x^3/3$ for x between $-1/2$ and $+1/2$.

5.2 SOLUTIONS TO EXERCISE 5.1

A Local Area Problem

(1) We are asked to compute $\displaystyle\int_0^\pi \frac{\sin(x)}{x}\,dx$. The first thing we do is write a little program in BASIC to get a feel for this function. Here is the program and its output. Study it carefully.

```
10   FOR X = .1 TO 3.1   STEP .1
20   PRINT X,SIN(X)/(X)
30   S = S + SIN(X)/X
40   NEXT X
50   PRINT "A = " S*.1
```

X	SIN(X)/X	X	SIN(X)/X
.1	.9983341	1.8	.5410264
.2	.9933466	1.9	.4980526
.3	.9850672	2	.4546486
.4	.9735459	2.1	.411052
.5	.9588511	2.2	.3674984
.6	.9410708	2.3	.3242197
.7	.920311	2.4	.281443
.8	.8966951	2.5	.239389
.9	.8703633	2.6	.1982699
1	.8414709	2.7	.158289
1.1	.8101885	2.8	.1196388
1.2	.7766993	2.9	.0825
1.3	.7411986	3.0	4.704026E-02
1.4	.7038926	3.1	1.341338E-02
1.5	.6649966	A = 1.802058	
1.6	.6247335		
1.7	.5833321	A = Approximate area = 1.8 square units	

First Try A Riemann Sum, Then A Power Series

Notice that the values of $\dfrac{\sin(x)}{x}$ have been computed at intervals of 0.1. If we sum these values, all of which are positive, and multiply by the interval width of 0.1, we obtain a Riemann sum approximation to the integral. If you think about it a bit, you will see that this Riemann sum is going to be a little bit smaller than the actual area we are seeking. In any case, we get a rough idea that the area is about 1.8 square units. Knowing this may keep us from making fools of ourselves in the calculations to follow. Now let's try and compute the same area using our infinite series expansion for $\int \dfrac{\sin(x)}{x}dx$. We had found from the tables in the appendix that

$$\int \frac{\sin(x)}{x}dx = x - \frac{x^3}{3\cdot 3!} + \frac{x^5}{5\cdot 5!} - \frac{x^7}{7\cdot 7!} + \frac{x^9}{9\cdot 9!}\cdots$$

Let's take the polynomial $F(x) = p_7(x)$ as our approximation to the integral. Here is a BASIC program to compute $F(\pi) = F(3.14)$. We shall learn later that in a series with decreasing terms that alternate in sign, the maximum error is less than the absolute value of the first term omitted. We have used that fact to estimate the error in this BASIC program's approximation to the integral. The value of the area as estimated by $F(3.14)$ is 1.84, which agrees pretty closely with our earlier Riemann sum approximation.

```
10   PRINT "ENTER X ":INPUT X
20   F3=3*2:F5=5*4*F3:F7=7*6*F5:F9=9*8*F7
30   P1=X:P3=X^3/(3*F3):P5=X^5/(5*F5):P7=X^7/(7*F7)
40   PRINT "F("X")=" P1-P3+P5-P7
50   PRINT "MAX ERROR IS ",X^9/(9*F9)

RUN
ENTER X
? 3.14
F( 3.14)= 1.843483
MAX ERROR IS    9.085761E-03
```

What If We Hadn't Found The Power Series In The Appendix?

In a certain sense, we lucked out in finding $\int \dfrac{\sin(x)}{x}dx$ expressed as an infinite series in our table. A more typical scenario would have been that we found only the series

$$\sin(x) = x - \frac{x^3}{3!} + \frac{x^5}{5!} - \frac{x^7}{7!} + \cdots$$

We then would have divided all of the terms of this series by x to obtain

$$\frac{\sin(x)}{x} = 1 - \frac{x^2}{3!} + \frac{x^4}{5!} - \frac{x^6}{7!} + \ldots$$

Integrating this series term by term gives the series for $\int \frac{\sin(x)}{x} dx$. We shall learn later how to justify these steps. For now let's be daring! After all, we wrote some BASIC programs to give us some independent evidence of the validity of these calculations. Generally, if these types of calculations are unjustified, they produce real garbage that is easily detected by a little common sense.

A Local Volume Problem

(2)   Using the methods of CHAPTER 4, we have that the volume is

$$V = \pi \int_0^\pi \frac{\sin^2(x)}{x^2} dx.$$

We begin with a short BASIC program that computes values of $\frac{\sin^2(x)}{x^2}$ at intervals of 0.1 and estimates the volume with a Riemann sum. The estimated volume is 4.3 cubic units. Here is the program and some output.

A Riemann Sum Approximation To The Local Volume Problem

```
10   FOR X = .1 TO 3.1 STEP .1
20   PRINT X, (SIN(X))^2/X^2
30   S = S + (SIN(X))^2/X^2
40   NEXT X
50   PRINT "INTEGRAL IS ABOUT "S*.1
60   PRINT "VOLUME IS ABOUT "3.14*S*.1
```

.1	.9966711	1.6	.3902919
.2	.9867376	1.7	.3402764
.3	.9703576	1.8	.2927095
.4	.9477916	1.9	.2480563
.5	.9193953	2	.2067054
.6	.8856142	2.1	.1689638
.7	.8469723	2.2	.135055
.8	.8040621	2.3	.1051184
.9	.7575323	2.4	7.921019E-02
1	.7080734	2.5	5.730708E-02

1.1	.6564054	2.6	3.931096E-02
1.2	.6032616	2.7	2.505542E-02
1.3	.5493753	2.8	1.431345E-02
1.4	.4954648	2.9	6.806249E-03
1.5	.4422204	3.0	2.212786E-03
		3.1	1.799188E-04
		INTEGRAL IS ABOUT	1.368151
		VOLUME IS ABOUT	4.295995

We Get A Chance To Compute A Power Series From Scratch

To apply power series methods to this problem, we need an infinite series for $\sin^2(x)$. No such series is given in the appendix. We could take the series for $\sin(x)$ and square it. A better approach for us at this point is to learn the general rule for forming power series expansions of a function about $x = 0$. The general method was described in CHAPTER 3, following the discussion of TAYLOR POLYNOMIALS. The general rule for any function $h(x)$ is

$$h(x) = h(0) + h^{(1)}(0)x + \frac{h^{(2)}(0)}{2!}x^2 + \frac{h^{(3)}(0)}{3!}x^3 + \ldots + \frac{h^{(n)}(0)}{n!}x^n + \ldots$$

In this expression, $h^{(n)}(x)$ is the nth derivative of $h(x)$. This n^{th} derivative evaluated at $x = 0$ is $h^{(n)}(0)$. Now let's compute some derivatives of $h(x) = \sin^2(x)$:

$$h^{(1)}(x) = 2\sin(x)\cos(x) = \sin(2x)$$

$$h^{(2)}(x) = 2\cos(2x)$$

$$h^{(3)}(x) = -2^2\sin(2x)$$

$$h^{(4)}(x) = -2^3\cos(2x)$$

$$h^{(5)}(x) = +2^4\sin(2x)$$

$$h^{(6)}(x) = +2^5\cos(2x)$$

$$h^{(7)}(x) = -2^6\sin(2x)$$

$$h^{(8)}(x) = -2^7\cos(2x)$$

$$h^{(9)}(x) = +2^8\sin(2x)$$

Do you see the pattern? Note that $f^{(k)}(0) = 0$ for $k = 0, 1, 3, 5, 7, \ldots$. Thus we obtain

$$\sin^2(x) = +\frac{2^1}{2!}x^2 - \frac{2^3}{4!}x^4 + \frac{2^5}{6!}x^6 - \frac{2^7}{8!}x^8 + \frac{2^9}{10!}x^{10} - \ldots$$

Now we divide this series for $\sin^2(x)$ by x^2 to obtain

$$\frac{\sin^2(x)}{x^2} = + 1 - \frac{2^3}{4!}x^2 + \frac{2^5}{6!}x^4 - \frac{2^7}{8!}x^6 + \frac{2^9}{10!}x^8 - \cdots$$

Integrating term by term, we obtain

$$\int \frac{\sin^2(x)}{x^2}dx = + x - \frac{2^3}{3\cdot 4!}x^3 + \frac{2^5}{5\cdot 6!}x^5 - \frac{2^7}{7\cdot 8!}x^7 + \frac{2^9}{9\cdot 10!}x^9 - \cdots$$

You should notice the powers of 2 that appear in numerators of the coefficients in this series. They are going to make convergence slow for any large values of x. Even for $x = 3.14$ it is necessary to take the following polynomial to get reasonable accuracy:

$$p_{13}(x) = x - \frac{2^3}{3\cdot 4!}x^3 + \frac{2^5}{5\cdot 6!}x^5 - \frac{2^7}{7\cdot 8!}x^7$$

$$+ \frac{2^9}{9\cdot 10!}x^9 - \frac{2^{11}}{11\cdot 12!}x^{11} - \frac{2^{13}}{9\cdot 14!}x^{13}.$$

The following BASIC program confirms this and computes directly $p_{13}(3.14)$. The volume estimated by this program is 4.46, which is pretty close to our Riemann sum method above. This program is a simple-minded and direct translation of the formula for $p_{13}(x)$ into BASIC. It would be much better numerically to compute the factorials F4, F6, etc., and the numbers P1, P3, etc., recursively within a loop. This is another story, however. Remember, we are just fooling around with series at this point.

A BASIC Program To Solve The Local Volume Problem

```
LIST
10   PRINT "ENTER X ":INPUT X
20   F4=24:F6=6*5*F4:F8=8*7*F6:F10=10*9*F8:F12=
     12*11*F10:F14=14*13*F12
30   P1=X:P3=(2*X)^3/(3*F4):P5=(2*X)^5/(5*F6):P7=
     (2*X)^7/(7*F8):P9=(2*X)^9/(9*F10):P11=(2*X)^(11/(11*F12):P13=
     (2*X)^13/(13*F14)
40   INTEGRAL = P1-P3+P5-P7+P9-P11+P13
50   PRINT "INTEGRAL IS ABOUT "INTEGRAL
60   PRINT "V("X") IS ABOUT "3.14*INTEGRAL
70   F16=16*14*F14
80   PRINT "MAXIMUM ERROR IS "(2*X)^15/(15*F16)
```

RUN
ENTER X
? 3.14
INTEGRAL IS ABOUT   1.420813
V(3.14) IS ABOUT   4.461352
MAXIMUM ERROR IS   3.181611E-03

Finally, A Local Arclength Problem!

(**3**)   Using the techniques of CHAPTER 4, we have

$$\text{ARCLENGTH} = \int_{-1/2}^{1/2} (1 + (y'(x))^2)^{1/2} dx = \int_{-1/2}^{1/2} (1 + x^4)^{1/2} dx$$

$$= 2 \int_{0}^{1/2} (1 + x^4)^{1/2} dx.$$

We look in the appendix and find that

$$(1 + x)^{1/2} = 1 + \frac{1}{2}x - \frac{1 \cdot 1}{2 \cdot 4}x^2 + \frac{1 \cdot 1 \cdot 3}{2 \cdot 4 \cdot 6}x^3 - \frac{1 \cdot 1 \cdot 3 \cdot 5}{2 \cdot 4 \cdot 6 \cdot 8}x^4 + \cdots$$

Substituting x^4 for x we obtain

$$(1 + x^4)^{1/2} = 1 + \frac{1}{2}x^4 - \frac{1 \cdot 1}{2 \cdot 4}x^8 + \frac{1 \cdot 1 \cdot 3}{2 \cdot 4 \cdot 6}x^{12}$$

$$- \frac{1 \cdot 1 \cdot 3 \cdot 5}{2 \cdot 4 \cdot 6 \cdot 8}x^{16} + \cdots$$

Integrating term by term, we find

$$\int (1 + x^4)^{1/2} dx = x + \frac{1}{2}\frac{x^5}{5} - \frac{1 \cdot 1}{2 \cdot 4}\frac{x^9}{9} + \frac{1 \cdot 1 \cdot 3}{2 \cdot 4 \cdot 6}\frac{x^{13}}{13} - \cdots$$

If we set $x = 1/2$ in this expression we should get a good approximation to half the arclength. Actually, we need only the first three terms of this series. With $x = 1/2$, we again have a series with decreasing terms of alternating sign. Thus the error is no worse than the first term omitted. The following little BASIC program shows us the error for omitting the term involving x^{13}:

A Basic Program For The Error

```
LIST
10   PRINT "INPUT X ":INPUT X
20   PRINT (3/(2*4*6*13))*X^(13)

RUN
INPUT X
? .5
 5.868765E-07
```

The next program computes the first three terms of the series evaluated at x = 1/2.

A Basic Program For The Arclength

```
LIST
10   PRINT "ENTER X ":INPUT X
20   P1 = X:P5 = X^5/2*5:P9 = X^9/2*4*9
30   PRINT "F("X") = "P1 + P5 − P9
40   PRINT "ARCLENGTH IS ABOUT "2*(P1 + P5 − P9)

RUN
ENTER X
? .5
F( .5) =   .5429688
ARCLENGTH IS ABOUT   1.085938
```

Something We've Swept Under The Rug: Intervals Of Convergence

The arclength is thus about 1.086. There is an important detail that we have swept under the rug thus far in our calculations with the series for $(1 + x)^{1/2}$. We found the series expansion for $(1 + x)^{1/2}$ in the math tables in the appendix under SERIES AND PRODUCTS. Just after the series expansion for $(1 + x)^{1/2}$ you see the notation $[x^2 < 1]$. This notation means that the series is valid only for values of x such that $x^2 < 1$. In other words, $|x| < 1$ or, equivalently, $-1 < x < +1$. The $\{x: -1 < x < +1\}$ is also denoted by $(-1, +1)$ and is called the "interval of convergence of the series." We will learn more about intervals of convergence (THEOREM 5.57). The main thing we need to note is that x = 1/2 is in the interval of convergence and so is $x^4 = (1/2)^4 = 1/16$. If you look at the series for sin(x) in the appendix, you will see that the interval of convergence is $[x^2 < \infty]$.

The series for sin(x) converges for all x. This means that no matter how large x is, if you take enough terms of the series for sin(x) you will eventually get a good approximation. The number of terms required may be very large, however! In the case of the series for $(1 + x)^{1/2}$, no matter how many terms you take, you won't get good approximations to values of x outside the interval of convergence (values such as $x = 2$ or $x = 3$, etc.). The general rule for integrating a series term by term to obtain a series that approximates the integral is that the limits of integration must both lie in the interval of convergence. Thus, in our case, the limits were 0 and 1/2, which both lie in the interval of convergence of $(1 + x)^{1/2}$ and hence of the series for $(1 + x^4)^{1/2}$, which also has interval of convergence $-1 < x < +1$. Does all this talk of intervals of convergence make you nervous about the validity of our calculation of the arclength of $x^3/3$ from $-1/2$ to $1/2$? If so that's a healthy sign. Let's check out our calculations with a little BASIC program:

Riemann Sum Approximation For Arclength

```
LIST
10   FOR X=0 TO .5 STEP .01
20   S=S+(1+X^4)^.5
30   NEXT X
40   PRINT "THE INTEGRAL IS ABOUT "2*S*.01

RUN
THE INTEGRAL IS ABOUT   1.026509
```

The Riemann sum approximation is close to the result obtained from the infinite series, so probably things are all right.

There is one thing left that we should do in connection with this arclength problem. We should derive the series for $(1 + x)^{1/2}$. Remember the general rule

$$y(x) = y(0) + \frac{y^{(1)}(0)}{1!}x + \frac{y^{(2)}(0)}{2!}x^2 + \frac{y^{(3)}(0)}{3!}x^3 \ldots$$

Computing some derivatives, we find $y^{(1)}(x) = (1/2)(1 + x)^{-1/2}$, $y^{(2)}(x) = (1/2)(-1/2)(1 + x)^{-3/2}$, $y^{(3)}(x) = (1/2)(-1/2)(-3/2)(1 + x)^{-5/2}$, $y^{(4)}(x) = (1/2)(-1/2)(-3/2)(-5/2)(1 + x)^{-7/2}, \ldots$. Substituting $x = 0$ into these expressions and putting the resulting numbers into the expansion for $y(x)$ gives the series in the appendix for $(1 + x)^{1/2}$. You should carry out this calculation carefully. Only the interval of convergence remains a mystery!

Now You Get To Try Some Local Calculus!

We now give a series of variations on EXERCISE 5.1. The problems in these variations correspond closely to their counterparts in EXERCISE 5.1, so you may want to check SOLUTIONS 5.2 periodically.

5.3   VARIATIONS ON EXERCISE 5.1

(1)   Find the area under the graph of $f(x) = \dfrac{\sin(x/2)}{x}$, $0 \leq x \leq \pi$. Sketch the graph of $f(x)$.

(2)   Find the volume of the solid of revolution obtained by revolving the curve $f(x) = \dfrac{\sin(x/2)}{x}$, $0 < x < \pi$, about the x-axis.

(3)   Find the arclength of the curve $y(x) = x^3/3$ for x between -0.9 and $+0.9$. Make sure the error is less than .001. This problem would be much harder if we asked for the arclength from -2 to $+2$. Do you see why?

Power Series Are Better Than Riemann Sums When Parameters Occur

5.4   VARIATIONS ON EXERCISE 5.1

(1)   Find the area under the graph of $f_\beta(x) = \dfrac{\sin(\beta x)}{x}$, $0 \leq x \leq \pi$, as a function of β, $0 \leq \beta \leq 1$. Sketch the graph of $f_\beta(x)$ for several values of β.

(2)   Find, as a function of β, the volume of the solid of revolution obtained by revolving the curve $f_\beta(x) = \dfrac{\sin(\beta x)}{x}$, $0 \leq x \leq \pi$, about the x-axis. Again, assume $0 \leq \beta \leq 1$.

(3)   Find the arclength of the curve $y(x) = (\alpha x)^3/3$, $0 \leq \alpha \leq 3/2$, for x between -0.5 and $+0.5$. Your answer should be a function of α such that the error is less than .001 for all values of α with $0 \leq \alpha \leq 3/2$.

A Different Nasty Function—Same Techniques

5.5   VARIATIONS ON EXERCISE 5.1

(1)   Find the area under the graph of $f(x) = \sin(x^2)$, $0 \leq x \leq (\pi)^{1/2}$. Sketch the graph of $f(x)$.

(2) Find the volume of the solid of revolution obtained by revolving the curve $f(x) = \sin(x^2)$, $0 \leq x \leq (\pi)^{1/2}$, about the x-axis.

(3) Find the arclength of the curve $y(x) = x^3$ for x between -0.5 and $+0.5$. Make sure the error is less than .001.

By now you should have some idea how infinite series can be used to extend the range of applicability of calculus. It's now time to be a little more systematic about our study of infinite series.

We Start With Sequences

5.6 DEFINITION: SEQUENCES Let $Z = \{0, 1, 2, \ldots\}$ denote the set of nonnegative integers. A function f whose domain D is a subset of Z and whose range is the set R of real numbers is called a real valued *sequence*. If D has infinitely many elements, then f is called an infinite sequence.

Playing The Envelope Game With Sequences

It's time for the ENVELOPE GAME again. Imagine you have an envelope. Inside is a sequence. What are you going to see when you open the envelope? Well, you might see some ordered pairs written on a piece of old yellow parchment: (0, 5.45), (1, 6.43), (5, 3.45). This would be the sequence with $D = \{0, 1, 5\}$. At 0, this sequence would have the value 5.45, at 1 the value 6.43, and 5 the value at 3.45. If the sequence is an infinite sequence, you won't see all of its values written down (obviously!). You might see something like (0, 0), (1, 2) (2, 4), (3, 6), (4, 8), ... The ". . . ," read "dot, dot, dot," is meant to imply that anyone should be able to figure out the general rule from what is given: (n, 2n). Another way to describe the same sequence is $D = Z$, $f(n) = 2n$. Another way is $D = Z$, $f(n) = a_n$ where $a_n = 2n$. Another way is $D = Z$, $a_n = 2n$. Another way is $a_n = 2n$, $n = 0, 1, 2,$ . . . . In all cases, we must be clear about D and about the rule which assigns to each element of D a real number.

The next definition will sound strange at first, but you will get used to it, if not learn to love it.

Epsilons And Limits

5.7 DEFINITION: LIMIT OF A SEQUENCE Let a_n, $n = 0, 1, 2, 3, \ldots$ be an infinite sequence. We say that a real number A is the limit of a_n as n goes to infinity and write

$$\lim_{n \to \infty} a_n = A$$

if for every real number $\epsilon > 0$ there exists an integer N_ϵ such that for all $n > N_\epsilon$, $|a_n - A| < \epsilon$.

You Give Me ϵ Then I Give You N_ϵ Such That . . .

Many students find DEFINITION 5.7 annoying. Consider, for example, the sequence $a_n = (2n + 1)/(n + 1)$. As n goes to infinity this sequence obviously approaches $A = 2$ as a limit. We don't need DEFINITION 5.7 to see that this is true, so why confuse the obvious? The reason that we need DEFINITION 5.7 is that it provides a necessary technical tool for discussing limits of sequences in general terms, apart from any specific examples. If DEFINITION 5.7 seems confusing to you, think of it as sort of a game. Imagine that you are in a room sitting at a desk. On the desk is a piece of paper with a sequence a_n, $n = 0, 1, 2, \ldots$, that you've announced converges to a number A. Every now and then, at random intervals, someone opens the door to the room and hands you a positive real number ϵ (like $\epsilon = .001$, for example). You have to give that person an integer N_ϵ such that for all $n > N_\epsilon$, $|a_n - A| < \epsilon$. If you can prove that you can do this for any ϵ that may be given to you, then that proves that A is the limit of the sequence a_n. For the sequence $a_n = (2n + 1)/(n + 1)$, if you are given $\epsilon > 0$, you can take N_ϵ to be any integer greater than or equal to ϵ^{-1}. If $n > N_\epsilon \geq \epsilon^{-1}$ then

$$|a_n - A| = \left| \frac{2n + 1}{n + 1} - 2 \right| = \left| \frac{-1}{n + 1} \right| < \frac{1}{n} < \frac{1}{N_\epsilon} \leq \epsilon.$$

This proves, using DEFINITION 5.7, the obvious fact that $(2n + 1)/(n + 1)$ converges to $A = 2$. If you take $\epsilon = .001$ in the above inequality, then $\epsilon^{-1} = 1000$ and $N_{.001}$ can be any integer greater than or equal to 1000. In fact, $N_{.001} = 1000$ works fine. Thus, for all $n > 1000$,

$$|a_n - A| = \left| \frac{2n + 1}{n + 1} - 2 \right| < .001.$$

It is more important that the beginning calculus student develop a strong intuitive feeling for limits than a technical ability to work with DEFINITION 5.7. A little awareness of the latter is all that we ask at this point. Actually, if you think about it a bit, DEFINITION 5.7 has a strong intuitive appeal. It says that if you claim that the sequence a_n approaches A, then, given *any* *level* of accuracy ϵ, which we think of as a small number, you must be able to specify an integer N_ϵ such that past N_ϵ the sequence gets *and stays* within that level of accuracy from A.

If It Doesn't Converge, It Diverges

If a sequence a_n has a limit in the sense of DEFINITION 5.7, it is called a *convergent sequence*. A sequence a_n, n = 0, 1, 2, . . . , that does not have a limit is called divergent. A simple example of a divergent sequence is the sequence $a_n = (-1)^n$. This sequence hops back and forth from $+1$ to -1. Given any $\epsilon < 2$, it is obviously impossible to find the N_ϵ demanded by DEFINITION 5.7. Another divergent sequence is the sequence $a_n = n$, n = 0, 1, 2, ... The A of DEFINITION 5.7 is specified to be a real number. This sequence $a_n = n$ will never get close and stay close to any real number A, because a_n just gets larger and larger as n gets larger and larger. We say that the sequence a_n is "unbounded."

5.8 DEFINITION A sequence a_n, n = 0, 1, 2, . . . , is *bounded* if there is a positive number B such that $|a_n| < B$ for all n.

5.9 THEOREM A convergent sequence is bounded.

Proof: Let a_n, n = 0, 1, 2, . . . , be a sequence with

$$\lim_{n \to \infty} a_n = A.$$

Taking $\epsilon = 1$ in DEFINITION 5.7, let N_1 be such that for $n > N_1$, $|a_n - A| < 1$. Let B be the maximum of $|a_0|, |a_1|, |a_2|, . . ., |a_{N_1}|, |A| + 1$. Then $|a_n| \leq B$ for all n. This completes the proof.

It's Equivalent To Its Contrapositive

THEOREM 5.9 says that if "a_n, n = 0, 1, 2, . . . , is convergent" then "a_n, n = 0, 1, 2, . . . , is bounded." This is equivalent to saying if "a_n, n = 0, 1, 2, . . . , is not bounded" then "a_n, n = 0, 1, 2, . . . , is divergent." These two statements are called *contrapositive* statements. If P and Q are propositions, then the statement "if P then Q" is equivalent to "if not Q then not P." In our example, P = "a_n, n = 0, 1, 2, . . . , is convergent" and Q = "a_n, n = 0, 1, 2, . . . , is bounded." Here is an example from real life: if "there is a cow in the barn" then "there is a mammal in the barn." The contrapositive statement, which is logically equivalent, is if "there is no mammal in the barn'' then ''there is no cow in the barn.'' Care must be taken with ''real life'' interpretations of the contrapositive.

The Converse Is Not The Same As The Contrapositive

You should pay careful attention to the distinction made in mathematical proofs between the ''contrapositive'' and the ''converse.'' The converse of the statement ''if P then Q'' is the statement ''if Q then P.'' These statements are definitely not logically equivalent. Each must be proved or disproved separately. In the statement if ''a_n, n = 0, 1, 2, . . ., is convergent'' then ''a_n, n = 0, 1, 2, . . ., is bounded,'' the converse is if ''a_n, n = 0, 1, 2, . . ., is bounded'' then ''a_n, n = 0, 1, 2, . . ., is convergent.'' This latter statement is false, as the example $a_n = (-1)^n$, n = 0, 1, 2, . . ., shows. If both the statement ''if P then Q'' and its converse ''if Q then P'' are true, then we say ''P if and only if Q'' is true. As an example, suppose we are talking about triangles in a trigonometry course and the lengths of the sides of a triangle T are denoted by $a \leqslant b \leqslant c$. Let P = ''T is a right triangle'' and Q = ''$a^2 + b^2 = c^2$.'' The statement ''if P then Q'' is the PYTHA-GOREAN THEOREM. The statement ''if Q then P'' is also a true theorem, a consequence of the law of cosines. Thus, the statement ''P if and only if Q'' is valid.

We now state some of the basic rules for operating with limits of sequences. These rules are a special case of the RULES FOR LIMITS (3.14).

Some More Rules And The Agony Of A Proof

5.10 THEOREM: RULES FOR LIMITS OF SEQUENCES Suppose that a_n, n = 0, 1, 2, . . ., and b_n, n = 0, 1, 2, . . ., are convergent infinite sequences. Let

$$\lim_{n \to \infty} a_n = A \text{ and } \lim_{n \to \infty} b_n = B.$$

Define sequences t_n, s_n, p_n, and q_n, n = 0, 1, 2, . . ., by $t_n = \alpha a_n$, where α is a real number, $s_n = a_n + b_n$, and $p_n = a_n b_n$. If $b_n \neq 0$ for all n, define $q_n = a_n/b_n$. Then

(1) $\lim_{n \to \infty} t_n = \alpha A$

(2) $\lim_{n \to \infty} s_n = A + B$

(3) $\displaystyle\lim_{n\to\infty} p_n = AB$

(4) $\displaystyle\lim_{n\to\infty} q_n = A/B$ if $B \neq 0$.

Proof: We give the proofs of (2) and (3). You should try to do similar proofs for (1) and (4). We use DEFINITION 5.7. To prove (2) we must show that given any $\epsilon > 0$ we can find an integer N_ϵ such that, for all $n > N_\epsilon$, $|s_n - (A + B)| = |a_n + b_n - A - B| < \epsilon$. We shall use the fact that for any real numbers x and y, $|x + y| \leq |x| + |y|$. Given ϵ, let $\rho = \epsilon/2$. From DEFINITION 5.7, we know we can find N_ρ such that, for all $n > N_\rho$, $|a_n - A| < \rho$. Likewise, we can find N_ρ (perhaps different than the one of the last sentence) such that, for all $n > N_\rho$, $|b_n - B| < \rho$. Take N_ϵ to be the larger of these two N_ρ. Thus, for $n > N_\epsilon$, we have

$$|s_n - (A + B)| = |a_n + b_n - A - B|$$

$$\leq |a_n - A| + |b_n - B| < \rho + \rho = \epsilon.$$

This completes the proof of (2). To prove (3), we must show that given any $\epsilon > 0$, we can find an integer N_ϵ such that, for all $n > N_\epsilon$, $|a_n b_n - AB| < \epsilon$. We have

$$|a_n b_n - AB| = |(a_n - A)b_n + (b_n - B)A|$$

$$\leq |a_n - A|\,|b_n| + |b_n - B|\,|A|.$$

By THEOREM 5.9, the sequence b_n is bounded by some number M, which we choose to be larger than $|A|$. Now, let $\rho = \epsilon/(2M)$. Then, by DEFINITION 5.7, there is an integer N_ρ such that, for all $n > N_\rho$, $|a_n - A| < \rho$. Likewise, there is an integer N_ρ (perhaps different from the one of the last sentence) such that, for all $n > N_\rho$, $|b_n - B| < \rho$. Take N_ϵ to be the larger of these two N_ρ. Then, for all $n > N_\epsilon$,

$$|a_n b_n - AB| \leq |a_n - A|\,|b_n| + |b_n - B|\,|A|$$

$$\leq |a_n - A|\,M + |b_n - B|\,M < \rho M + \rho M = \epsilon.$$

This completes the proof of (3) of THEOREM 5.10.

Look again at THEOREM 5.9 and DEFINITION 5.8. A sequence a_n, $n = 0, 1, 2, \ldots$, is "bounded" by B if $|a_n| < B$ for all n. We can now refine that idea a bit.

Upper And Lower Bounds And A Fundamental Axiom

5.11   DEFINITION   Let a_n, $n = 0, 1, 2, \ldots$, be a sequence. We say that the real number U is an *upper bound* for the sequence if $a_n < U$ for all n. We say the real number L is a *lower bound* for the sequence if $a_n > L$ for all n. A real number LU is the *least upper bound* of the sequence a_n if it is an upper bound for a_n and if, for any real number $x < LU$, x is not an upper bound of a_n. A real number GL is the *greatest lower bound* of the sequence a_n if it is a lower bound for a_n and if, for any real number $x > GL$, x is not a lower bound of a_n.

Let's look at a couple of examples of DEFINITION 5.11. As a first example, take $f(x) = -(x - 3)^2 + 4$. Define $a_n = f(n)$, $n = 0, 1, 2, \ldots$. Then, since the maximum value of $f(x)$ is $+4$ at $x = +3$, any number $U > 4$ is an upper bound for this sequence. The number 4 is the least upper bound because any number $x < 4$ has $a_3 > x$ and hence x is not an upper bound for the sequence a_n. As a second example, take $g(x) = 4x^2/(x^2 + 5)$ and define $b_n = g(n)$, $n = 0, 1, 2, \ldots$. By computing $g'(x) = 40x/(x^2 + 5)^2$ we see that $g(x)$ is increasing and hence the sequence b_n satisfies $b_n < b_{n+1}$ for all $n = 0, 1, 2, \ldots$. It is easy to see the limit as n approaches infinity of b_n is 4. Thus 4 is an upper bound. In fact, 4 is the least upper bound. See if you can convince yourself why this is true! It is a special case of THEOREM 5.13 below.

It is a basic axiom of the real number system that every sequence of real numbers that has an upper bound U must have a unique least upper bound LU. Likewise, every sequence of real numbers that has a lower bound must have a unique greatest lower bound GL. For certain types of sequences, these unique bounds are also the limits.

Bounded Monotone Sequences Always Converge

5.12   DEFINITION   A sequence a_n, $n = 0, 1, 2, \ldots$, is nondecreasing if $a_n \leq a_{n+1}$ for $n = 0, 1, 2, \ldots$. A sequence a_n, $n = 0, 1, 2, \ldots$, is nonincreasing if $a_n \geq a_{n+1}$ for $n = 0, 1, 2, \ldots$.

5.13   THEOREM   Every bounded nondecreasing sequence converges to its least upper bound LU. Every bounded nonincreasing sequence converges to its greatest lower bound GL.

Proof: Suppose that a_n, n = 0, 1, 2, . . ., is nondecreasing and bounded. In particular, a_n must be bounded above and hence must have a least upper bound LU. Given $\epsilon > 0$, we note that the real number x = LU $- \epsilon$ is not an upper bound for the sequence a_n by the definition of the least upper bound LU. Thus, there is an integer m such that $a_m > $ LU $- \epsilon$. Of course, $a_m \leq$ LU as LU is an upper bound for the whole sequence a_n. If we take N_ϵ = m, then, by the fact that the sequence a_n is nondecreasing, for all n > N_ϵ, LU $- \epsilon < a_m \leq a_n \leq$ LU. In other words, for n > N_ϵ, $|a_n - $ LU$| < \epsilon$. Thus, by DEFINITION 5.7, LU is the limit of the sequence a_n. This completes the proof of THEOREM 5.13 (the proof of the second statement of the threorem is directly analogous).

Diverging To Plus Or Minus Infinity

In DEFINITION 5.7, we defined what it meant for a sequence to converge to a limit A, where A is a real number. We stated that a sequence that did not converge was called a *divergent sequence*. Divergent sequences can be bounded, such as the sequence $(-1)^n$, n = 0, 1, 2, . . ., or unbounded, such as a_n = n, n = 0, 1, 2, . . ., or $a_n = -$n, n = 0, 1, 2, . . ., or $a_n = (-1)^n$n, n = 0, 1, 2, ... In the sequence a_n = n, the terms get steadily larger and larger without any upper bound. In the sequence $a_n - -$n, the terms get steadily smaller and smaller without any lower bound. In the first case, we say that the sequence "diverges to plus infinity" or "tends to plus infinity." In the second case, we say the sequence "diverges to minus infinity" or "tends to minus infinity." Although it is technically an abuse of the notation to do so, we sometimes write

$$\lim_{n \to \infty} a_n = +\infty \qquad \text{and} \qquad \lim_{n \to \infty} a_n = -\infty$$

to describe these two situations. The sequence $(-1)^n$n doesn't tend to either plus infinity or minus infinity. It hops back and forth between larger and larger positive and negative values. We simply call this sequence an "unbounded divergent sequence." You should be getting sophisticated enough by now to understand the following definition:

5.14 DEFINITION Let a_n, n = 0, 1, 2, . . ., be a sequence. We say that a_n *diverges to plus infinity* if for all x > 0, there exists an integer N_x such that for all n > N_x, a_n > x. Similarly, we say that a_n diverges to minus infinity if for all x < 0, there exists an integer N_x such that for all n > N_x, a_n < x.

The intuitive idea of DEFINITION 5.14 is that if the sequence a_n diverges to plus infinity, then given any number x, which we imagine to be very large, we can always find some point in the sequence beyond which *all* of the terms are larger than x.

Some Practice With Sequences

In the next exercise, you are asked to discuss the divergence or convergence of certain sequences. This means that you should state whether or not the sequence diverges or converges. If it converges, try to find the limit. If it diverges, state whether or not the sequence is bounded. If the sequence is not bounded, state whether or not the sequence diverges to plus infinity or to minus infinity or neither.

5.15 EXERCISES

(1) Discuss the convergence or divergence of the following sequences:

(a) $\dfrac{2n^3 + 3n + 1}{3n^3 + 2}$, n = 0, 1, 2, . . .

(b) $\dfrac{-n^3 + 1}{2n^2 + 3}$, n = 0, 1, 2, . . .

(c) $\dfrac{(-n)^{n+1} + 1}{n^n + 1}$, n = 0, 1, 2, . . .

(d) $\dfrac{n^n}{e^n}$, n = 1, 2, . . .

(e) $\dfrac{(-n)^3 + 1}{n^3 + \ln(n)}$, n = 0, 1, 2, . . .

(f) $\cos(\pi n)$, n = 0, 1, 2, . . .

(g) $\dfrac{\log_2(n)}{\log_3(n)}$, n = 2, 3, . . .

(h) $\dfrac{\log_2(n)}{n^{0.1}}$, $n = 1, 2, \ldots$

(i) $\dfrac{\log_2(\log_2(n))}{\log_2(n)}$, $n = 2, 3, \ldots$

(j) $\cos(n)$, $n = 0, 1, 2, \ldots$

(k) $\left(1 + \dfrac{1}{n}\right)^n$, $n = 1, 2, \ldots$

(2) By using a computer to generate these sequences, discuss their convergence or divergence:

(a) $\cos(n^2)$, $n = 0, 1, 2, \ldots$

(b) s_n, $n = 1, 2, \ldots$, where $s_{n+1} = (1/n)^{2^{-n}} s_n$ and $s_1 = 1$

(c) s_n, $n = 1, 2, \ldots$, where $s_{n+1} = \left(1 + \dfrac{1}{n}\right)^{2^{-n}} s_n$ for $n = 2, 3,$ $\ldots$, and $s_1 = 1$

Study The Solution, Then Modify The Original

We now discuss the solutions to EXERCISE 5.15. Remember, after learning the solution, go back to the original problem, change it a little bit, and rework it. After doing this for all problems in EXERCISE 5.15, you should move on to the VARIATIONS ON EXERCISE 5.15.

5.16 SOLUTIONS TO EXERCISE 5.15

(1)(a) This sequence is a ratio of two polynomials in n (i.e., a rational function of n). If we divide numerator and denominator by the highest power of n that appears in the denominator, we can write the same sequence, except for the first term, as

$$\frac{2 + (3/n^2) + (1/n^3)}{3 + (2/n^3)}, n = 1, 2, \ldots$$

In this form, it is obvious that the sequence converges to the limit 2/3 as n goes to infinity.

(1)(b)   Again, we have a rational function of n. Divide by the highest power of n that appears in the denominator to obtain

$$\frac{-n + (1/n^2)}{2 + (3/n)}, \; n = 1, 2, \ldots$$

In this form, it is evident that the sequence diverges to minus infinity.

(1)(c)   Again, we divide by the highest power of n in the denominator to obtain

$$\frac{-n(-1)^n + n^{-n}}{1 + n^{-n}}.$$

This is an unbounded, divergent sequence that oscillates between very large negative and positive values.

(1)(d)   This is the same as the sequence $(n/e)^n$, which obviously diverges to plus infinity.

(1)(e)   Dividing numerator and denominator by n^3 gives

$$\frac{-1 + 1/n^3}{1 + (\ln(n))/n^3}.$$

If you happen to know that the limit as n tends to infinity of $(\ln(n)/n^3$ is zero, then you will see that the limit of this expression is obviously -1. To understand this fact, we use L'HOPITAL'S RULE 3.15 to write

$$\lim_{n \to \infty} \frac{\ln(n)}{n^3} = \lim_{x \to \infty} \frac{\ln(x)}{x^3} = \lim_{x \to \infty} \frac{1}{3x^3} = 0.$$

(1)(f)   This is the sequence $+1, -1, +1, \ldots$ which is bounded, divergent.

(1)(g)   If we write $\log_3(n) = \log_3(2)\log_2(n)$, we see that every term in the sequence is $(\log_3(2))^{-1}$ and thus the sequence converges to this constant.

(1)(h)   Using L'HOPITAL'S RULE again, we obtain

$$\lim_{n \to \infty} \frac{\log_2(n)}{n^{0.1}} = \lim_{x \to \infty} \frac{\log_2(x)}{x^{0.1}} = \lim_{x \to \infty} \frac{1}{(0.1)\ln(2)x^{0.1}} = 0.$$

Thus the limit of this sequence is zero. If you worked part (b) of EXERCISE 3.20(3) you will recall the useful fact that the limit as x goes to infinity of $(\ln(x))/x^a$ is zero for any $a > 0$.

(1)(i)   This limit is zero. The reason is again the fact that the limit as x goes to infinity of $(\ln(x))/x^a$ is zero for any $a > 0$. In this case, we take $a = 1$ and use the fact that as x goes to infinity so does $\log_2(x)$.

(1)(j)   This problem is a little trickier than it seems. If you write the following little BASIC program

```
10   PRINT COS(N)
20   N=N+1
30   GOTO 10
```

and watch the numbers stream out, it is clear that cos(n) does not converge. This is the correct intuition. The function cos(x) is +1 at all $x = 2\pi n$, n an integer. At all real numbers of the form $x = 2\pi n + \pi$, cos(x) is −1. Think about the intervals $[2\pi n - .5, 2\pi n + .5]$. In every such interval there must be an integer m. This means that $\cos(m) > \cos(.5)$. In every interval of the form $[2\pi n + \pi - .5, 2\pi n + \pi + .5]$ there must be an integer m' for which $\cos(m') < \cos(\pi - 0.5) = -\cos(0.5)$. Thus, for infinitely many integers n, $\cos(n) > \cos(.5)$ and for infinitely many integers n, $\cos(n) < -\cos(0.5)$. This proves what common sense tells us, the sequence cos(n), $n = 0, 1, 2, \ldots$, does not converge. It is bounded and divergent.

(1)(k)   Let's write a little BASIC program to check this sequence out.

```
10   PRINT (1+ 1/N)^N
20   N=N+1
30   GOTO 10
       .
       .
       .
     2.717815
     2.717536
     2.717995
     2.717692
     2.718148
```

Amazing! It looks like this sequence is converging to the number $e = 2.718$ . . .. The easiest way to see this is to investigate the limit as n goes to infinity of $\ln\left(1 + \dfrac{1}{n}\right)^n = n\ln\left(1 + \dfrac{1}{n}\right)$. Using good old L'HOPITAL'S RULE again gives

$$\lim_{x\to\infty} x\ln(1 + x^{-1}) = \lim_{x\to\infty} \frac{\ln(1 + x^{-1})}{x^{-1}} = \lim_{x\to\infty} \frac{1}{1 + x^{-1}} = 1.$$

Thus, for the sequence $a_n = \left(1 + \dfrac{1}{n}\right)^n$, we have shown that the limit($\ln(a_n)$) = 1. This implies that limit(a_n) = e^1 = e. In general, if limit($\ln(a_n)$) = A > 0 then limit(a_n) = e^A. Formally, this is because the function $\ln(x)$ is one-to-one (FIGURE 2.13) and continuous (DEFINITION 3.12). It is also intuitively obvious if you take a look at the graph of $\ln(x)$ shown in FIGURE 2.13.

Now The Solutions To Exercise 5.15(2)

(2)(a)   The method used to analyze problem 1(j) above doesn't work here. The cos(x) function is evaluated at the integer points 1, 4, 16, . . ., n^2, ... If $\cos(n^2)$ were to have a limit A as n goes to infinity, then the points of the sequence n^2, n = 1, 2, . . ., would have to eventually cluster about points in the set $\{x : \cos(x) = A\}$. This seems highly unlikely, and can, with a little more effort than we want to make at this point, be proved not to happen. Here is a BASIC program with some sample output from the screen after the program has run for several minutes.

```
10   PRINT COS(N*N)
20   N=N+1
30   GOTO 10
          .
          .
          .
      .1177025
     −.9925256
      .9545957
      .9154154
      .6455998
     −.5313178
     −.9999428
     −.5312158
      .6454156
      .9155609
      .9544519
     −.992599
      .1169841
     −.3111826
     −.3815438
     −.7837191
     −.8679878
```

It is apparent that this sequence is bounded but divergent.

(2)(b)   A sequence s_n defined in this manner is said to be "defined re-cursively." Note that the ratio s_{n+1}/s_n is less than 1. Thus the sequence s_n is nonincreasing and, since 0 is a lower bound for the sequence, it must converge to its greatest lower bound. The following program, when allowed to run a bit, produces the output shown.

```
10   S = 1:N = 1
20   S = S*(1/N)^(2^ − N)
30   PRINT S
40   N = N + 1
50   GOTO 20
       .
       .
       .
     .6017975
     .6017975
     .6017975
     .6017975
     .6017975
     .6017975
     .6017975
```

It seems that .6017975 is the approximate limit.

(2)(c)   This sequence is again one that is defined recursively. The ration s_{n+1}/s_n is greater than one, so the sequence is nondecreasing. If it is bounded above, then it must converge to its least upper bound. Using techniques that we shall develop later in this chapter, it will be easy for us to show that this sequence is bounded above. In any case, the following program plus output tells us that this sequence is bounded above and gives us a good approximation to the limit.

```
10   S = 1:N = 1
20   S = S*(1 + (1/N))^(1/(2^N))
30   PRINT S
40   N = N + 1
50   GOTO 20
     1.414213
     1.565085
     1.622389
     1.645175
     1.654575
     1.658565
     1.660296
```

1.66106
1.661402
1.661556
1.661627
1.661659
1.661674
1.661681
1.661685
1.661686
1.661687
1.661687
1.661688
1.661688
1.661688
1.661688

Now Try The Variations

We now begin VARIATIONS on EXERCISE 5.15. Remember that corresponding problems use similar methods. Many of these problems have "general ideas" behind them. By the time you have worked the last variation of each type, you should try to articulate these ideas. Don't worry about being wrong in stating these general ideas, just try to make sense!

5.17   VARIATIONS ON EXERCISE 5.15

(1)   Discuss the convergence or divergence of the following sequences:

(a) $\dfrac{4n^5 + 3n^4 + 1}{7n^5 + 2}$, $n = 0, 1, 2, \ldots$

(b) $\dfrac{-n^3 + 1}{2n^4 + 3}$, $n = 0, 1, 2, \ldots$

(c) $\dfrac{(-n)^n + n^{n-1}}{n^n - n^{n-1}}$, $n = 2, \ldots$

(d) $n^n/(1 + n^{-1})^{n^2}$, $n = 1, 2, \ldots$

(e) $\dfrac{(-n)^3 + 1}{n^3 + (\ln(n))^5}$, $n = 0, 1, 2, \ldots$

(f) $\cos\left(\pi\,\dfrac{n^2 + 1}{n + 2}\right)$, $n = 0, 1, 2, \ldots$

(g) $\dfrac{\log_2(n^5)}{\log_3(n)}$, $n = 2, 3, \ldots$

(h) $\dfrac{\log_2(n)}{n^{0.001}}$, $n = 1, 2, \ldots$

(i) $\dfrac{\log_2(\log_2(n))}{\log_2(n)}$, $n = 2, 3, \ldots$

(j) $\cos\left(\dfrac{n^2 + 1}{n + 2}\right)$, $n = 0, 1, 2, \ldots$

(k) $\left(1 + \dfrac{2}{n}\right)^n$, $n = 1, 2, \ldots$

(2) By using a computer to generate these sequences, discuss their convergence or divergence:

(a) $\sin((44n^3 + n + 1)/(7n^2 + 1))$, $n = 0, 1, 2, \ldots$

(b) s_n, $n = 1, 2, \ldots$, where $s_{n+1} = \left(1 + \dfrac{1}{n}\right)^{(\ln(n))^{-1}} s_n$ for $n = 2$, $3, \ldots$, and $s_1 = 1$

(c) s_n, $n = 1, 2, \ldots$, where $s_{n+1} = \left(\dfrac{n + 1}{2n + 1}\right)^{\frac{(-1)^n}{n}} s_n$ and $s_1 = 1$

5.18 VARIATIONS ON EXERCISE 5.15

(1) Discuss the convergence or divergence of the following sequences:

(a) $\dfrac{4(\ln(n))^5 + 3(\ln(n))^4 + 1}{7(\ln(n))^5 + 2}$, $n = 1, 2, \ldots$

(b) $\dfrac{-n^{-3} + 1}{2n^{-4} + 3}$, $n = 1, 2, \ldots$

(c) $\dfrac{(-n)^{n-1} + n^n}{n^n - n^{n-1}}$, $n = 2, \ldots$

(d) $(1 + n)^{1/n}$, $n = 1, 2, \ldots$

(e) $\dfrac{(-n\ln(n))^3 + 1}{n^3 + (\ln(n))^5}$, $n = 1, 2, \ldots$

(f) $\sin\left(\dfrac{\pi^2 n^2 + 1}{\pi n + 2}\right)$, $n = 0, 1, 2, \ldots$

(g) $\dfrac{\log_2(n^5 + \log_2(n))}{\log_3(n)}$, $n = 2, 3, \ldots$

(h) $\dfrac{\ln(n)}{n^{(\ln(n))^{-1}}}$, $n = 2, 3, \ldots$

(i) $\dfrac{\log_2(\log_2(n))}{\log_3(\log_3(n))}$, $n = 4, 5, \ldots$

(j) $\sin(2n)$, $n = 0, 1, 2, \ldots$

(k) $\left(1 + \dfrac{\ln(n)}{n}\right)^n$, $n = 1, 2, \ldots$

(2) By using a computer to generate these sequences, discuss their convergence or divergence:

(a) $\cos((n^3 + 1)/(n + 1))$, $n = 0, 1, 2, \ldots$

(b) s_n, $n = 1, 2, \ldots$, where $s_{n+1} = (1/n^p)^{q^{-n}} s_n$ and $s_1 = 1$ for $p = 1, 2, 3$ and $q = 2, 3, 4$.

(c) s_n, $n = 1, 2, \ldots$, where $s_{n+1} = \left(\dfrac{\log_2(n)}{\log_2(qn)}\right)^{p^{-n}} s_n$, $n = 2, 3, 4,$ $\ldots$, and $s_1 = 1$ for $p = 2, 3, 4$ and $q = 2, 4, 8$.

5.19   VARIATIONS ON EXERCISE 5.15

(1)   Discuss the convergence or divergence of the following sequences:

(a) $\dfrac{4a_n^5 + 3a_n^4 + 1}{7a_n^5 + 2}$, $n = 0, 1, 2, \ldots$, where $a_n = (n^2 + 1)/(2n^2 + 1)$

(b) $\dfrac{-\sin^3 n + 1}{2\sin^4 n + 3}$, $n = 0, 1, 2, \ldots$

(c) $\dfrac{(-n)^n + n^n}{n^n - n^{n-1}}$, $n = 2, \ldots$

(d) $(2 + \sin(n))^{1/n}$, $n = 1, 2, \ldots$

(e) $\dfrac{(-n\ln(n))^3 + 2n^3}{n^3 + 2(n\ln(n))^3}$, $n = 1, 2, \ldots$

(f) $\sin\left(\dfrac{\pi^2 n^2 + n}{\pi n + 2}\right)$, $n = 0, 1, 2, \ldots$

(g) $\dfrac{\log_2(n^5 + 2^n)}{n}$, $n = 2, 3, \ldots$

(h) $\dfrac{\ln(n)}{n^{(\ln(\ln(n)) - 1)}}$, $n = 4, 5, \ldots$

(i) $\dfrac{\log_2(\log_2(\log_2(n)))}{\log_3(\log_3(\log_3(n)))}$, $n = 28, 29, \ldots$ The general rule?

(j) $\sin(3n)$, $n = 0, 1, 2, \ldots$ The general rule?

(k) $\left(1 + \dfrac{\ln(n)}{n}\right)^{n/\ln(n)}$, $n = 2, 3, \ldots$

(2)   By using a computer to generate these sequences, discuss their convergence or divergence:

(a) s_n, n = 1, 2, 3, . . ., where $s_{n+1} = \dfrac{(2n)^2}{(2n - 1)(2n + 1)} s_n$, n = 1, 2, 3, . . ., and $s_1 = 2$.

(b) $\sin(s_n n)$, n = 1, 2, 3, . . ., where s_n is as in (a).

(c) $\sin(s_n \lfloor \ln(\ln(n + 3)) \rfloor)$, n = 1, 2, 3, . . ., where s_n is as in (a). The function $\lfloor x \rfloor$ is the "greatest integer function" or "floor function" and is called INT in BASIC. For example, INT(4.7) = 4, INT(3.14) = 3, etc.

An Infinite Series Is A Sequence Of Partial Sums

We are now ready to start our discussion of infinite series. Let's start with an old friend, an infinite *sequence* a_k, k = 0, 1, 2, ... Remember, this is nothing more than a function from the domain {0, 1, 2, . . .} to the real numbers. The value of the function at k is denoted by a_k. If we have another sequence b_k, k = 0, 1, 2, . . ., then we say the two sequences are the same or are "equal" if $a_k = b_k$ for all k, k = 0, 1, 2, ... Suppose that for each integer n, n = 0, 1, 2, . . ., we form the sum $s_n = a_0 + a_1 + . . . + a_n$. This defines a new sequence s_n, n = 0, 1, 2, . . ., called the *infinite series with terms from the infinite sequence* a_k, k = 0, 1, 2, ... The sequence s_n is also called the *sequence of partial sums* of the sequence a_k, k = 0, 1, 2, ... If s_n, n = 0, 1, 2, . . ., is the sequence of partial sums of a_k, k = 0, 1, 2, . . ., and t_n, n = 0, 1, 2, . . ., is the sequence of partial sums of b_k, k = 0, 1, 2, . . ., then to say the sequence (s_n) equals the sequence (t_n) means that $s_n = t_n$, n = 0, 1, 2, ... In particular, $s_0 = t_0$, so $a_0 = b_0$. In general, for n > 0, $a_n = s_n - s_{n-1} = t_n - t_{n-1} = b_n$, and hence $a_n = b_n$ for all n. Thus, two sequences can give rise to the same sequence of partial sums if and only if these two sequences are themselves equal.

Infinite Sums—Two Interpretations

We now confront another notational artifact of calculus. The infinite series with terms from the infinite sequence a_k, k = 0, 1, 2, . . ., is denoted by

$$\sum_{k=0}^{\infty} a_k.$$

So, nothing wrong with that! Unfortunately, if $s_n = a_0 + a_1 + . . . + a_n$ is the nth partial sum, we also have the notation

$$\sum_{k=0}^{\infty} a_k = \lim_{n \to \infty} s_n = A$$

which says that the limit of the sequence of partial sums is A. In other words, the same infinite summation notation is used to designate both the sequence and its limit, two very different things. Usually, there is enough additional information floating around in any given discussion to avoid confusion.

Equal Series . . . Equal Limits, It's Not The Same

In general, if you want to say that two infinite series are equal, you say something like

"The infinite series $\sum_{k=0}^{\infty} a_k$ equals the infinite series $\sum_{k=0}^{\infty} b_n$."

This would mean that $a_k = b_k$ for all k. On the other hand, the statement

$$\sum_{k=0}^{\infty} a_k = \sum_{k=0}^{\infty} b_k,$$

by itself usually means that the limits of the two series are equal and does not mean that $a_k = b_k$ for all k.

As a specific example, let $a_k = (1/2)^k$, $k = 0, 1, 2, \ldots$, and let $b_0 = 0$, $b_k = (2/3)^k$, $k = 1, 2, \ldots$. These "geometric series" are studied in high school algebra or precalculus courses and both converge to the number 2. Thus, we write

$$\sum_{k=0}^{\infty} a_k = \sum_{k=0}^{\infty} b_k = 2.$$

Every Sequence Is A Sequence Of Partial Sums

Before studying examples of series, there is one other general remark to be made. It seems from the definition that infinite series are a special class of infinite sequences. This is true in the sense that an infinite series s_n, $n = 1, 2, \ldots$, is specified as the sequence of partial sums of some infinite sequence a_n, $n = 0, 1, 2, \ldots$. Because of this, we discuss infinite series in terms of this underlying sequence a_n and its properties, giving the theory of infinite series a special notational and conceptual flavor. You should realize, however, that any sequence b_n, $n = 0, 1, 2, \ldots$, can be regarded as an infinite series. Just define $a_0 = b_0$ and $a_n = b_n - b_{n-1}$, $n = 1, 2, \ldots$, and the sequence $b_n = s_n$, where $s_{\hat{n}}$, $n = 0, 1, 2, \ldots$, is the sequence of partial sums of a_n, $n = 0, 1, 2, \ldots$. This is the type of remark that interests mathematicians and bores everyone else. It does come in handy in some specific problems concerning sequences and series, so at least take note of it.

5.20 EXAMPLE: GEOMETRIC SERIES Let $a_k = r^k$, $k = 0, 1, 2$, $\ldots$, where r is any real number. The infinite series with terms from this sequence is the sequence of partial sums $s_n = 1 + r + \ldots + r^n$, $n = 0$, $1, 2, \ldots$. We know from precalculus courses that $s_n = (1 - r^{n+1})/(1 - r)$ if $r \neq 1$. If $r = 1$ then $s_n = n + 1$. It is obvious from this formula that the sequence s_n diverges if $|r| \geq 1$ and converges to $1/(1 - r)$ if $|r| < 1$. Thus, we write

$$(*) \quad \sum_{k=0}^{\infty} r^k = \frac{1}{1 - r} \text{ if } |r| < 1.$$

As examples of geometric series, take $r = (1/2)$ and $r = (2/3)$ to get

$$\sum_{k=0}^{\infty} (1/2)^k = \frac{1}{1 - (1/2)} = 2 \text{ and } \sum_{k=0}^{\infty} (2/3)^k = \frac{1}{1 - (2/3)} = 3.$$

If the first term is missing from a geometric series $(*)$ then the sum of the series is $r/(1 - r)$ instead of $1/(1 - r)$. Thus, for example,

$$\sum_{k=1}^{\infty} (2/3)^k = \frac{(2/3)}{1 - (2/3)} = 2.$$

Our next example has some very special properties that are important to understanding series.

5.21 EXAMPLE: HARMONIC AND ALTERNATING HARMONIC SERIES Let $a_k = 1/k$, $k = 1, 2, \ldots$. The series of partial sums $s_n = 1 + (1/2) + (1/3) + \ldots + (1/n)$, $n = 1, 2, \ldots$, is called the *harmonic series*. This series diverges. One way to see this is to write the series as

$$1 + [(1/2)] + [(1/3) + (1/4)] + [(1/5) + (1/6) + (1/7) + (1/8)] + \ldots$$

where the general term in square brackets is

$$[(1/(2^n + 1)) + \ldots + (1/2^{n+1})].$$

Each term in square brackets is greater than or equal to 1/2 and there are infinitely many such terms, thus the series must diverge.

An important related series is the *alternating harmonic series* where $a_k = (-1)^{k-1}(1/k)$, $k = 1, 2, \ldots$. The partial sums of this series are of the form $s_n = 1 - (1/2) + (1/3) - (1/4) + \ldots + (-1)^{n-1}(1/n)$. This series converges. To see why intuitively, imagine that you are standing in a room with your back against a wall. Imagine that you step forward 1 meter, then backward 1/2 meter, then forward 1/3 meter, etc. After n such steps, your distance from the wall is the value of the partial sum s_n. By the time you are stepping forward one millimeter, etc., an observer in the room (who by now has decided you are nuts) would conclude that you are standing still. In other words, you have converged. This argument works just as well for any step sizes, as long as they are alternating forward and backward, of decreasing size, and tend to zero. In the case of the alternating harmonic series, your final distance from the wall is ln(2) meters. Check it out on your computer. See also EXERCISE 5.70(16).

Alternating Series

From the discussion of EXAMPLE 5.21, we have the following theorem:

5.22   THEOREM   Let a_k, $k = 1, 2, \ldots$, be a sequence of positive numbers such that $a_k \geq a_{k+1}$ for $k = 1, 2, \ldots$. If $\lim\limits_{k \to \infty} a_k = 0$ then

$$\sum_{k=1}^{\infty} (-1)^{k-1} a_k$$

converges and

$$\left| \sum_{k=1}^{\infty} (-1)^{k-1} a_k - \sum_{k=1}^{n} (-1)^{k-1} a_k \right| \leq |a_{n+1}|$$

Proof: The intuitive idea of the proof was discussed in connection with EXAMPLE 5.21. Let s_{2n}, $n = 1, 2, \ldots$, be the sequence of even partial sums. We write

$$s_{2n} = (a_1 - a_2) + (a_3 - a_4) + \ldots + (a_{2n-1} - a_{2n}).$$

Each of the terms in parenthesis is nonnegative. Thus the sequence s_{2n} is nondecreasing. The sequence s_{2n} is also bounded above by a_1. By THEOREM 5.13, this sequence converges to its least upper bound, LU. Similarly, define the sequence s_{2n+1}, $n = 0, 1, \ldots$, of odd partial sums. This sequence can be written

$$s_{2n+1} = a_1 - (a_2 - a_3) - (a_4 - a_5) - \ldots - (a_{2n} - a_{2n+1}).$$

This sequence is obviously nonincreasing and bounded below by zero. Thus, it converges to its greatest lower bound, GL. But, $\left|s_{2n+1} - s_{2n}\right| = a_{2n+1}$ tends to zero as n tends to infinity and hence GL = LU is the limit of the sequence s_n, n = 1, 2, 3, ... This proves that the series converges. The ideas of EXAMPLE 5.21 explain the error estimate.

Absolute Convergence

There is the germ of another important idea in EXAMPLE 5.21.

5.23   DEFINITION   Let s_n, n = 0, 1, 2, . . ., be the series with terms a_k, k = 0, 1, 2, ... Let t_n be the series with terms $\left|a_k\right|$, k = 0, 1, 2, ... If the series t_n converges, then the series a_n is said to *converge absolutely* or to be an *absolutely convergent series*

Another way that DEFINITION 5.23 is stated is

$$\sum_{k=0}^{\infty} a_k \text{ converges absolutely if } \sum_{k=0}^{\infty} \left|a_k\right| \text{ converges.}$$

It may not be obvious to you that a series that converges absolutely must converge. We shall explain why below. The converse is false, in that a series that converges need not converge absolutely. This is the case with the alternating harmonic series, which converges but does not converge absolutely because the harmonic series diverges.

Convergent Series—A Thought Algorithm

It's time to refresh our intuition about infinite series. Remember, we start with a sequence a_k, k = 0, 1, 2, ... These are the terms of the infinite series s_n, n = 0, 1, 2, . . ., where $s_n = a_0 + a_1 + \cdots + a_n$. Imagine that we have written a computer program to evaluate these numbers s_n. We start the computer program running and watch the screen. The numbers s_n stream onto the screen. At first we are thrilled. The numbers are changing rapidly. But after awhile they begin to change only in the 10th decimal place and we become bored. This is the sort of thing that happens when the series is converging to some number A. We may never see A on the screen, only numbers close to A. Here is a definition of convergence of infinite series that corresponds to our computer intuition.

Alternative Definition Of Convergence

5.24 DEFINITION Let s_n, $n = 0, 1, 2, \ldots$, be the sequence of partial sums of the sequence a_k, $k = 0, 1, 2, \ldots$. The sequence s_n converges if for every $\epsilon > 0$ there exists an integer N_ϵ such that for all $q \geq p > N_\epsilon$, $|a_p + \ldots + a_q| < \epsilon$.

In our imaginary computer experiment, the ϵ is a small number, like 10^{-9}, where, if the numbers are changing by an amount less than that number, we lose interest. The number N_ϵ corresponds to the number of terms in the sum that have been computed when this begins to happen. After this begins to happen, we can watch the computer add any amount of additional terms from any p to any q and the sum won't change by much. The advantage of DEF-INITION 5.24 is that it doesn't explicitly mention the limit of the series. Of course, we have already given the definition of the convergence of a series in terms of the convergence of the sequence of partial sums. This uses DEF-INITION 5.7. Technically this means that DEFINITION 5.24 is really a little theorem. We won't worry about this technicality.

Absolute Convergence Implies Convergence

Just to give DEFINITION 5.24 a try, let's prove that a series that converges absolutely must converge.

5.25 THEOREM If a series converges absolutely then it converges.

Proof: Let t_n, $n = 0, 1, 2, \ldots$, be the sequence of partial sums of the sequence $|a_0|, |a_1|, |a_2|, \ldots$. We assume that t_n converges. Thus, given any $\epsilon > 0$, there exists an N_ϵ such that for $q \geq p > N_\epsilon$, $|a_p| + \ldots + |a_q| < \epsilon$. But, $|a_p + \ldots + a_q| \leq |a_p| + \ldots + |a_q|$ and hence, for $q \geq p > N_\epsilon$, $|a_p + \ldots + a_q| < \epsilon$. By DEFINITION 5.24, this proves that the sequence s_n of partial sums of a_k, $k = 0, 1, 2, \ldots$, converges. This completes the proof.

Series Have Tails . . .

5.26 DEFINITION Let a_k, $k = 0, 1, 2, \ldots$, be an infinite sequence. For any integer $t = 0, 1, 2, \ldots$, we consider the sequence $a_t, a_{t+1}, a_{t+2}, \ldots$. This sequence will be called the "t^{th} tail sequence of the sequence a_k." The sequence of partial sums $s_n^t = a_t + a_{t+1} + \ldots + a_{t+n}$, $n = 0, 1, 2,$

. . ., is the "t^{th} tail series" of the series $s_n = a_0 + \ldots + a_n$, $n = 0, 1, 2, \ldots$.

In the infinite sum notation, we say that

"the series $\displaystyle\sum_{k=t}^{\infty} a_k$ is the tth tail of the series $\displaystyle\sum_{k=0}^{\infty} a_k$"

In this context, we are using the infinite sum notation to specify the series or sequence of partial sums and not the limit. When we say

$$\sum_{k=0}^{\infty} a_k = a_0 + a_1 + \ldots + a_{t-1} + \sum_{k=t}^{\infty} a_k$$

we are using the infinite sum notation to denote the limits of these respective series. Again, although this notation is not the greatest, it is concise and doesn't usually lead to confusion.

It's Enough To Test The Tail . . .

The next result is simple, but extremely useful for testing series for convergence.

5.27 THEOREM If any tail series of a series converges then the series converges. Conversely, if any tail series diverges then the series diverges.

Proof: Suppose that a_k, $k = 0, 1, 2, \ldots$, are the terms of the series. To say that the tth tail series converges means that given any $\epsilon > 0$ there exists N_ϵ such that for all $q \geq p > N_\epsilon$, $|a_p + a_{p+1} + \ldots + a_q| < \epsilon$. This is exactly what is required by DEFINITION 5.24 for the whole series to converge. Conversely, if the tail series diverges then there is some $\epsilon > 0$ such that for every integer N there exists $q \geq p > N$ such that $|a_p + \ldots + a_q| \geq \epsilon$. This shows that the conditions of DEFINITION 5.24 are not valid when applied to the whole series and thus the whole series diverges.

As an example of the way THEOREM 5.27 is used, consider the series with terms $a_k = (-1)^k 100/|k - 99.5|$, $k = 1, 2, \ldots$. We would like to apply THEOREM 5.22, but the hypothesis of this theorem doesn't quite apply. In particular, we don't have $|a_k| \geq |a_{k+1}|$ for $k = 1, 2, \ldots$. The sequence $|a_k|$ is increasing until $k = 99$, then $|a_{99}| = |a_{100}| = 200$. After that, the sequence $|a_k|$ does satisfy the conditions of THEOREM 5.22, and hence the

t^{th} tail series converges for t = 100. By THEOREM 5.27, the whole series converges. We are using $|a_k|$ for what we called a_k in THEOREM 5.22.

Here is another useful result that follows from DEFINITION 5.24.

A Result About Products Of Sequences

5.28 THEOREM Suppose that the series with terms a_k, k = 0, 1, 2, . . ., converges absolutely and let b_k, k = 0, 1, 2, . . ., be any bounded sequence. Then the series with terms $a_k b_k$, k = 0, 1, 2, . . ., converges absolutely.

Proof: Let M be such that $|b_k| < M$ for all k. By hypothesis, the series $|a_0| + |a_1| + . . . + |a_n| + . . .$, converges. Given any $\epsilon > 0$, there exists an N such that for all $q \geqslant p > N$, $|a_q| + . . . + |a_p| < \epsilon/M$. This implies that $|a_q b_q| + . . . + |a_p b_p| < \epsilon$. Thus, by DEFINITION 5.24, the series with terms $a_k b_k$ converges absolutely. This completes the proof.

We'll learn later that the series with terms $a_k = 1/k^p$, k = 1, 2, 3, . . ., converges if $p > 1$ and the convergence is obviously absolute since the terms are all positive. Thus, if we take p = 3, then the series $1 + 2^{-3} + 3^{-3} + 4^{-3} + . . .$ converges absolutely. If we multiply the terms of this series by the divergent, unbounded sequence $b_k = k$, then we still obtain an absolutely convergent series, $1 + 2^{-2} + 3^{-2} + 4^{-2} + ...$ This shows that the converse of THEOREM 5.28 is false in that absolute convergence of the series with terms $a_k b_k$ does not imply that the sequence b_k is bounded.

The Comparison Test For Convergence

One of the easiest ways to determine the absolute convergence of a series is to compare it with another series known to converge absolutely. The idea is stated in the following corollary.

5.29 COROLLARY (COMPARISON TEST) Suppose that the series with terms a_k, k = 0, 1, 2, . . ., converges absolutely. Let c_k, k = 0, 1, 2, . . ., be a sequence such that $|c_k| \leqslant M|a_k|$, k = 0, 1, 2, . . ., where M is some positive real number. Then, the series with terms c_k, k = 0, 1, 2, . . ., converges absolutely.

Proof: We apply THEOREM 5.28. Define a sequence b_k by $b_k = 0$ if $a_k = 0$ and, otherwise, $b_k = c_k/a_k$. This sequence c_k is bounded by M. Thus, by THEOREM 5.28, the sequence $a_k b_k = c_k$ converges absolutely. This completes the proof.

The Comparison Test For Divergence

Another way to state the COMPARISON TEST is that if the series with terms $|a_k|$, $k = 0, 1, 2, \ldots$, diverges and if $C|c_k| \geq |a_k|$, for $k = 0, 1, 2, \ldots$ and $C > 0$, then the series with terms $|c_k|$, $k = 0, 1, 2, \ldots$, diverges. To see this, we note that if the series with terms $|c_k|$ converged then, by setting $M = C^{-1}$ in COROLLARY 5.29, we would have that, contrary to assumption, the series with terms $|a_k|$ would converge. In this application of COROLLARY 5.29, the roles of the a_k and c_k are reversed.

By now you are thinking "ENOUGH THEORY!" You are right. It's time to put this stuff to work. One of the best ways to learn a subject is to teach it. Our goal in what follows is to show you how to be a good instructor in the subject of infinite series, by teaching you how to make up problems for your classmates to solve.

Now You Learn To Make Up Infinite Series Problems

5.30   PROBLEMS BASED ON THE INTEGRAL TEST   Suppose that we have a continuous function $f(x)$ such that $f(x) > 0$ and $f(x)$ is nonincreasing for $x > t > 0$. Let a_k, $k = 0, 1, 2, \ldots$, be a sequence such that $|a_k| \leq f(k)$, $k \geq t$. Then the "integral test" states

$$\sum_{k=0}^{\infty} a_k \text{ converges absolutely if } \int_{t}^{\infty} f(x)dx < \infty.$$

On the other hand, if $|a_k| \geq f(k)$ for $k \geq t$ then

$$\sum_{k=0}^{\infty} |a_k| \text{ diverges if } \int_{t}^{\infty} f(x)dx = \infty.$$

The reason that these results are true is shown in FIGURE 5.31. There we see that the area under the curve is related to the sum of the tth tail of the series with terms a_k in such a way that both either converge or diverge.

FIGURE 5.31 The Integral Test

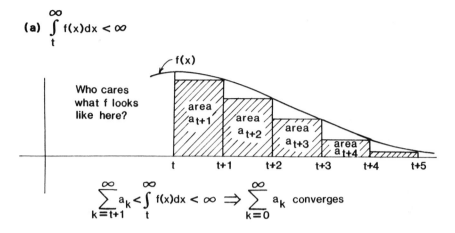

(a) $\int\limits_{t}^{\infty} f(x)dx < \infty$

Who cares what f looks like here?

area a_{t+1}

area a_{t+2}

area a_{t+3}

area a_{t+4}

f(x)

t   t+1   t+2   t+3   t+4   t+5

$$\sum_{k=t+1}^{\infty} a_k < \int_{t}^{\infty} f(x)dx < \infty \Rightarrow \sum_{k=0}^{\infty} a_k \text{ converges}$$

(b) $\int\limits_{t}^{\infty} f(x)dx = \infty$

Who cares what f looks like here?

area a_t

area a_{t+1}

area a_{t+2}

area a_{t+3}

f(x)

t   t+1   t+2   t+3   t+4   t+5

$$\sum_{k=t}^{\infty} a_k > \int_{t}^{\infty} f(x)dx = \infty \Rightarrow \sum_{k=0}^{\infty} a_k \text{ diverges}$$

Look For Positive Nonincreasing Functions

To construct problems based on the integral test, all we need to do is find functions $f(x)$ that eventually, past some real number t, become positive and nonincreasing. We then evaluate the integral from t to infinity of $f(x)$. If it diverges we can construct a divergent series, if it converges we can construct a convergent series.

For example, let's take $f(x) = x^{-1}$. The integral

$$\int \frac{dx}{x} = \ln(x) \text{ so } \int_{1}^{\infty} \frac{dx}{x} = \lim_{x\to\infty} \ln(x) - 0 = +\infty.$$

Thus, the series

$$\sum_{k=1}^{\infty} \frac{1}{k}$$

diverges to infinity. We already knew this.

Let's try a function $f(x) = (x\ln(x))^{-1}$ that goes to zero a little bit faster than x^{-1}. Perhaps the related series will converge. We compute

$$\int \frac{dx}{x\ln(x)} = \ln(\ln(x)) \text{ so } \int_2^{\infty} \frac{dx}{x\ln(x)} = \lim_{x \to \infty} \ln(\ln(x)) - \ln(\ln(2)) = +\infty$$

Thus, the series

$$\sum_{k=2}^{\infty} \frac{1}{k\ln(k)} = +\infty$$

is a divergent series also.

Trying once again along these same lines, we discover that

$$\int \frac{dx}{x(\ln(x))^2} = \frac{-1}{\ln(x)} \text{ so } \int_2^{\infty} \frac{dx}{x(\ln(x))^2} < \infty.$$

Thus we obtain the interesting fact that

$$\sum_{k=2}^{\infty} \frac{1}{k(\ln(k))^2}$$

converges.

Some Tricks For Making Difficult Problems

In EXERCISE 5.32, you will be asked to construct four problems based on the integral test and exchange these problems with your classmates. If you want to make these problems appear more difficult than they really are, here is a useful trick. If you have a series with terms a_k that converges absolutely and $\tilde{a}_k$ is any sequence such that the sequence $b_k = \tilde{a}_k/a_k$ is a bounded sequence, then the series with terms $\tilde{a}_k$ also converges absolutely. This is just a restatement of THEOREM 5.28. For example, take

$$a_k = \frac{1}{k(\ln(k))^2} \text{ and } \tilde{a}_k = \frac{3k + 1}{(k^2 + 1)(\ln(k))^2}, k = 2, 3, \ldots.$$

It is easy to check that $\lim_{k \to \infty} (\tilde{a}_k/a_k) = 3$ and hence the sequence $b_k = \tilde{a}_k/a_k$, $k = 2, 3, \ldots$, is bounded. Thus, the series

$$\sum_{k=2}^{\infty} \frac{3k + 1}{(k^2 + 1)(\ln(k))^2}$$

converges. To make this look even more messy, the expression $\ln(k)$ could be replaced by someting like $\ln((k^2 + 1)/(k + 2))$. The same sort of trick can be played with divergent sequences.

Check The Tables And Chapter 4 For Integrals

In finding integrals to use in constructing convergent and divergent series, don't forget to browse through the problems from CHAPTER 4 and the TABLE OF INTEGRALS in the back of the book. The TABLE OF INTEGRALS contains a section called MISCELLANEOUS DEFINITE INTEGRALS that has some interesting integrals. Two such integrals are

$$\int_0^{\infty} \ln\left(\frac{e^x + 1}{e^x - 1}\right) dx \text{ and } \int_0^1 \ln\left(\frac{1 + x}{1 - x}\right) \frac{dx}{x} .$$

The second integral can be transformed into the first by the substitution $x = e^{-v}$. Using the first integrand, we compute

$$\frac{d}{dx} \ln\left(\frac{e^x + 1}{e^x - 1}\right) = \frac{-2e^x}{e^{2x} - 1} .$$

Thus the function $\ln((e^x + 1)/(e^x - 1))$ is positive and decreasing for $x > 0$. According to our integral table, the integral of this function from 0 to infinity is $\pi^2/4$. In particular, the integral from 1 to infinity is finite and hence

$$\sum_{k=1}^{\infty} \ln\left(\frac{e^k + 1}{e^k - 1}\right)$$

converges.

The integral test, as we have used it thus far, is based on integrals where the upper limit is infinity. The second integral mentioned above

$$\int_0^1 \ln\left(\frac{1 + x}{1 - x}\right) \frac{dx}{x}$$

has upper limit 1. As remarked above, by making the change of variable $x = e^{-v}$, this integral can be transformed into the integral of the previous paragraph. Instead, let's make the change of variable $x = v^{-1}$. Then $dx = -v^{-2}dv$ and the integral becomes

$$-\int_{\infty}^0 \ln\left(\frac{1 + v^{-1}}{1 - v^{-1}}\right) \frac{dv}{v} = \int_0^{\infty} \ln\left(\frac{v + 1}{v - 1}\right) \frac{dv}{v} .$$

This means that the series

$$\sum_{k=2}^{\infty} \frac{1}{k} \ln \left(\frac{k + 1}{k - 1} \right)$$

converges.

As one final example of the integral test, consider the integral

$$\int x^{-p} dx = \frac{x^{-p+1}}{-p + 1} \text{ if } p > 1 \text{ so } \int_1^{\infty} x^{-p} dx = \frac{1}{p - 1} \text{ if } p > 1.$$

This means that the series

$$\sum_{k=1}^{\infty} k^{-p}$$

converges if $p > 1$. Otherwise, by the integral test again, the series diverges.

Transforming Integrals Into Series

So now it's your turn to make up some exercises for your classmates based on the integral test. BE MEAN!

5.32 EXERCISES Make up four exercises based on the integral test to exchange with your classmates.

5.33 PROBLEMS BASED ON THE COMPARISON TEST Suppose the infinite series

$$\sum_{k=0}^{\infty} a_k$$

converges absolutely. COMPARISON TEST 5.29 says that if c_k, $k = 0, 1,$ $2, \ldots,$ is a sequence such that $|c_k| \leq M|a_k|$ for some positive real number M and $k = 0, 1, 2, \ldots,$ then the series

$$\sum_{k=0}^{\infty} c_k$$

also converges absolutely. The usual way that the inequality $|c_k| \leq M|a_k|$ is established is by showing that

$$\lim_{k \to \infty} \frac{|c_k|}{|a_k|}$$

exists. For example, the rational function

$$c_k = \frac{3k + 3}{k^2 + 5} = \frac{3 + (3/k)}{k + (5/k)}$$

is very close to $a_k = 3/k$ for large values of k. In fact, the ratio c_k/a_k approaches 1 as k tends to infinity. The series with terms $3/k$ diverges and thus so does the series with terms c_k.

On the other hand, the series with terms c_k defined by

$$c_k = \frac{3k + 3}{k^3 + 5} = \frac{3 + (3/k)}{k^2 + (5/k)}$$

has terms that look like $3/k^2$ for large values of k. The series with terms $3/k^2$ converges and therefore so does the series

$$\sum_{k=0}^{\infty} \frac{3k + 3}{k^3 + 5}.$$

Thus, if you know that a series $a_0 + a_1 + \ldots$ converges absolutely, and you define a series c_k, $k = 0, 1, 2, \ldots$, such that the limit $|c_k/a_k|$ exists, then the series $c_0 + c_1 + \ldots$ converges absolutely.

Two Series Behave The Same If Their Terms Behave The Same

5.34 EXERCISES Make up four exercises based on the comparison test to exchange with your classmates.

5.35 PROBLEMS BASED ON THE ROOT TEST The method called the ROOT TEST for convergence is a special case of the comparison test just discussed. We take $a_k = r^k$ so that the series $a_0 + a_1 + \ldots$ is the geometric series. If b_k, $k = 0, 1, 2, \ldots$, is any sequence with limit $\lim_{k \to \infty} |b_k| < 1$, then there is some $r < 1$ and some integer t such that, for all $k \geq t$, $|b_k| < r$. Thus $|b_k|^k < r^k$ for $k \geq t$, and hence

$$\sum_{k=t}^{\infty} |b_k|^k \text{ and thus } \sum_{k=0}^{\infty} |b_k|^k$$

converges. For example, take

$$b_k = \frac{2k^2 + k + 5}{3k^2 - k + 6}.$$

The limit of the sequence b_k is $2/3$. It then follows that the series

$$\sum_{k=0}^{\infty} \left(\frac{2k^2 + k + 5}{3k^2 - k + 6} \right)^k$$

converges.

Another example is gotten by taking $b_k = (1 - (1/k))^k$, $k = 1, 2, \ldots$. The limit of b_k as k tends to infinity is e^{-1}. Thus, the series

$$\sum_{k=1}^{\infty} (1 - (1/k))^{k^2}$$

converges. If the limit as k tends to infinity of $|b_k|$ is greater than 1, then the sequence $|b_k|^k$ doesn't tend to zero, and consequently

$$\sum_{k=0}^{\infty} |b_k|^k$$

diverges. If the limit of $|b_k|$ is equal to 1, then the series

$$\sum_{k=0}^{\infty} |b_k|^k$$

can either converge or diverge. For example, $b_k = k^{(-1/k)}$ and $b_k = k^{(-2/k)}$ both tend to 1 as k tends to infinity. In the first case, the series

$$\sum_{k=0}^{\infty} |b_k|^k$$

diverges and in the second case it converges.

Root Test: Let limit $|b_k| = \beta$; $\Sigma |b_k|^k$ Converges If $\beta < 1$, Diverges If $\beta > 1$

There is a very common class of series whose convergence or divergence can be determined by the root test. These series, which look sort of like the geometric series, have the following form:

$$\sum_{k=0}^{\infty} f(k)r^k, \text{ r a real number, } f(k) > 0.$$

If we think of $f(k)r^k = |b_k|^k$, then $|b_k| = (f(k))^{1/k}r$. In most examples, $f(k)$ is chosen such that the limit as k goes to infinity of $f(k)^{1/k}$ is 1. Here are some examples of such $f(k)$: $f(k) = k$, $f(k) = k^2$, $f(k) = k^p$, p any real number, $f(k) = 3k^3 + 2k^2 + k + 1$, $f(k)$ any positive valued polynomial in k, f(k) any positive valued rational function of k. In most cases, the way you show that $f(k)^{1/k}$ tends to 1 is by showing that its logarithm, $(1/k)\ln(f(k))$ tends to zero as k goes to infinity. Here is a series that converges, constructed by this method

$$\sum_{k=0}^{\infty} \left(\frac{9k^5 + 5k}{k^2 + 3}\right)^3 (1/2)^k.$$

In this series, the (1/2) can be replaced by any function of k that has limit 1/2. For example, $(2k + 1)/(4k + 3)$ could be used in place of the 1/2.

The name ROOT TEST that is given to this method comes from the fact that the limit condition is put on the numbers $|b_k|$, which are the kth roots of the terms $|b_k|^k$ that appear in the series. This method is also called CAUCHY'S TEST. If the series is initially written $\sum_{k=0}^{\infty} a_k$, then the root test is applied to $|b_k| = |a_k|^{1/k}$. The series converges absolutely if $\alpha = \text{limit } |a_k|^{1/k} < 1$, diverges absolutely if $\alpha > 1$.

Root Test: Let limit $|a_k|^{1/k} = \alpha;$ $\Sigma |a_k|$ Converges If $\alpha < 1$, Diverges If $\alpha > 1$

5.36 EXERCISES Make up four exercises based on the root test to exchange with your classmates.

5.37 PROBLEMS BASED ON THE RATIO TEST The method we now discuss is called the RATIO TEST for convergence. This method is again based on comparison with the geometric series. It is a less powerful method than the root test discussed previously. If we have a series with terms a_k, k = 0, 1, 2, . . ., then we may consider

$$\lim_{k \to \infty} \frac{|a_{k+1}|}{|a_k|} = \rho.$$

The RATIO TEST states that if $\rho < 1$ then the series converges absolutely, and if $\rho > 1$ then the series $|a_k|$, k = 0, 1, 2, . . ., diverges. It is not too difficult to show that if the limit of $|a_{k+1}|/|a_k|$ is ρ then the limit of $|a_k|^{1/k}$ is also ρ. The converse is not true. For example, if we take

$$a_k = 2^{-k-(-1)^k}$$

then

$$\lim_{k \to \infty} \frac{|a_{k+1}|}{|a_k|} = \lim_{k \to \infty} 2^{-1+2(-1)^k}$$

which does not exist. On the other hand,

$$\lim_{k \to \infty} |a_k|^{1/k} = \lim_{k \to \infty} 2^{-1-(-1)^k/k} = 1/2$$

shows that the series converges.

The Ratio Test Is Less Powerful But Sometimes More Useful

The ratio test is sometimes more useful than the root test in dealing with series whose terms involve factorials. Consider the series

$$\sum_{k=1}^{\infty} \frac{(k!)^2}{(2k)!} .$$

For this series, we obtain

$$\lim_{k \to \infty} \frac{|a_{k+1}|}{|a_k|} = \lim_{k \to \infty} \frac{(k+1)^2}{(2k+1)(2k+2)} = 1/4.$$

Thus the series converges. Applying the root test involves taking the limit of $(k!)^{2/k}$ and $((2k)!)^{1/k}$. The best way to do this is to use a formula called STIRLING'S FORMULA, which approximates $k!$ by $(2\pi k)^{1/2}(k/e)^k$. Making this substitution for $k!$ and the corresponding substitution of $(2\pi(2k))^{1/2}(2k/e)^{2k}$ for $(2k)!$ in $a_k = (k!)^2/(2k)!$, taking the kth root, and then taking the limit also gives 1/4. The ratio test is much easier in this example!

It is easy to see why the ratio test is valid for determining convergence. Suppose that the limit of $|a_{k+1}|/|a_k|$ is less than 1. Then there is some positive $r < 1$ and some integer t, such that for all $k > t$, $|a_{k+1}|/|a_k| < r$. In particular, $|a_{t+1}| < r|a_t|$, $|a_{t+2}| < r^2|a_t|$, . . ., $|a_{t+k}| < r^k|a_t|$, ... The series with terms $|a_t|r^k$, $k = 1, 2, . . .$, is the geometric series and converges. Thus

$$\sum_{k=t}^{\infty} a_k \text{ converges and hence } \sum_{k=0}^{\infty} a_k \text{ converges.}$$

Ratio Test: Let limit $(|a_{k+1}|/|a_k|) = \alpha;$ $\Sigma |a_k|$ Converges If $\alpha < 1$, Diverges If $\alpha > 1$

5.38 EXERCISES Work as many of EXERCISE 5.36 as you can using the ratio test.

Conditional Convergence—You Can Get Anything You Want

We have already studied, in EXAMPLE 5.21, the alternating harmonic series

$$\sum_{k=1}^{\infty} (-1)^{k-1}(1/k).$$

This series does not converge absolutely as the harmonic series $1 + (1/2) + (1/3) + \ldots$ diverges. In general, we have the following definition:

5.39 DEFINITION A series that converges but does not converge absolutely is called a *conditionally convergent* series.

Here are some things to be aware of in connection with conditionally convergent series. If we take the conditionally convergent series $1 - (1/2) + (1/3) - \ldots$ and extract the subseries consisting of every other term, we obtain the divergent series $1 + (1/3) + (1/5) \ldots$. For absolutely convergent series, any subseries also converges. If that doesn't seem very interesting to you, a related fact, pointed out by the mathematician Riemann, states that, given any number, by rearranging the terms of a conditionally convergent series you can make the resulting series converge to exactly that number. We won't have any use for Riemann's result, but it's fun to contemplate and can be proved in a way understandable to the beginner. If, on the other hand, you rearrange the terms of an absolutely convergent series, the new series converges to the same value as the old series.

Another thing you can do to create a new series from a given series $a_0 + a_1 + \ldots$ is to insert parentheses: $(a_0 + a_1) + (a_2 + a_3 + a_4) + (a_5 + a_6) + \ldots$, for example. If the series $a_0 + a_1 + \ldots$ converges, conditionally or absolutely, it doesn't matter, then this new series converges to the same thing. If all of the a_k are nonnegative, then the divergence of the series a_k implies the divergence of the new parenthesized series. For divergent series with both positive and negative terms, be careful. For example, the series $1 - 1 + 1 - 1 \ldots$ is divergent, but $(1 - 1) + (1 - 1) + \ldots$ obviously converges to zero.

Conditional Convergence—Dirichlet's Test

5.40 PROBLEMS BASED ON DIRICHLET'S TEST AND ABEL'S TEST Let a_k, $k = 0, 1, 2, \ldots$, be a sequence of nonnegative numbers such that $a_{k+1} \le a_k$, $k = 0, 1, 2, \ldots$, and the limit of the sequence a_k is zero. In other words, a_k is a nonincreasing sequence which tends to zero. Let b_k, $k = 0, 1, 2, \ldots$, be a sequence with the property that the corresponding sequence of partial sums, $s_n = b_0 + \ldots + b_n$, $n = 0, 1, 2, \ldots$, is bounded. Then the series

$$\sum_{k=0}^{\infty} a_k b_k$$

converges. This result is called DIRICHLET'S TEST. Of course, by replacing a_k by $-a_k$, we see that the result is valid if the a_k are nondecreasing and tend to zero.

THEOREM 5.22 was an example of a class of series of this type. If we take $b_k = (-1)^k$ then clearly the partial sums of this sequence are bounded. This gives the class of series in THEOREM 5.22. In constructing convergent series by the Dirichlet method, the challenge is to think up interesting sequences, b_k, $k = 0, 1, 2, \ldots$, with bounded partial sums. You can also simply make up sequences with bounded partial sums. For example $+1$, $+2$, $+3$, -1, -2, -3, $+1$, $+2$, $+3$, -1, -2, -3, $\ldots$ has its partial sums bounded. In the paragraph just prior to EXAMPLE 5.20, we pointed out that every sequence is the sequence of partial sums of some series. Thus, if we start with any bounded sequence c_k, $k = 0, 1, 2, \ldots$, and form the sequence $b_0 = c_0$, $b_k = c_k - c_{k-1}$, $k - 1, 2, \ldots$, then the sequence b_k is a sequence with bounded partial sums, $s_n = c_n$.

Another class of interesting examples can be gotten by taking $b_k = \sin(k\tau + \gamma)$, $k = 0, 1, 2, \ldots$, where γ is any real number and τ is any real number that is not an integral (including zero) multiple of 2π. That the partial sums $b_0 + b_1 + \ldots + b_n$, $n = 0, 1, 2, \ldots$, form a bounded sequence is an easy result in complex analysis. Of course, the sequence $\cos(k\tau + \gamma)$, $k = 0, 1, 2, \ldots$, also has bounded partial sums. Feel free to use these sequences in the EXERCISES 5.41. This means that the following series converge conditionally (remember EXERCISE 5.15 (1-j)).

$$\sum_{k=2}^{\infty} \sin(k)/\ln(k) \qquad \sum_{k=1}^{\infty} \cos(\pi k + 1)/k \qquad \sum_{k=0}^{\infty} \sin(\pi k)/k^{1/2}.$$

Abel's Test

There is a useful variation on Dirichlet's method, called ABEL'S TEST, where instead of just having the sequence of partial sums of the sequence bounded, we make the stronger assumption that the series $b_0 + b_1 + b_2 + \ldots$ actually converges. In this case we can take the sequence a_k, $k = 0, 1,$

2, . . ., to be any nonincreasing sequence or nondecreasing sequence which has a limit. In other words, this "monotonic" sequence tends to a limit, but the limit need not be zero as in DIRICHLET'S TEST. For example, $a_k = 1 + 1/k, k = 1, 2, 3, . . .$, would be such a sequence. ABEL'S TEST states that under these conditions, the series $a_0 b_0 + a_1 b_1 + . . .$ converges. In the general statement of Dirichlet's method or Abel's method, we take $k = 0$, 1, 2, . . ., but the sequences a_k and b_k can start at any value of k and these results are still valid. Remember, the fact that "the tail series of the series converges implies that the series converges" is a result for all convergent series, conditional or absolute. As an application of Abel's method, the series

$$\sum_{k=2}^{\infty} (-1)^k (1 + 1/k)(1/\ln(k))$$

converges.

Both DIRICHLET'S TEST and ABEL'S TEST deal with series formed by taking term-by-term products of sequences where various conditions were put on the sequences. In the case of absolute convergence, we had available the very powerful THEOREM 5.28 which said that the termwise product of a bounded sequence and an absolutely convergent series gave rise to another absolutely convergent series. This theorem is clearly false for conditionally convergent series. As an example, multiply the alternating harmonic series

$$\sum_{k=1}^{\infty} (-1)^{k-1}(1/k)$$

termwise by the bounded sequence $(-1)^{k-1}, k = 1, 2, . . .$.

Dirichlet's And Abel's Test—You Make Up The Problems

5.41 EXERCISES Make up four exercises, using either DIRICHLET'S TEST or ABEL'S TEST, to exchange with your classmates.

Basic Arithmetic Rules For Series

We now give some EXERCISES on infinite series, followed by the SO-LUTIONS and VARIATIONS on these exercises. As usual, after reading the solution to a problem, change the original problem slightly and rework it. Then go on to the variations. One thing to keep in mind above all else is the basic fact that an infinite series is a sequence of partial sums. You have had some good solid practice with sequences to fall back on when stuck! In particular, RULES FOR LIMITS OF SEQUENCES 5.10 apply. You should translate these rules into series notation. For example, if $a_0 + a_1 + \ldots +$ $a_k + \ldots$ and $b_0 + b_1 + \ldots + b_k + \ldots$ are two convergent series (conditional or absolute, it makes no difference) and α is any real number, then we have

(1)
$$\sum_{k=0}^{\infty} \alpha a_k = \alpha \sum_{k=0}^{\infty} a_k$$

(2)
$$\sum_{k=0}^{\infty} (a_k + b_k) = \sum_{k=0}^{\infty} a_k + \sum_{k=0}^{\infty} b_k.$$

How would you define multiplication of two series in terms of their partial sums? Give an example.

Exercises On Series

5.42 EXERCISES

(1) Apply the root test, the ratio test, and the comparison test to each of the following series:

(a) $\displaystyle\sum_{k=1}^{\infty} \frac{(k + 1)(k + 2)(k + 3)}{k!}$

(b) $\displaystyle\sum_{k=1}^{\infty} \frac{2^{k/2}}{k^2 + k + 1}$

(c) $\displaystyle\sum_{k=1}^{\infty} \frac{k + 1}{2k + 3}$

(d) $\displaystyle\sum_{k=0}^{\infty} \frac{k^5}{5^k}$

(2) Apply the comparison test to each of the following series:

(a) $\displaystyle\sum_{k=0}^{\infty} \frac{1}{k^2 - 150}$

(b) $\displaystyle\sum_{k=5}^{\infty} \frac{1}{2^k - k^2}$

(c) $\displaystyle\sum_{k=2}^{\infty} \frac{1}{(k^3 - k^2 - 1)^{1/2}}$

(d) $\displaystyle\sum_{k=1}^{\infty} \frac{(k + 1)^{1/2} - (k - 1)^{1/2}}{k}$

(3) Discuss the convergence of the following series:

(a) $\displaystyle\sum_{k=1}^{\infty} \frac{(-1)^k}{k} (1 + 2^{-2} + 3^{-2} + \ldots + k^{-2})$

(b) $\displaystyle\sum_{k=1}^{\infty} \frac{(-1)^k}{k} \left(\frac{1}{1} + \frac{1}{2} + \ldots + \frac{1}{k} \right)$

(4) The geometric series is the series $1 + r + r^2 + \ldots + r^k + \ldots$ where r is any real number. The terms of this series are of the form $r^{f(k)}$ where $f(k) = k$. Discuss the following generalization of the geometric series in which $f(k) = k$ is replaced by $f(k) = (\ln(k))^\beta$, β a real number:

$$\sum_{k=1}^{\infty} r^{(\ln(k))^\beta}.$$

(5) Discuss whether or not the following series converge or diverge. If convergent, specify whether the series is conditionally convergent or absolutely convergent and explain why.

(a) $\displaystyle\sum_{k=0}^{\infty} (-1)^k \operatorname{arccot}(k)$

(b) $\displaystyle\sum_{k=0}^{\infty} \frac{\sin(k)}{|k - 99.5|}$

(c) $\displaystyle\sum_{k=1}^{\infty} \frac{(-1)^k}{k}\cos(1/k)$

(d) $\displaystyle\sum_{k=0}^{\infty} \frac{-9k^2 - 5}{k^3 + 1} \sin(k)$

Study The Solution—Change And Rework The Problem

5.43   SOLUTIONS TO EXERCISE 5.42

(1)(a)   We are asked to apply three methods, the root test, the ratio test, and the comparison test to this problem. We'll do all three methods, but first let's think a bit. If a_k denotes the kth term of this series, then the numerators of a_k is a polynomial of degree 3 and the denominator is an expression that grows faster than any polynomial in k. If the denominator were only the degree 5 polynomial $p(k) = k(k - 1)(k - 2)(k - 3)(k - 4)$, then by comparison with the series with terms k^{-2} the series would converge. The actual denominator, which is $p(k)(k - 5)!$, is bigger than $p(k)$ for $k > 6$, thus the series converges by comparison with the series with terms k^{-2}. The comparison test is the easy way to go here and would, when you have had a little practice, tell you at a glance that the series converges.

To apply the root test, we would compute $(a_k)^{1/k}$. Using $k! = (2\pi k)^{1/2}(k/e)^k$, approximately, and using the fact that $(bk + c)^{1/k}$ tends to 1 as k tends to infinity for any real numbers $b > 0$ and c, we get that $(a_k)^{1/k}$ tends to zero. Thus, the series converges by the root test.

By the ratio test, we must compute a_{k+1}/a_k. We get

$$\frac{a_{k+1}}{a_k} = \frac{k + 4}{(k + 1)^2}$$

which tends to zero as k tends to infinity. Thus, the series converges by the ratio test.

(1)(b) Remember, it is an immediate consequence of DEFINITION 5.24, that if a series with terms a_k converges then the terms a_k must tend to zero as k tends to infinity. The contrapositive is that if a_k does not tend to zero then the series with terms a_k does not converge. In this example, the terms a_k tend to infinity. Thus, this series is divergent. If you like, this can be thought of as comparison with the divergent series with each term 1.

To apply the root test, we compute easily that the limit of $(a_k)^{1/k}$ as k tends to infinity is $2^{1/2}$. The series diverges by the root test.

To apply the ratio test, we compute a_{k+1}/a_k equals

$$2^{1/2}\, \frac{k^2 + k + 1}{(k + 1)^2 + (k + 1) + 1}$$

which again tends to $2^{1/2}$ and again implies divergence.

(1)(c) If a_k denotes the kth term of this series, then it is obvious that a_k tends to 1/2 as k tends to infinity. The series diverges. The root test and the ratio test both yield limits of 1 and hence no conclusion can be drawn from them. Check this out by computing the limits in both cases.

(1)(d) An exponential function a^x, $a > 1$, grows much faster than any polynomial of any degree. In particular, $p(x)/a^x$ goes to zero as x tends to infinity, for any fixed polynmial p(x) and any $a > 1$. We can write the term $c_k = k^5 5^{-k}$ as $(k^5 5^{-k/2})5^{-k/2}$. The expression $k^5 5^{-k/2}$ tends to zero as k tends to infinity and thus the series converges by comparison with the geometric series with terms $a_k = 5^{-k/2}$ (take $r = 5^{-1/2}$ in the geometric series). Stop now and reread 5.33 PROBLEMS BASED ON THE COMPARISON TEST. Using that notation, we have c_k/a_k tends to zero and hence the series with terms c_k converges.

The root test is easy for this problem. The numerator of $(c_k)^{1/k}$ is $k^{5/k}$ and the denominator is $5^{k/k} = 5$. The numerator tends to 1 and the denominator, of course, tends to 5. Thus the limit of $(c_k)^{1/k}$ is 1/5 and the series converges.

By the ratio test, we get that $c_{k+1}/c_k = (1/5)(k + 1)^5/k^5$ which again tends to 1/5. It is a theorem that if the limit of $(c_k)^{1/k}$ exists then the limit of c_{k+1}/c_k exists and is the same. The converse, as we noted in 5.37 PROBLEMS BASED ON THE RATIO TEST, is false.

Solutions To (2) Of EXERCISE 5.42

(2)(a)   For large values of k, $(k^2 - 150)^{-1}$ "behaves like" k^{-2}, so this series with terms $(k^2 - 150)^{-1}$ converges absolutely by comparison with the series with terms k^{-2}. To be more precise about what we mean by "behaves like," let $c_k = 1/(k^2 - 150)$ and let $a_k = 1/k^2$. The sequence c_k/a_k converges to 1 and is consequently a bounded sequence. The series with terms a_k converges absolutely and thus, by the discussion of 5.33 PROBLEMS BASED ON THE COMPARISON TEST, the series with terms $c_k = 1/(k^2 - 150)$ converges absolutely.

(2)(b)   This problem is another example of the idea discussed in connection with the solution to problem (b) of (1). The exponential function 2^k grows much faster than k^2. The terms of this series behave like 2^{-k} and hence this series converges absolutely by comparison with the geometric series $1 + 2^{-1} + 2^{-2} + \ldots$. To apply the discussion of 5.33 PROBLEMS BASED ON THE COMPARISON TEST directly, we let $c_k = (2^k - k^2)^{-1}$ and $a_k = 2^{-k}$. The ratio c_k/a_k tends to 1 as k tends to infinity. The series with terms a_k converges absolutely and hence so does the series with terms c_k.

(2)(c)   In the polynomial $k^3 - k^2 - 1$, the term k^3 is dominant for large k. Thus $(k^3 - k^2 - 1)^{1/2}$ behaves like $k^{3/2}$ for large k, and the series of this problem converges absolutely by comparison with the series with terms $k^{-3/2}$. In applying the discussion of 5.33 PROBLEMS BASED ON THE COMPARISON TEST, take $a_k = k^{-3/2}$ and $c_k = 1/(k^3 - k^2 - 1)^{1/2}$. The sequence c_k/a_k converges to 1 and the series with terms a_k converges absolutely.

(2)(d)   Multiply the numerator and denominator of the kth term of this series by $[(k + 1)^{1/2} + (k - 1)^{1/2}]$ to get the same series with the kth term now written $2k^{-1}[(k + 1)^{1/2} + (k - 1)^{1/2}]^{-1}$. This term behaves like $4k^{-3/2}$ for large k and hence the series of this problem converges absolutely. We leave it to you to be more precise about "behaves like" for this problem.

Solutions To (3) Of Exercise 5.42

(3)(a) We are going to apply ABEL'S TEST with

$$a_k = 1 + 2^{-2} + 3^{-2} + \ldots + k^{-2}.$$

We showed, using the integral test, that the series

$$1 + 2^{-2} + 3^{-2} + \ldots + k^{-2} + \ldots$$

converged. Hence, the a_k, which are the partial sums of this series, form a nondecreasing sequence with a limit. Taking $b_k = (-1)^k (1/k)$ and applying ABEL'S TEST gives that the series of this problem converges.

(3)(b) This problem looks a lot like part **(a)** except that the sequence of terms $s_k = 1 + (1/2) + \ldots + (1/k)$ diverges, being the partial sums of the harmonic series. This looks bad for ABEL'S TEST. From the integral test, we know that s_k is approximately $\ln(k)$. Since $\ln(k)/k$ goes to zero, there may be hope for DIRICHLET'S TEST. Define $a_k = s_k/k$. The fact that $\ln(k)/k$ goes to zero implies that a_k goes to zero. Here's a computer program to compute the a_k.

```
10   K=1:SK=0
20   SK-SK+1/K
30   PRINT K,SK/K
40   K=K+1
50   GOTO 20
```

k	a_k
1	1
2	.75
3	.6111111
4	.5208334
5	.4566667
6	.4083334
7	.3704082
8	.3397322
9	.3143298
10	.2928968
11	.2745343
12	.2586009
13	.2446257
14	.2322545
15	.2212153
16	.2112956
17	.2023266

18	.1941727
19	.1867231
20	.179887
21	.1735885
22	.1677643
23	.1623605
24	.1573316
25	.1526383
26	.1482469
27	.144128
28	.1402561
29	.1366088
30	.1331662

It looks like a_k is monotonically decreasing to zero. By DIRICHLET'S TEST or by THEOREM 5.22, this means that the series of this problem converges. Actually, we can show by direct computation that the a_k are monotonically decreasing to zero. First, write

$$a_k - a_{k+1} = \frac{1}{k}(1 + (1/2) + \ldots + (1/k))$$

$$- \frac{1}{k+1}\left(1 + (1/2) + \ldots + (1/k) + (1/(k+1))\right).$$

Now write this expression as

$$\frac{1}{k(k+1)}\left((k+1)(1 + (1/2) + \ldots + (1/k))\right.$$

$$\left. - k(1 + (1/2) + \ldots + (1/k) + (1/(k+1)))\right).$$

Simplifying this expression gives

$$a_k - a_{k+1} = \frac{1}{k(k+1)}\left((1 + (1/2) + \ldots + (1/k)) - \frac{k}{k+1}\right).$$

This shows that $a_k - a_{k+1}$ is greater than zero for all k since there are k terms in the sum $1 + (1/2) + \ldots + (1/k)$ and each term is greater than $1/(k+1)$. That a_k tends to zero follows by comparison with $\ln(k)/k$.

(4) The exponents $(\ln(k))^\beta$ are all positive real numbers but are not integers unless $\beta = 0$, so we should require $r \geq 0$ to avoid complex numbers. If $r \geq 1$ then the series diverges for all values of β as any positive power of a

number greater than or equal to 1 is still greater than or equal to 1. Think of the graph of x^ϵ for $x \geq 1$ where $\epsilon > 0$ is small. Thus, we assume $0 \leq r < 1$. If $\beta = 0$ then $(\ln(k))^\beta = 1$ for all k and the series diverges unless $r = 0$. If $\beta < 0$ then the sequence of exponents $(\ln(k))^\beta$ tends to 0 and, unless $r = 0$, the terms of the series tend to 1 which implies divergence. Remember, any series diverges if its terms don't converge to zero. So far, nothing but bad news! If this series converges, we must have $0 \leq r < 1$ and $\beta > 0$. Of course, if $r = 0$ the series converges, so let's assume $0 < r < 1$ and $\beta < 0$. Write $r = e^{-\alpha}$ where $\alpha > 0$. Then we have

$$r^{(\ln(k))^\beta} = e^{-\alpha(\ln(k))^\beta} = e^{-(\ln(k))\alpha(\ln(k))^{\beta-1}} = k^{-\alpha(\ln(k))^{\beta-1}}.$$

Clearly, if $0 < \beta < 1$ then $\beta - 1$ is negative and this series diverges (its terms become larger than k^{-1}). If $\beta = 1$ then this series converges only if $\alpha > 1$. If $\beta > 1$ then this series converges for all $\alpha > 0$. Why? Because for $\beta > 1$ the expression $\alpha(\ln(k))^{\beta-1}$ tends to infinity as k tends to infinity. As soon as $\alpha(\ln(k))^\beta$ is bigger than 2, say, we can compare the tail of our series with the series k^{-2}, which we know converges.

To summarize, we have that the series

$$\sum_{k=1}^{\infty} r^{(\ln(k))^\beta} \text{ where } r > 0 \text{ and } \beta \text{ is a real number}$$

converges if $\beta = 1$ and $0 < r < e^{-1}$ or if $\beta > 1$ and $0 < r < 1$. In all other cases, it diverges. Remember this fact! Its very useful for comparison tests.

Do you see how this result contains our previous result about the convergence of the series with terms k^{-p} derived just prior to EXERCISE 5.32? Write

$$\sum_{k=1}^{\infty} k^{-p} = \sum_{k=1}^{\infty} (e^{-p})^{\ln(k)}.$$

By the result we have just obtained, with $r = e^{-p}$, this series converges if $e^{-p} < e^{-1}$ and diverges otherwise. In other words, it converges if $p > 1$ and diverges otherwise.

(5)(a) The graph of the function arccot(x) is shown in FIGURE 2.39. The sequence arccot(k) converges monotonically to zero. Thus, the series of this problem converges by DIRICHLET'S TEST with $a_k = $ arccot(k) and $b_k = (-1)^k$. But, does this series converge absolutely? In the TABLE OF INTEGRALS, we find that the integral of arccot(x) is $x\text{arcctn}(x) + (1/2)\ln(1 + x^2)$. Thus the series with terms arccot(k) diverges by the integral test. The series with terms $(-1)^k\text{arcctn}(k)$ is conditionally convergent.

(5)(b)   This looks like DIRICHLET'S TEST again with b_k = sin(k) and a_k = $|k - 99.5|^{-1}$. The DIRICHLET TEST calls for the a_k to be nonincreasing (or nondecreasing) with limit zero. The a_k of this example in fact increase until k = 99. After that, the sequence is nonincreasing and converges to zero. But that's good enough. That means, by the DIRICHLET TEST, that the tail series $a_{99}b_{99}$ + $a_{100}b_{100}$ + . . . converges. This implies that the whole series converges. But does it converge conditionally or absolutely? Read again the solution to EXERCISE 5.15 (1-j) and you will see that, for the same reason presented there, this series does not converge absolutely.

(5)(c)   This is an application of ABEL'S TEST with a_k = $(-1)^k/k$ and b_k = cos(1/k). As k tends to infinity, cos(1/k) is nondecreasing and tends to 1. The a_k are the terms of the alternating harmonic series, which converges. This series does not converge absolutely, by comparison with the harmonic series.

(5)(d)   Again, we have an application of DIRICHLET'S TEST with a_k = $(-9k^2 - 5)/(k^3 + 1)$ and b_k = sin(k). The sequence a_k is nondecreasing with limit zero and the sequence b_k has bounded partial sums.

This completes the solutions to EXERCISE 5.42. It's time for the variations. Remember to refer to the corresponding problem in EXERCISE 5.42 if you get stuck. Also, don't forget your computer! With simple programs, often involving no more than five or six lines of code, you can gain much valuable information about the infinite series in these problems.

5.44   VARIATIONS ON EXERCISE 5.42

(1)   Apply the root test, the ratio test, and the comparison test to each of the following series:

(a) $\displaystyle\sum_{k=1}^{\infty} \frac{(k - 1)(k - 2)(k - 3)}{k!}$

(b) $\displaystyle\sum_{k=1}^{\infty} \frac{2^{k/2}}{k^{200} + k + 1}$

(c) $\displaystyle\sum_{k=1}^{\infty} \frac{k^2 + 1}{2k^2 + 3}$

(d) $\displaystyle\sum_{k=0}^{\infty} \frac{k^{500}}{(1.01)^k}$

(2) Apply the comparison test to each of the following series:

(a) $\displaystyle\sum_{k=0}^{\infty} \frac{1}{k^{1.0001} - 1.50}$

(b) $\displaystyle\sum_{k=1}^{\infty} \frac{1}{2^k - 3k^{2000}}$

(c) $\displaystyle\sum_{k=2}^{\infty} \frac{1}{(k^3 - k^2 - 1)^{1/4}}$

(d) $\displaystyle\sum_{k=1}^{\infty} \frac{(k+1)^{1/4} - (k-1)^{1/4}}{k}$

(3) Discuss the convergence of the following series

(a) $\displaystyle\sum_{k=2}^{\infty} \frac{(-1)^k}{k} \left(\frac{1}{2(\ln(2))^2} + \frac{1}{3(\ln(3))^2} + \ldots + \frac{1}{k(\ln(k))^2} \right)$

(b) $\displaystyle\sum_{k=1}^{\infty} \frac{(-1)^k}{k} \left(\frac{(\ln(1))^2}{1} + \frac{(\ln(2))^2}{2} + \ldots + \frac{(\ln(k))^2}{k} \right)$

(4) The geometric series is the series $1 + r + r^2 + \ldots + r^k + \ldots$ where r is any real number. The terms of this series are of the form $r^{f(k)}$ where $f(k) = k$. Discuss the following generalization of the geometric series in which $f(k) = k$ is replaced by $f(k) = k^\gamma$, γ a real number:

$$\sum_{k=1}^{\infty} r^{k^\gamma} \text{ where } r > 0.$$

(5) Discuss whether or not the following series converge or diverge. If convergent, specify whether the series is conditionally convergent or absolutely convergent and explain why.

(a) $\displaystyle\sum_{k=1}^{\infty} (-1)^k \ln(k) \operatorname{arccot}(k)$

(b) $\displaystyle\sum_{k=0}^{\infty} \frac{\sin(k)}{\ln|k - 99.5|}$

(c) $\displaystyle\sum_{k=1}^{\infty} \frac{(-1)^k}{k}\cos(k)$

(d) $\displaystyle\sum_{k=0}^{\infty} \frac{-9k^2 - 5}{k^3 + 1}\sin(k)\cos(k)$

5.45   VARIATIONS ON EXERCISE 5.42

(1) Apply the root test, the ratio test, and the comparison test to each of the following series:

(a) $\displaystyle\sum_{k=4}^{\infty} \frac{(k + 1)(k + 2)(k + 3)}{k(k - 1)(k - 2)(k - 3)}$

(b) $\displaystyle\sum_{k=1}^{\infty} \frac{k^2}{2^{\ln(k)}}$

(c) $\displaystyle\sum_{k=1}^{\infty} \frac{\ln(k) + (1/k)}{(\ln(k))^3 + 3}$

(d) $\displaystyle\sum_{k=1}^{\infty} \frac{k^{\ln(k)}}{5^k}$

(2) Apply the comparison test to each of the following series:

(a) $\displaystyle\sum_{k=1}^{\infty} \frac{1}{(\ln(k))^2 - 150}$

(b) $\displaystyle\sum_{k=5}^{\infty} \frac{1}{(\ln(\ln(k))^k}$

(c) $\displaystyle\sum_{k=2}^{\infty} \frac{1}{(k^3 - k^2 - 1)^{1/2.99}}$

(d) $\displaystyle\sum_{k=1}^{\infty} \frac{(k + 1)^{1/200} - (k - 1)^{1/200}}{k}$

(3) Discuss the convergence of the following series. In each case give a theoretical reason for convergence and divergence and write a program to evaluate the series to back up your conclusion. The series expansion for e^x is given in the appendix, if you think that might help.

(a) $\sum_{k=1}^{\infty} (-1)^k(e^{-1/k} - 1 + 1/k)$

(b) $\sum_{k=1}^{\infty} (e^{-1/k} - 1 + 1/k)$

(4) Discuss convergence of the following series in terms of the real numbers α and β:

$$\sum_{k=1}^{\infty} k^{\alpha(\ln(k))^\beta}$$

(5) Discuss whether or not the following series converge or diverge. If convergent, specify whether the series is conditionally convergent or absolutely convergent and explain why.

(a) $\sum_{k=0}^{\infty} (-1)^k(\operatorname{arccot}(k))^\beta$, $\beta < 1$

(b) $\sum_{k=0}^{\infty} \dfrac{\sin(k) - \cos(k)}{|k - 99.5|}$

(c) $\sum_{k=1}^{\infty} \cos(k)\sin(1/k)$

(d) $\sum_{k=0}^{\infty} \dfrac{-9k^2 - 5}{k^3 + 1} (\sin(k)\cos(k/2) + \cos(k)\sin(k/2))$

5.46   VARIATIONS ON EXERCISE 5.42

(1) Apply the root test, the ratio test, *or* the comparison test to each of the following series:

(a) $\sum_{k=1}^{\infty} \dfrac{k^{\sqrt{k}}}{k!}$

(b) $\sum_{k=1}^{\infty} \dfrac{(k/4)^k}{k!}$

(c) $\displaystyle\sum_{k=1}^{\infty} \frac{k!}{k(k+1)\ldots(2k-1)}$

(d) $\displaystyle\sum_{k=1}^{\infty} \frac{(k!)^{2k}}{(k^2)!}$

(2)   Apply the comparison test to each of the following series:

(a) $\displaystyle\sum_{k=1}^{\infty} \frac{(2k)!}{k^2(k^2-1)\ldots(k^2-k+1)}$

(b) $\displaystyle\sum_{k=5}^{\infty} \left(\frac{1}{\ln(\ln(k))}\right)^{\ln(k)}$

(c) $\displaystyle\sum_{k=5}^{\infty} \left(\frac{1}{\ln(k)}\right)^{\ln(\ln(k))}$

(d) $\displaystyle\sum_{k=2}^{\infty} \frac{k^{\ln(k)}}{(\ln(k))^k}$

(3)   Discuss the convergence of the following series in terms of the real numbers α and β:

$$\sum_{k=4}^{\infty} (\ln(k))^{\alpha(\ln(k))^\beta}$$

(4)   Discuss convergence of the following series in terms of the real number β:

$$\sum_{k=4}^{\infty} r^{(\ln(\ln(k))^\beta}, \ r > 0$$

(5)   Discuss whether or not the following series converge or diverge. If convergent, specify whether the series is conditionally convergent or absolutely convergent and explain why.

(a) $\displaystyle\sum_{k=0}^{\infty} (-1)^k(\operatorname{arccot}(k))^\beta, \ \beta > 1$

(b) $\displaystyle\sum_{k=0}^{\infty} \frac{\sin(k) - \cos(k)}{\ln(|k - 99.5|)}$

(c) $\displaystyle\sum_{k=1}^{\infty} (\sin(k) - \sin(k + (1/k)))$

(d) $\displaystyle\sum_{k=0}^{\infty} \ln(1 + (1/k))\sin(2k + \pi)$

Sequences Of Functions

Up to now, we have been concentrating mainly on sequences and series where the terms are real numbers. We began this chapter, however, looking at "power series" approximations to functions. We now return to the idea of series of functions. The idea is simple; we consider a sequence $a_k(x)$, $k = 0, 1, 2, \ldots$, where each $a_k(x)$ is a real valued function of x. For example, $a_k(x)$ could be $\sin(kx)$ or $a_k(x)$ could be x^k. The case $a_k(x) = x^k$ is shown in FIGURE 5.47. Notice that for any particular values of x, say $x = 1/2$, the sequence $a_k(x)$, in this case $a_k(1/2) = (1/2)^k$, becomes a sequence of real numbers. We are experts on such sequences by now, so most of the hard work needed to study sequences of functions is done.

We know that when we study functions, we have to know the domains of the functions. In the sequence of functions $a_k(x)$ of FIGURE 5.47, we have

FIGURE 5.47   The Sequence of Functions x^k, $k = 1, 2, \ldots$

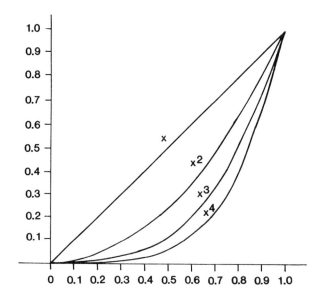

been vague about this. Let's consider two cases. First let's take the domain of $a_k(x)$ to be the interval $[0,1/2] = \{x : 0 \leqslant x \leqslant 1/2\}$ for all $k = 1, 2, 3,$ .... For each x in the interval $[0,1/2]$, the sequence of numbers $a_k(x)$ converges to zero. We say that the function $a(x) = 0$ is the "limit of the sequence of functions $a_k(x)$ on the interval $[0,1/2]$."

x^k, $k = 0, 1, 2, \ldots$, Converges Uniformly On $[0,1/2]$

For the second case, let's take the domains of the $a_k(x)$ to be the interval $[0,1) = \{x : 0 \leqslant x < 1\}$. It is still the case that for each x in $[0,1)$, $a_k(x)$ converges to zero. Thus the function $a(x) = 0$ on $[0,1)$ is still the limit of the sequence $a_k(x)$. But, there is something interesting and very important about the difference between these two examples. In the first case, where the common domain is the interval $[0,1/2]$, given any $\epsilon > 0$, if we choose N such that $1/2^N < \epsilon$, then for all $k > N$ and all x in $[0,1/2]$, $|a_k(x)| < \epsilon$. The key phrase here is "and all x in $[0,1/2]$." A glance at FIGURE 5.47 will explain why this is true. Each $a_k(x)$ has $a_k(1/2)$ as its maximum value over the interval $[0,1/2]$. Thus, if we make this maximum value small, all other values of $a_k(x)$ on the interval $[0,1/2]$ will be even smaller!

x^k, $k = 0, 1, 2, \ldots$, Does Not Converge Uniformly On $[0,1)$

But what if we try to play the same game with the common domain of the sequence taken to be the interval $[0,1)$? It won't work. Even though the sequence of functions $a_k(x)$ converges to the function $a(x) = 0$ at each x in this interval, we can't claim that given any $\epsilon > 0$ there is an N such that, for all $k > N$ and all x in $[0,1)$, $|a_k(x)| < \epsilon$. Do you see why? Look at FIGURE 5.47. Given $\epsilon = 1/2$, for example, and any k, no matter how large, there will always be some x in the interval $[0,1)$, perhaps very close to 1, with $a_k(x) > \epsilon$. This example leads to the following *extremely important* definition.

The Definition Of Uniform Convergence

5.48   DEFINITION   Let $a_k(x)$ be a sequence of real valued functions defined on a common domain D. Suppose that for each x in D, the sequence $a_k(x)$ converge to $a(x)$. If, given any $\epsilon > 0$, it is possible to choose N_ϵ such that for all $k > N_\epsilon$ *and for all* x *in* D, $|a_k(x) - a(x)| < \epsilon$, then we say that $a_k(x)$ converges *uniformly* to $a(x)$ on D.

FIGURE 5.49   A Uniformly Convergent Sequence

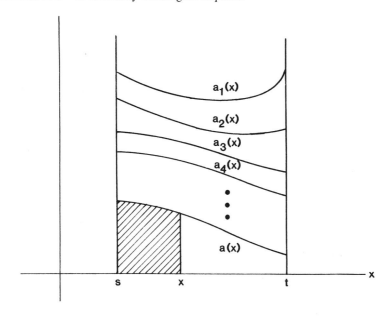

The Intuitive Idea Of Uniform Convergence

We now want to understand the intuitive idea behind DEFINITION 5.48. Look at FIGURE 5.49. There you see a sequence of functions, $a_k(x)$, $k = 1, 2, 3, 4, \ldots$, which we imagine converging uniformly to a limit function $a(x)$ on the interval [s,t]. The graph of the function $a(x)$ is shown by a black line which has a certain thickness (or else you couldn't see it). Let's call the thickness of this line ϵ. According to DEFINITION 5.48, there is an integer N such that for all $k > N$, $|a_k(x) - a(x)| < \epsilon$. Intuitively, this means that for all $k > N$, the lines representing the graphs of the functions $a_k(x)$ completely disappear into the black line that represents the graph of $a(x)$. This is the idea behind the following theorem:

The Limit Of The Integrals Is The Integral Of The Limit

5.50   THEOREM   Let $a_k(x)$, $k = 0, 1, 2, \ldots$, be a sequence of real valued continuous functions that converge uniformly on the interval [s,t] to a function $a(x)$. Define functions $A_k(x)$, $k = 0, 1, 2, \ldots$, and $A(x)$ by

$$A_k(x) = \int_s^x a_k(t)dt, \ k = 0, 1, 2, \ldots, \text{ and } A(x) = \int_s^x a(t)dt.$$

Then the sequence of functions $A_k(x)$, $k = 0, 1, 2, \ldots$, converges uniformly to $A(x)$ on [s,t].

The Intuitive Idea Of Theorem 5.50

To understand the meaning of THEOREM 5.50, we refer again to FIGURE 5.49. The shaded area under the graph of the function a is the area under that function and above the interval [s,x]. This area is the integral.

$$\int_s^x a(t)dt$$

which is a function of its upper limit x as x varies in the integral [s,t]. For notational simplicity, let's define, as in THEOREM 5.50,

$$A(x) = \int_s^x a(t)dt \text{ and } A_k(x) = \int_s^x a_k(t)dt.$$

We know that, for all $k > N$, the graphs of the functions $a_k(x)$ disappear into the black line representing the graph of $a(x)$. This means that, for all $k > N$ and all x in [s,t], the difference $|A_k(x) - A(x)|$ is less than the area of the black line used to draw the graph of the function a from $x = s$ to $x = t$. Theoretically, this black line can be made as thin as we wish to imagine it. This is the intuitive meaning behind uniform convergence of the sequence $A_k(x)$ to $A(x)$ for x in [s,t]. In more advanced courses in analysis, students learn a precise statement and proof of THEOREM 5.50. They study "pathologies" of the functions $a_k(x)$. The functions might be discontinuous, so badly so that they don't have integrals in some sense, etc. We don't have to worry about such strange things in our brief introduction to sequences of functions. Knowing the intuitive idea behind the very useful THEOREM 5.50 is 99 percent of what is required to use it intelligently.

One way to paraphrase THEOREM 5.50 is "For uniformly convergent sequences of functions, the integral of the limit is the limit of the integrals." There is a corresponding result that, loosely stated, says: "For uniformly convergent sequences of functions, the derivative of the limit is the limit of the derivatives." We have to be more careful about this statement. Here is a more precise version which concerns the sequence of derivative functions $a_k'(x)$ of a sequence of functions $a_k(x)$.

The Limit Of The Derivative Is The Derivative Of The Limit

5.51 THEOREM Let $a_k(x)$, $k = 0, 1, 2, \ldots$, be a sequence of real valued functions that converge uniformly on the interval [s,t] to a function a(x). If the sequence of derivative functions $a_k'(x)$, k = 0, 1, 2, . . ., converges uniformly to a function b(x) on [s,t], then b(x) is the derivative function a'(x) for s < x < t.

There are technical difficulties with THEOREM 5.51 that must be studied in more advanced courses. The main difference that we should be aware of between THEOREM 5.51 and THEOREM 5.50 is that in THEOREM 5.51, the uniform convergence of the derivative sequence must be verified in each case and does not follow generally from uniform convergence of the original sequence $a_k(x)$. It is not too hard to imagine why this is the case in terms of FIGURE 5.49. Imagine for k > N that all of the functions $a_k(x)$ have disappeared into the black line that represents the graph of a(x), s < x < t. As we have noted already, the integral of these "disappearing functions" can't differ from each other by more than the area of the black line. But we could imagine that these functions inside the black line could still be very wiggly, which means that their derivatives might oscillate wildly and differ a lot from each other. This is why the uniform convergence of the derivative functions must be a part of the hypothesis in THEOREM 5.51.

Examples Of Theorems 5.50 And 5.51

Let's test out the ideas of THEOREM 5.50 and THEOREM 5.51 on the sequence $a_k(x) - x^k$ of FIGURE 5.47. In this example, we have uniform convergence on any interval [0,t] where t < 1. The integral $A_k(x) = x^{k+1}/(k + 1)$. THEOREM 5.50 says that these functions $A_k(x)$ converge uniformly on [0,t] to the integral A(x) of the function a(x) = 0. Of course, A(x) = 0 also and it is obviously true that the $A_k(x)$ converge uniformly on [0,t] to the zero function. The derivative functions $a_k'(x) = kx^{k-1}$ can easily be shown to converge uniformly to the function a'(x) = 0 on [0,t] if t < 1. This is because the maximum of $a_k'(x)$ on the interval [0,t] occurs at t and has value kt^{k-1}. If $0 \le t < 1$ then this sequence converges to zero. This verifies THEOREM 5.51 for the sequence x^k on [0,t].

To see what can happen if we don't have uniform convergence, consider the sequence $b_k = kx^{k-1}$, k = 1, 2, . . ., on the interval [0,1). The functions kx^{k-1} were called $a_k'(x)$ in the previous paragraph, but we change the name to $b_k(x)$ to emphasize the new domain [0,1). On [0,1), the sequence $b_k(x)$ converges to b(x) = 0 at each x, but the convergence is not uniform (why?). The integrals

$$B_k(x) = \int_0^x b_k(t)dt = x^k$$

do not converge uniformly to the zero function on $[0,1)$. In particular, the area under each curve $b_k(x) = kx^{k-1}$, for $0 < x < 1$, is 1. The area under the limit function $b(x)$ is 0. Thus, the limit of the integrals need not be the integral of the limit for non-uniformly convergent sequences.

We now must consider infinite series of functions. Just as with sequences of real numbers, we start with a sequence $a_k(x)$, $k = 0, 1, 2, \ldots$, of functions defined on a common domain D and consider the sequence $s_n(x) = a_0(x) + a_1(x) + \ldots + a_n(x)$, $n = 0, 1, 2, \ldots$, of partial sums with terms from this sequence. This sequence of partial sums is the *infinite series* with terms from the sequence $a_k(x)$, $k = 1, 2, \ldots$. Just as before, we use without serious harm, the notation

$$\sum_{k=0}^{\infty} a_k(x)$$

to mean both the series and its limit function, depending on the discussion.

If the sequence of partial sums $s_n(x)$ converges uniformly to a limit on the domain D, then we say that the series with terms from the sequence $a_k(x)$, $k = 0, 1, 2, \ldots$, is *uniformly convergent*. The following definition, technically a theorem in a more advanced course, corresponds to DEFINITION 5.24.

Uniform Convergence Of Series

5.52 DEFINITION (UNIFORM CONVERGENCE OF SERIES) The infinite series $a_0(x) + a_1(x) + \ldots$ converges *uniformly* on the domain D if, given any $\epsilon > 0$, there exists an N_ϵ such that for all $q \geqslant p > N_\epsilon$ and for all x in D, $|a_p(x) + \ldots + a_q(x)| < \epsilon$.

The key phrase in DEFINITION 5.52 is "for all x in D." Through our practice with infinite series of numbers, we have already acquired the technical skills needed to deal with infinite series of functions. Our main concern with series of functions will be how to deal with the important issue of uniform convergence. The next theorem gives a simple but very useful test for uniform convergence.

5.53 THEOREM (WEIERSTRASS'S M TEST) Let $a_0(x) + a_1(x) + \ldots$ be an infinite series of functions defined on a domain D. Let M_k be a sequence of numbers such that $|a_k(x)| \leqslant M_k$ for all x in D. If the series $M_0 + M_1 + \ldots$ converges then the series $a_0(x) + a_1(x) + \ldots$ converges uniformly on D.

Proof: Given any $\epsilon > 0$, we can choose N_ϵ such that for all $q \geq p > N_\epsilon$, $|M_p + \ldots + M_q| = M_q + \ldots + M_p < \epsilon$. But $|a_q(x) + \ldots + a_p(x)| \leq |a_p(x)| + \ldots + |a_q(x)| \leq M_q + \ldots + M_p$ for all x in D. Thus the sequence $a_0(x) + a_1(x) + \ldots$ converges uniformly on D by DEFINITION 5.52. This completes the proof.

You can see from the proof of THEOREM 5.53 that, in fact, the series of absolute values, $|a_0(x)| + |a_1(x)| + \ldots$, converges uniformly on D. The series with terms $a_k(x)$, $k = 0, 1, 2, \ldots$, is thus "absolutely uniformly convergent." Notice also that if the sequence M_k, $k = 0, 1, 2, \ldots$, is replaced by a sequence of functions $M_k(x)$, $k = 0, 1, 2, \ldots$, which converge uniformly on D and satisfy $|a_k(x)| \leq M_k(x)$ for all x in D, the proof is still valid.

As a simple example of the M-test, consider the series

$$\sum_{k=1}^{\infty} \frac{x^k}{k^\alpha}, \alpha > 1.$$

Let $M_k = \dfrac{1}{k^\alpha}$. Take the domain D of the functions x^k/k^α to be $[-1, +1]$. On the interval $[-1, +1]$, $|x^k/k^\alpha| \leq M_k$. Thus the convergence of the series with terms M_k, $k = 1, 2, \ldots$, implies the uniform convergence of the series of functions with terms x^k/k^α on the interval $[-1, +1]$.

The analogs THEOREMS 5.50 and 5.51, which concern integrals and derivatives of sequences of functions, follow immediately. The series with terms from the sequence of functions $a_k(x)$, $k = 0, 1, 2, \ldots$, is, by definition, the sequence of partial sums $s_n(x)$, $n = 0, 1, 2, \ldots$, of this sequence. Suppose the domain D of these functions is the interval [s,t]. Suppose that the sequence $s_n(x)$ converges uniformly on [s,t] to s(x). By definition, this means that the series with terms $a_k(x)$, $k = 1, 2, \ldots$, converges uniformly on [s,t]. Define functions $S_n(x)$ and $S(x)$ on [s,t] by

$$S_n(x) = \int_s^x s_n(t)dt \quad \text{and} \quad S(x) = \int_s^x s(t)dt.$$

By THEOREM 5.50, the sequence $S_n(x)$ converges uniformly to $S(x)$ on the interval [s,t].

The previous paragraph contains all of the ideas needed to apply THEOREM 5.50 to series. It is necessary, however, to fool around with the notation a bit to make sure you can recognize the various possible ways of saying the same thing. For example, we can write

$$S(x) = \int_s^x \left(\sum_{k=0}^{\infty} a_k(t) \right) dt \text{ and}$$

$$S_n(x) = \int_s^x s_n(t)dt = \int_s^x \left(\sum_{k=0}^{n} a_k(t) \right) dt = \sum_{k=0}^{n} \int_s^x a_k(t)dt.$$

The last of the above expressions for $S_n(x)$ is a common way of thinking about $S_n(x)$. Starting with the sequence of functions $a_k(x)$, $k = 0, 1, 2, \ldots$, we form the new sequence of functions

$$\int_s^x a_k(t)dt \text{ , } k = 0, 1, 2, \ldots.$$

This sequence is the sequence of integrals of the terms of the series $a_0(x) + a_1(x) + \ldots$. Doing this is called "term by term integration of the series." $S_n(x)$ is the sequence of partial sums of these integrals. Here is a common way to state THEOREM 5.50 applied to series when using this type of notation.

Term-By-Term Integration

5.54 THEOREM Suppose that the series of functions of x

$$\sum_{k=0}^{\infty} a_k(x)$$

converges uniformly on the interval [s,t]. Then the series of functions of x

$$\sum_{k=0}^{\infty} \left(\int_s^x a_k(t)dt \right)$$

converges uniformly on [s,t], and

$$\sum_{k=0}^{\infty} \left(\int_s^x a_k(t)dt \right) = \int_s^x \left(\sum_{k=0}^{\infty} a_k(t) \right) dt.$$

The analog of THEOREM 5.51, concerning term-by-term differentiation is as follows:

Term-By-Term Differentiation.

5.55 THEOREM Suppose that both of the series

$$\sum_{k=0}^{\infty} a_k(x) \quad \text{and} \quad \sum_{k=0}^{\infty} a'_k(x)$$

converge uniformly on [s,t]. The for x in [s,t],

$$\frac{d}{dx} \sum_{k=0}^{\infty} a_k(x) = \sum_{k=0}^{\infty} a'_k(x).$$

We began this chapter by working with infinite series of functions called power series. We didn't really know what we were doing, but it was fun. In particular, we did term-by-term integration on power series. For example, at one point we had the power series

$$\frac{\sin(x)}{x} = 1 - \frac{x^2}{3!} + \frac{x^4}{5!} - \frac{x^6}{7!} + \dots$$

We integrated this power series term by term. We knew it was o.k. to do this because we checked our answer on the computer by other means (a Riemann sum). The limits of integration that we used were from 0 to π. According to THEOREM 5.54, integration term by term is all right if this series converges uniformly on the interval $[0,\pi]$. Not only must the series converge uniformly on $[0,\pi]$, but it must converge to $\sin(x)/x$ for us to get the correct answer.

When a series does or does not converge to a given function is another matter that we will take up below. To show that the series for $\sin(x)|x$ converges uniformly on $[0,\pi]$, we can use the M-test (THEOREM 5.53) with $M_k = \pi^k/(k + 1)!$ Each term in the series satisfies $|x^k/(k + 1)!| \le M_k$ for $0 < x < \pi$. The series with terms M_k, $k = 0, 2, 4, \dots$, is easily seen to converge (root test, ratio test, or comparison test) and hence the given series for $\sin(x)/x$ converges uniformly on $[0,\pi]$. The same argument shows that it converges uniformly on any interval $[-c, +c]$ for any real number $c > 0$.

Power Series: $\Sigma \, c_k(x - a)^k$

In general, a power series is a series of functions $a_k(x)$, $k = 0, 1, 2, \dots$, where $a_k(x) = c_k(x - a)^k$ and the c_k, called the "coefficients of power series," are real numbers and a is a real number.

5.56   DEFINITION (Power Series)   A series of functions of the form

$$\sum_{k=0}^{\infty} c_k(x - a)^k$$

where a and c_k, $k = 0, 1, 2, \ldots$ are real numbers, is called a *power series*.

Suppose we try to apply the root test to the power series with terms $c_k(x - a)^k$. We compute the limit

$$\lim_{k \to \infty} |c_k(x - a)^k|^{1/k} = \lim_{k \to \infty} |c_k|^{1/k} |x - a| = L|x - a|.$$

The number L is the limit as k goes to infinity of $|c_k|^{1/k}$. The series converges if $|x - a| < L^{-1}$ and diverges if $|x - a| > L^{-1}$. The number $R = L^{-1}$ is called the "radius of convergence" of the power series. Except for some technical details (the limit of $|c_k|^{1/k}$ might not exist), we have proved the following theorem.

5.57   THEOREM   Let

$$\sum_{k=0}^{\infty} c_k(x - a)^k$$

be a power series and let L be the limit of $|c_k|^{1/k}$. The number $R = L^{-1}$ is called the radius of convergence of the power series. The power series converges absolutely for all numbers x such that $|x - a| < R$ and diverges for $|x - a| > R$. For $|x - a| = R$ the power series may either converge or diverge, depending on the particular case.

For example, if $c_k = 2^{-k}$ then $L = 1/2$. Thus the radius of convergence of the power series

$$\sum_{k=0}^{\infty} 2^{-k}(x - a)^k$$

is 2. If $c_k = 1/k!$, then $L = 0$. In this case, we say that the radius of convergence is infinite. The power series

$$\sum_{k=0}^{\infty} (x - a)^k/k!$$

converges for all x.

The Proof Of Taylor's Theorem

The next result concerns power series approximations to functions. We shall prove a theorem called "TAYLOR'S THEOREM." This theorem is very useful and has an unforgettably simple proof based on integration by parts. Recall that, just prior to EXERCISE 4.26, we introduced the following tabular notation or "box notation" for describing integration by parts:

$f(x)$	$g(x)$
$f'(x)$	$g'(x)$

$$\int f(x)g'(x)dx = f(x)g(x) - \int f'(x)g(x)dx$$

In words, the integral of the product of the entries on either diagonal is the product of the entries in the top row minus the integral of the product of the entries on the other diagonal.

Suppose we have a function $f(x)$ that is "smooth" in the sense that it has lots of derivatives. Most functions we use in calculus are like that. They can be differentiated over and over again. The functions $\sin(x)$, $\tan(x)$, $\cos(x)$, $\ln(x)$, e^x, $\cosh(x)$, $\arcsin(x)$ $(1 + x)^{1/2}$, etc., have this property for many values of x.

Let b be any real number, and consider the following sequence of boxes with their corresponding integration by parts formulas:

$f^{(1)}(x)$	$b - x$
$f^{(2)}(x)$	-1

$$\int f^{(1)}(x)d(b - x) = f^{(1)}(x)(b - x) + \int f^{(2)}(x)(b - x)d(b - x)$$

$f^{(2)}(x)$	$\dfrac{(b - x)^2}{2!}$
$f^{(3)}(x)$	$-(b - x)$

$$\int f^{(2)}(x)(b-x)d(b-x) = f^{(2)}(x)\frac{(b-x)^2}{2!} + \int f^{(3)}(x)\frac{(b-x)^3}{3!}d(b-x)$$

$f^{(3)}(x)$	$\dfrac{(b - x)^3}{3!}$
$f^{(4)}(x)$	$-\dfrac{(b - x)^2}{2!}$

$$\int f^{(3)}(x)\,\frac{(b-x)^2}{2!}\,d(b-x) \;=\; f^{(3)}(x)\,\frac{(b-x)^3}{3!} + \int f^{(4)}(x)\,\frac{(b-x)^3}{3!}\,d(b-x)$$

.
.
.

(general case)

$f^{(n)}(x)$	$\dfrac{(b - x)^n}{n!}$
$f^{(n+1)}(x)$	$-\dfrac{(b-x)^{n-1}}{(n-1)!}$

$$\int f^{(n)}(x)\,\frac{(b-x)^{n-1}}{(n-1)!}\,d(b-x) \;=\; f^{(n)}(x)\,\frac{(b-x)^n}{n!} + \int f^{(n+1)}(x)\,\frac{(b-x)^n}{n!}\,d(b-x).$$

In the above sequence of boxes, focus your attention on the boxes, not the formulas. It's easier to remember that way, and the formulas follow automatically. In these formulas, the minus sign that usually appears in the last integral of the integration by parts formula is incorporated into the $d(b - x)$. Starting with the first formula, replace the last integral in each formula with the expression given for it in the next formula to get

$$\int f^{(1)}(x)\,d(b - x) \;=\;$$

$$f^{(1)}(x)(b - x) + f^{(2)}(x)\,\frac{(b - x)^2}{2!} + \ldots + f^{(n)}(x)\,\frac{(b - x)^n}{n!}$$

$$+ \int f^{(n+1)}(x)\,\frac{(b - x)^n}{n!}\,d(b - x).$$

This expression is an exact statement about indefinite integrals. It looks useless at first glance. However, now let's compute the definite integral

$$\int_b^a f^{(1)}(x)d(b-x) = f(b) - f(a).$$

The above formula now becomes

$$f(b) - f(a) =$$

$$f^{(1)}(a)(b-a) + f^{(2)}(a)\frac{(b-a)^2}{2!} + \ldots + f^{(n)}(a)\frac{(b-a)^n}{n!}$$

$$+ \int_b^a f^{(n+1)}(x)\frac{(b-x)^n}{n!}d(b-x).$$

By writing $d(b-x) = dx$ and reversing the order of the limits of integration, this same formula can be written

$$f(b) = f(a) + f^{(1)}(a)(b-a) + f^{(2)}(a)\frac{(b-a)^2}{2!} + \ldots + f^{(n)}(a)\frac{(b-a)^n}{n!}$$

$$+ \int_a^b f^{(n+1)}(x)\frac{(b-x)^n}{n!}dx.$$

The above formula is not an approximation. It is an identity valid for any function f with $n + 1$ continuous derivatives in some interval that contains a and b. In this identity, you can imagine the number a as fixed and b as a variable. When thinking of the identity in this way, it is common to replace the symbol b by x and write

$$f(x) = f(a) + f^{(1)}(a)(x-a) + f^{(2)}(a)\frac{(x-a)^2}{2!} + \ldots + f^{(n)}(a)\frac{(x-a)^n}{n!}$$

$$+ \int_a^x f^{(n+1)}(t)\frac{(x-t)^n}{n!}dt.$$

We have changed the variable from x to t in the definite integral, to avoid confusion with x, the new name for b. We introduce the following standard terminology.

Taylor Polynomial

5.58 DEFINITION Let $f(x)$ be a function with $n + 1$ derivatives at a point a. The polynomial

$$T_n(x) = f(a) + f^{(1)}(a)(x-a) + f^{(2)}(a) \frac{(x-a)^2}{2!} + \ldots + f^{(n)}(a) \frac{(x-a)^n}{n!}$$

is called the "Taylor polynomial of degree n of f at a."

Putting these ideas together, we have the following important theorem.

5.59 THEOREM (Taylor's Theorem) Let f be a function with continuous derivative function $f^{(n+1)}$ in the interval $I = (\alpha, \beta) = \{x : \alpha < x < \beta\}$. For any a and x in I

$$f(x) = T_n(x) + \int_a^x f^{(n+1)}(t) \frac{(x-t)^n}{n!} \, dt \text{ where}$$

$$T_n(x) = f(a) + f^{(1)}(a)(x-a) + f^{(2)}(a) \frac{(x-a)^2}{2!} + \ldots + f^{(n)}(a) \frac{(x-a)^n}{n!}$$

is the TAYLOR POLYNOMIAL of degree n of f at a.

The expression

$$R_{n+1}(x,a) = \int_a^x f^{(n+1)}(t) \frac{(x-t)^n}{n!} \, dt$$

is called the *remainder in the Taylor polynomial approximation to* $f(x)$. This integral represents the amount $f(x)$ differs from its Taylor polynomial of degree n at a. If for fixed x and a, the sequence $R_n(x,a)$ tends to zero as n tends to infinity, then we write

$$f(x) = \sum_{k=0}^{\infty} f^{(k)}(a) \frac{(x-a)^k}{k!} \, .$$

The power series

$$\sum_{k=0}^{\infty} f^{(k)}(a) \frac{(x-a)^k}{k!}$$

is called the "Taylor series of f at a."

There are two different questions that can be asked about the Taylor series of f at a. The first question is "For what values of x does the Taylor series of f at a converge?" The second question is "For what values of x does the Taylor series of f at a converge to f(x)?" For example, in the math tables in the appendix we find the statement

$$\sin(x) = x - \frac{x^3}{3!} + \frac{x^5}{5!} - \frac{x^7}{7!} + \ldots \qquad [x^2 < \infty]$$

This statement conveys the information that the Taylor series of sin(x) at a = 0 is

$$x - \frac{x^3}{3!} + \frac{x^5}{5!} - \frac{x^7}{7!} + \ldots$$

and, by the statement $[x^2 < \infty]$, that this series converges to sin(x) for all x. The function $z(x) = e^{-1/x^2}$, defined to be zero at x = 0, has derivatives of all orders, $z^{(n)}(x)$, for all values of x, including x = 0 (you should show this to be true). At 0, $z^{(n)}(0) = 0$, for all n = 0, 1, 2, ... Thus the Taylor series of this function at a = 0 is the identically zero series. The Taylor series of this function at a = 0 obviously converges for all x but is equal to z(x) only at x = 0. Using z(x), you can construct other curious examples of Taylor series. The function g(x) = sin(x) + z(x) has the same Taylor series as sin(x) at a = 0. This Taylor series converges everywhere but converges to g(x) only at x = 0. The function h(x) = z(x)sin(x) + sin(x) has the same Taylor series as sin(x) at a = 0 but converges to h(x) only at x = πk, k = 0, ±1, ±2, ...

Estimating The Remainder In Taylor's Theorem

In regard to the second question, "For what values of x does the Taylor series of f at a converge to f(x)?", it is helpful to have some techniques for showing convergence of the series of remainders $R_n(x,a)$. One useful technique supposes that we know an upper bound for $f^{(n+1)}(t)$ on the interval [a,x].

5.60   REMAINDER ESTIMATE   Suppose that $\left|f^{(n+1)}(t)\right| \leq B_{n+1}$ for all t in the interval [a,x]. Then

$$\left|R_{n+1}(x,a)\right| \leq B_{n+1} \frac{|x - a|^{n+1}}{(n + 1)!} .$$

Proof: We see from the form of $R_{n+1}(x,a)$ given above as a definite integral that $|R_{n+1}(x,a)|$ is less than or equal to the upper bound B_{n+1} times the area between the curve $|x - t|^n/n!$ and the interval $[a,x]$ on the x-axis. This area is exactly $|x - a|^{n+1}/(n + 1)!$ which proves the assertion

There is another common way of expressing the remainder, $R_{n+1}(x,a)$, called LAGRANGE'S FORM OF THE REMAINDER.

5.61   LAGRANGE'S FORM OF THE REMAINDER   There exists some number c between a and x such that

$$R_{n+1}(x,a) = f^{(n+1)}(c) \frac{(x - a)^{n+1}}{(n + 1)!}.$$

The good news about Lagrange's form for the remainder is that it is not an approximation but an equality. The bad news is that we hardly ever are able to find the mysterious number c. By considering all possibilities for c, we can get back to estimating the remainder, as in REMAINDER ESTIMATES 5.60.

As an application of REMAINDER ESTIMATES 5.60, consider the Taylor series for sin(x) at 0:

$$\sum_{k=0}^{\infty} (-1)^k \frac{x^{2k+1}}{(2k + 1)!}.$$

By the ratio test, or the root test together with Stirling's formula, we see easily that this series converges absolutely for all x. In fact, from Weierstrass's M test, it converges uniformly on every interval $[\alpha,\beta]$, where $\alpha < \beta$ are real numbers. The higher order derivatives of sin(x) are all either $+\sin(x)$, $-\sin(x)$, $+\cos(x)$, or $-\sin(x)$. Consequently, the bound $B_{n+1} = 1$ works for all n. Thus, the sequence of remainders, $|R_{n+1}(x,0)|$, is less than or equal to $|x|^{n+1}/(n + 1)!$, which converges to zero for all x. This shows that the Taylor series for sin(x) at 0 converges to sin(x) for all x.

There are three basic operations on general power series that we now must take a look at.

5.62   BASIC OPERATIONS ON A POWER SERIES

(1)   Starting with $\sum_{k=0}^{\infty} c_k(x - a)^k$ form $\sum_{k=1}^{\infty} kc_k(x - a)^{k-1}$ by differentiating term by term.

(2)   Starting with $\displaystyle\sum_{k=0}^{\infty} c_k(x - a)^k$ form $\displaystyle\sum_{k=0}^{\infty} \frac{c_k}{(k + 1)}(x - a)^{k+1}$ by integrating term by term.

(3)   Starting with $\displaystyle\sum_{k=1}^{\infty} c_k(x - a)^k$ form $\displaystyle\sum_{k=1}^{\infty} c_k(x - a)^{k-1}$ by dividing each term by $x - a$.

5.63   THEOREM   Each of the power series of 5.62 BASIC OPERA-TIONS ON A POWER SERIES has the same radius of convergence $R = L^{-1}$ where

$$L = \lim_{k \to \infty} |c_k|^{1/k}.$$

Proof: In case (1) we compute $\displaystyle\lim_{k \to} \left| k c_k(x - a)^{k-1} \right|^{1/k} =$

$$\left(\lim_{k \to \infty} (k^{1/k})\right) \left(\lim_{k \to \infty} |c_k|^{1/k}\right) \left(\lim_{k \to \infty} |x - a|^{(k-1)/k}\right) = (1)(L)(|x - a|).$$

Thus, in case (1), the second series converges if $|x - a| < L^{-1} = R$ and diverges if $|x - a| > R$.

In case (2), using the fact that $\displaystyle\lim_{k \to} (k + 1)^{-1/k} = 1$ and

$$\lim_{k \to \infty} |x - a|^{(k+1)/k} = |x - a|,$$

we again get R as the radius of convergence of the second series.

In case (3), we need only that

$$\lim_{k \to \infty} |x - a|^{(k-1)/k} = |x - a|$$

to get R as the radius of convergence. This completes the proof of THEOREM 5.63.

To complete our understanding of BASIC OPERATIONS ON A POWER SERIES we need the next easy result.

Uniform And Absolute Convergence Of Taylor's Series

5.64   THEOREM   If R is the radius of convergence of the power series

$$\sum_{k=0}^{\infty} c_k(x - a)^k$$

and $0 < r < R$, then this series converges uniformly on the interval $[a - r, a + r]$.

Proof: Apply the Weierstrass M test with $M_k = |c_k| \, r^k$.

It obviously follows from THEOREM 5.64 that a power series converges uniformly on any interval $[\alpha,\beta]$, where α and β are contained in the interval of convergence of the power series.

It's time to worry about notation again. If you see a statement like this

$$\text{``Consider the power series } c(x) = \sum_{k=0}^{\infty} c_k(x - a)^k \ldots\text{''}$$

then the symbol $c(x)$ stands for the power series itself. That is, $c(x)$ stands for the sequence of partial sums of the sequence $c_k(x - a)^k$, $k = 0, 1, 2,$ . . . . If, on the other hand, you see a statement like this

$$\text{``Consider the function } c(x) = \sum_{k=0}^{\infty} c_k(x - a)^k \text{ where } |x - a| < R \ldots\text{''}$$

the $c(x)$ stands for the function defined by the limit of the power series for values of x such that $|x - a|$ is less than the radius of convergence R. There is usually little chance of confusing these two different meanings for $c(x)$.

We now summarize the above discussion with an important theorem for working with power series.

Differentiation And Integration Of Power Series

5.65   THEOREM   Let R be the radius of convergence of the power series with terms $c_k(x - a)^k$, $k = 0, 1, 2, \ldots$, and define the function

$$c(x) = \sum_{k=0}^{\infty} c_k(x - a)^k, \quad |x - a| < R.$$

Then

(1) $$c'(x) = \sum_{k=1}^{\infty} k c_k(x - a)^{k-1} \text{ for } |x - a| < R$$

(2) $$\int_a^x c(x) \, dx = \sum_{k=0}^{\infty} \frac{c_k}{k + 1} (x - a)^{k+1} \text{ for } |x - a| < R$$

and, if $c(a) = 0$,

(3) $$\frac{c(x)}{x - a} = \sum_{k=1}^{\infty} c_k(x - a)^{k-1} \text{ for } |x - a| < R.$$

Proof: Choose $r < R$ such that $a - r < x < a + r$. All series of the theorem converge uniformly on the interval $[a - r, a + r]$. Then (1) follows from THEOREM 5.51 and (2) follows from THEOREM 5.50. The uniform convergence of the series in (2) is a consequence of THEOREM 5.50, although we have verified uniform convergence directly by THEOREMS 5.63 and 5.64. In the case of (3), $c(a) = 0$ means that $c_0 = 0$. Thus, (3) follows by multiplying both sides by $(x - a)$ and using the basic arithmetic rule (1) for series stated just prior to EXERCISE 5.42.

In the beginning of this chapter, we started with the series for $\sin(x)$.

$$\sin(x) = x - \frac{x^3}{3!} + \frac{x^5}{5!} - \ldots$$

We then divide by x to obtain

$$\frac{\sin(x)}{x} = 1 - \frac{x^2}{3!} + \frac{x^4}{5!} - \ldots$$

Next we integrate term by term to obtain

$$\int \frac{\sin(x)}{x} \, dx = 1 - \frac{x^3}{3 \, 3!} + \frac{x^5}{5 \, 5!} - \ldots$$

All of the above operations are now justified by THEOREM 5.65, and all series have the same radius of convergence ($R = \infty$) as the original series.

The operation of differentiation as in THEOREM 5.65(1) can be applied repeatedly, always obtaining a series with the same radius of convergence as the original. When you compute $c^{(n)}(x)$ and evaluate at $x = a$, you obtain $c^{(n)}(a) = n!c_n$. Thus, if a function is defined by a power series in its radius of convergence, such as $c(x)$ was in THEOREM 5.65, that power series is the Taylor series of that function.

Addition And Multiplication Of Power Series

The same argument used to prove THEOREM 5.65(3) shows that if you take any polynomial $p(z) = a_0 + a_1 z + a_2 z^2 + \ldots + a_n z^n$ and form the product

$$p(x - a)c(x) = p(x - a) \sum_{k=0}^{\infty} c_k(x - a)^k$$

and collect terms on the right according to powers of $(x - a)$, then you obtain the power series for $p(x - a)c(x)$ and that power series converges for $|x - a| < R$.

More generally, you can take any two power series

$$c(x) = \sum_{k=0}^{\infty} c_k(x - a)^k \text{ and } d(x) = \sum_{k=0}^{\infty} d_k(x - a)^k$$

and add them term by term or take their product, expressing the answer as a sum of powers of $(x - a)$. These operations can be done purely "formally" without regard to intervals of convergence. If, however, r is the minimum of the two radii of convergence for these two power series, then $c(x)$ and $d(x)$ are functions defined for $|x - a| < r$. In that case, the sum of the two series converges to the function $c(x) + d(x)$ for $|x - a| < r$ and the product converges to $c(x)d(x)$ for $|x - a| < r$. In either case the actual radius of convergence of the sum series or product series might be larger than r. For example, if the series for $c(x)$ had radius of convergence 1 and you subtract that series from itself, you get the zero series, which has infinite radius of convergence.

As an example of taking the product of two series, consider

$$\sin(x) = \sin(a) + \cos(a)(x - a) - \sin(a)(x - a)^2/2! + \ldots$$

and

$$e^x = e^a + e^a(x - a) + e^a(x - a)^2/2! + \ldots$$

The product

$$e^x\sin(x) = e^a\sin(a) + (e^a\sin(a) + e^a\cos(a))(x - a) + e^a\cos(a)(x - a)^2 + \ldots .$$

Instead of doing such a calculation, you can always compute the Taylor series directly. For $f(x) = e^x\sin(x)$, we find that $f^{(1)}(x) = e^x\sin(x) + e^x\cos(x)$ and $f^{(2)}(x) = 2e^x\cos(x)$. Thus $f^{(1)}(a) = e^a\sin(a)$ and $f^{(2)}(a) = 2e^a\cos(a)$ and we obtain the same series directly from the definition of the Taylor series.

Composing Power Series

Finally, there are the operations of division of power series and composition of power series. Both operations usually involve tedious operations. Modern algebraic symbol manipulation software is a great help with regard to these and all operations on power series. To simplify matters, we shall stick to power series in x. Taylor series expansions of functions about $a = 0$ are called MACLAURIN SERIES. Let's compute the Maclaurin series of the composition $e^{\sin(x)}$. We have

$$e^x = 1 + x + x^2/2! + x^3/3! + \ldots$$

and

$$\sin(x) = x - x^3/3! + x^5/5! - \ldots.$$

Both series converge for all x. The composition $e^x \sin(x)$ is thus

$$1 + (x - x^3/3! + x^5/5! - \ldots) + (x - x^3/3! + x^5/5! - \ldots)^2/2!$$

$$+ (x - x^3/3! + x^5/5! - \ldots)^3/3! + \ldots$$

Taking powers and collecting terms, we get

$$e^{\sin(x)} = 1 + x + x^2/2! - 3x^4/4! - \ldots$$

What happens with radii of convergence under composition of power series can be quite tricky. A little common sense will avoid the worst blunders. If you are going to compose c(x), which converges for $|x| < R_1$ and d(x) which converges for $|x| < R_2$ and want the power series for c(d(x)) to converge at x = t, then you better have both $|t| < R_2$ and $|d(t)| < R_1$. In our example, both power series had infinite radius of convergence. In combinatorial mathematics, composition of power series plays an important role even when the series don't converge. One very common type of composition of series is to compose a series, such as that of sin(x), with a power of x, such as x^2, to obtain a series such as $\sin(x^2) = x^2 - x^6/3! + \ldots$. Here, c(x) = sin(x) and $d(x) = x^2$.

Dividing Power Series

Dividing one series c(x) by d(x) to obtain the series for c(x)/d(x) is directly analogous to polynomial division and every bit as tedious. Of course, you must take care not to divide by zero. Here we divide the series for $\sin(x^2)$ by cos(x) to obtain a few terms of the series for $\sin(x^2)/\cos(x)$.

$$x^2 + \frac{x^4}{2} + \frac{x^6}{24} + \ldots$$

$$1 - \frac{x^2}{2} + \frac{x^4}{24} - \ldots \overline{\bigg)\; x^2 \qquad\qquad - \frac{x^6}{6} \qquad\qquad + \ldots}$$

$$x^2 - \frac{x^4}{2} + \frac{x^6}{24}$$

$$\frac{x^4}{2} - \frac{5x^6}{24} + \ldots$$

$$\frac{x^4}{2} - \frac{x^6}{4} + \ldots$$

$$\frac{x^6}{24} + \ldots$$

$$\frac{x^6}{24} + \ldots$$

$$0 + \ldots$$

Thus, computed to three terms, we have

$$\frac{\sin(x^2)}{\cos(x)} = x^2 + \frac{x^4}{2} + \frac{x^6}{24} + \ldots \qquad -\frac{\pi}{2} < x < \frac{\pi}{2}.$$

Limits Of Functions

In DEFINITION 3.11 we defined the notion of a limit of a function $f(x)$ at $x = t$ in terms of limits from the left and limits from the right at t. Using sequences, we can be more precise about these definitions. Let x_k, $k = 0$, $1, 2, \ldots$, be an infinite sequence. If the limit as k goes to infinity of x_k is t, we use the shorthand notation $x_k \to t$. If also $x_x < t$ for all k then we write $x_k \to t-$ and if $x_k > t$ for all k, we write $x_k \to t+$. We have the following definition:

5.66 DEFINITION We say that the "limit from the left of $f(x)$ at t is A" and write

$$\lim_{x \to t-} f(x) = A$$

if for every sequence $x_k \to t-$ the sequence $f(x_k) \to A$. Similarly, we define the "limit from the right of $f(x)$ at t is A" and write

$$\lim_{x \to t+} f(x) = A.$$

Combining these ideas, if the limit from the left *and* the right of $f(x)$ at t is A, then we say that "the limit of $f(x)$ at t is A" and write

$$\lim_{x \to t} f(x) = A.$$

The Epsilon—Delta Definition Of A Limit

The definition of "the limit of f(x) at t is A" can be stated directly as "for every sequence $x_k \to t$, $f(x_k) \to t$." An alternative definition of "the limit of f(x) at t is A" is to say that "For every $\epsilon > 0$ there exists a $\delta > 0$ such that for all x such that $|x - t| < \delta$, $|f(x) - A| < \epsilon$." This definition is called "the epsilon-delta definition of the limit." It is equivalent to the definition of "the limit of f(x) at t is A" given in DEFINITION 5.66. You should be getting to the point now where you can prove that these two definitions are equivalent. Give it a try!

Continuous Functions

Corresponding to DEFINITION 3.12, we have

5.67 DEFINITION We say that f(x) is *left continuous at t* if

$$\lim_{x \to t-} f(x) = f(t).$$

Similarly, we define "f(x) is *right continuous at t*." If f(x) is both left continuous and right continuous at t then we say that f(x) is *continuous at t*.

A direct statement that f(x) is continuous at t is "for every sequence $x_k \to t$, $f(x_k) \to f(t)$." Intuitively, continuous functions at t have nice smooth graphs at the point (t,f(t)). The function $f(x) = +1$ if $x \geq 0$ and $f(x) = -1$ if $x < 0$ is discontinuous at $x = 0$. It is right continuous at $x = 0$ but not left continuous. At $x = 0$, the graph of this function, which you should sketch, has a "jump discontinuity" at $x = 0$. Imagine a sequence of functions $f_n(x)$ converging uniformly to this function f(x) on $[-1, +1]$. As in FIGURE 5.49, there is some N such that for all $n > N$, the graphs of the functions $f_n(x)$ have disappeared into the line representing the graph of f(x). Thus, for $n > N$, the functions $f_n(x)$ must also be discontinuous at $x = 0$. The idea is that if a sequence of functions $f_n(x)$ converges uniformly to a function f(x) that is discontinuous at t, then all but possibly a finite number of the $f_n(x)$ must also be discontinuous at t. The contrapositive to this statement is as follows:

Uniform Convergence And Continuity

5.68 THEOREM Let $f_n(x)$ be a sequence of functions that converge uniformly to f(x) on some interval $[a - r, a + r]$. If all but possibly a finite number of the functions $f_n(x)$ are continuous at t, $a - r < t < a + r$, then $f(x)$ is continous at t.

Proof: We must show that for any sequence $x_k \to t$, $f(x_k) \to f(t)$. The values of x_k may be assumed to be in the interval $(a - r, a + r)$. This is equivalent to showing that for any $\epsilon > 0$, there is some K such that for all $k > K$, $|f(t) - f(x_k)| < \epsilon$. We write

$$|f(t) - f(x_k)| = |f(t) - f_n(t) + f_n(t) - f_n(x_k) + f_n(x_k) - f(x_k)|$$

$$\leq |f(t) - f_n(t)| + |f_n(t) - f_n(x_k)| + |f_n(x_k) - f(x_k)|.$$

By uniform convergence, we can, by picking n large enough, make sure that $|f_n(x) - f(x)| < \epsilon/3$ for all x in the interval, in particular for $x = t$ and $x = x_k$, $k = 0, 1, \ldots$. Fix such an n and choose it to be such that $f_n(x)$ is continuous at t (only finitely many f_n are not continuous at t). Now, choose K such that for all $k > K$, $|f_n(t) - f_n(x_k)| < \epsilon/3$. Thus, for all $k > K$, $|f(t) - f(x_k)| < \epsilon$. This completes the proof.

Limits And Uniformly Convergent Series

In THEOREM 5.68, the word "continuous" can be replaced by "right continuous" or "left continuous." The next corollary follows from THEOREM 5.68 by replacing the function $f_n(x)$ by the partial sums $s_n(x)$.

5.69 COROLLARY If all of the functions in the sequence $a_k(x)$, $k = 0, 1, 2, \ldots$ are continuous at t and the series $a_0(x) + a_1(x) + \ldots$ converges uniformly on an interval containing t, then

$$\lim_{x \to t} \sum_{k=0}^{\infty} a_k(x) = \sum_{k=0}^{\infty} a_k(t).$$

The most useful applications of COROLLARY 5.69 are to power series. Consider the Maclaurin series

$$\sum_{k=0}^{\infty} c_k x^k, \quad |x| < R$$

where R is the radius of convergence. If $|t| < R$ then, by uniform convergence on $[-r, +r]$ where $|t| < r < R$, we have

$$\lim_{x \to t} \sum_{k=0}^{\infty} c_k x^k = \sum_{k=0}^{\infty} c_k t^k.$$

Abel's Limit Theorem

If the series converges absolutely for $x = R$, then, by the M test, the series converges uniformly on $[-R, R]$. Thus,

$$\lim_{x \to R} \sum_{k=0}^{\infty} c_k x^k = \sum_{k=0}^{\infty} c_k R^k.$$

It is an interesting result, due to Abel, that the above equality is valid even if the series

$$\sum_{k=0}^{\infty} c_k R^k$$

converges only conditionally. This result can be proved from the ABEL'S TEST that we studied in 5.40 PROBLEMS BASED ON DIRICHLET'S TEST and ABEL'S TEST. Here is a sequence of steps leading to an interesting fact. You should be able to justify each step, the last of which uses Abel's limit theorem.

$$\frac{1}{1 + t} = 1 - t + t^2 - t^3 + \ldots \qquad |t| < 1$$

$$\frac{1}{1 + t^2} = 1 - t + t^2 + t^4 - t^6 + \ldots \qquad |t| < 1$$

$$\arctan(x) = \int_0^x \frac{dt}{1 + t^2} = x - \frac{x^3}{3} + \frac{x^5}{5} - \frac{x^7}{7} + \ldots \, |x| < 1$$

$$\frac{\pi}{4} = \lim_{x \to 1} \arctan(x) = 1 - \frac{1}{3} + \frac{1}{5} - \frac{1}{7} + \ldots$$

The last identity, which can be written

$$\pi = 4 \left(1 - \frac{1}{3} + \frac{1}{5} - \frac{1}{7} + \ldots \right)$$

is called Gregory's series for π. Try it on your computer!

5.70   EXERCISES

(1)   Verify that all series marked " $->$ " in the SERIES AND PRODUCTS section of the MATH TABLES in the Appendix are correct to at least four nonzero terms.

(2) Verify that all series marked "$->>$" in the SERIES AND PROD-UCTS section of the MATH TABLES in the Appendix are correct to at least four nonzero terms.

(3) Verify that all series marked "$*$" in the SERIES AND PRODUCTS section of the MATH TABLES in the Appendix are correct to at least four nonzero terms.

(4) Compute the TAYLOR SERIES expansions of e^x, $\sin(x)$, and $\cos(x)$ about $x = a$. Show your work clearly.

(5) Find the radius of convergence R of the following series from EXER-CISE 5.42. Discuss convergence at R and $-R$ if $R < \infty$.

(a) $\displaystyle\sum_{k=1}^{\infty} \frac{(k + 1)(k + 2)(k + 3)}{k!} x^k$

(b) $\displaystyle\sum_{k=1}^{\infty} \frac{2^{k/2}}{k^2 + k + 1} x^k$

(c) $\displaystyle\sum_{k=1}^{\infty} \frac{k + 1}{2k + 3} x^k$

(d) $\displaystyle\sum_{k=0}^{\infty} \frac{k^5}{5^k} x^k$

(e) $\displaystyle\sum_{k=0}^{\infty} \frac{x^k}{k^2 - 150}$

(f) $\displaystyle\sum_{k=5}^{\infty} \frac{x^k}{2^k - k^2}$

(g) $\displaystyle\sum_{k=2}^{\infty} \frac{1}{(k^3 - k^2 - 1)^{1/2}} x^k$

(h) $\displaystyle\sum_{k=1}^{\infty} \frac{(k + 1)^{1/2} - (k - 1)^{1/2}}{k} x^k$

(i) $\displaystyle\sum_{k=1}^{\infty} \frac{(-x)^k}{k}(1 + 2^{-2} + 3^{-2} + \cdots + k^{-2})$

(j) $\displaystyle\sum_{k=1}^{\infty} \frac{(-x)^k}{k}\left(\frac{1}{1} + \frac{1}{2} + \cdots + \frac{1}{k}\right)$

(k) $\displaystyle\sum_{k=0}^{\infty} (-1)^k \operatorname{arccot}(k)x^k$

(l) $\displaystyle\sum_{k=0}^{\infty} \frac{\sin(k)}{|k - 99.5|}x^k$

(m) $\displaystyle\sum_{k=1}^{\infty} \frac{(-x)^k}{k}\cos(1/k)$

(n) $\displaystyle\sum_{k=0}^{\infty} \frac{-9k^2 - 5}{k^3 + 1}\sin(k)x^k$

(6) Find the radius of convergence R of the following series from VARI-
ATIONS 5.44. Discuss convergence at R and $-R$ if $R < \infty$.

(a) $\displaystyle\sum_{k=1}^{\infty} \frac{(k - 1)(k - 2)(k - 3)}{k!}x^k$

(b) $\displaystyle\sum_{k=1}^{\infty} \frac{2^{k/2}}{k^{200} + k + 1}x^k$

(c) $\displaystyle\sum_{k=1}^{\infty} \frac{k^2 + 1}{2k^2 + 3}x^k$

(d) $\displaystyle\sum_{k=0}^{\infty} \frac{k^{500}}{(1.01)^k}x^k$

(e) $\displaystyle\sum_{k=0}^{\infty} \frac{x^k}{k^{1.0001} - 150}$

(f) $\displaystyle\sum_{k=1}^{\infty} \frac{x^k}{2^k - 3k^{2000}}$

(g) $\displaystyle\sum_{k=2}^{\infty} \frac{x^k}{(k^3 - k^2 - 1)^{1/4}}$

(h) $\displaystyle\sum_{k=1}^{\infty} \frac{(k + 1)^{1/4} - (k - 1)^{1/4}}{k} x^k$

(i) $\displaystyle\sum_{k=2}^{\infty} \frac{(-x)^k}{k} \left(\frac{1}{2(\ln(2))^2} + \frac{1}{3(\ln(3))^2} + \cdots + \frac{1}{k(\ln(k))^2} \right)$

(j) $\displaystyle\sum_{k=1}^{\infty} \frac{(-x)^k}{k} \left(\frac{(\ln(1))^2}{1} + \frac{(\ln(2))^2}{2} + \cdots + \frac{(\ln(k))^2}{k} \right)$

(k) $\displaystyle\sum_{k=1}^{\infty} (-x)^k \ln(k) \operatorname{arccot}(k)$

(l) $\displaystyle\sum_{k=0}^{\infty} \frac{\sin(k)}{\ln|k - 99.5|} x^k$

(m) $\displaystyle\sum_{k=1}^{\infty} \frac{(-x)^k}{k} \cos(k)$

(n) $\displaystyle\sum_{k=0}^{\infty} \frac{-9k^2 - 5}{k^3 + 1} \sin(k)\cos(k)x^k$

(7) Find the radius of convergence R of the following series from VARI-
ATIONS 5.45. Discuss convergence at R and $-$R if R $< \infty$.

(a) $\displaystyle\sum_{k=4}^{\infty} \frac{(k + 1)(k + 2)(k + 3)}{k(k - 1)(k - 2)(k - 3)} x^k$

(b) $\displaystyle\sum_{k=1}^{\infty} \frac{k^2}{2^{\ln(k)}} x^k$

(c) $\displaystyle\sum_{k=1}^{\infty} \frac{\ln(k) + (1/k)}{(\ln(k))^3 + 3} x^k$

(d) $\displaystyle\sum_{k=1}^{\infty} \frac{k^{\ln(k)}}{5^k} x^k$

(e) $\displaystyle\sum_{k=1}^{\infty} \frac{x^k}{(\ln(k))^2 - 150}$

(f) $\displaystyle\sum_{k=5}^{\infty} \frac{x^k}{(\ln(\ln(k)))^k}$

(g) $\displaystyle\sum_{k=2}^{\infty} \frac{x^k}{(k^3 - k^2 - 1)^{1/2.99}}$

(h) $\displaystyle\sum_{k=1}^{\infty} \frac{(k + 1)^{1/200} - (k - 1)^{1/200}}{k} x^k$

(i) $\displaystyle\sum_{k=1}^{\infty} (-1)^k (e^{-1/k} - 1 + 1/k) x^k$

(j) $\displaystyle\sum_{k=1}^{\infty} (e^{-1/k} - 1 + 1/k) x^k$

(k) $\displaystyle\sum_{k=0}^{\infty} (-1)^k (\text{arccot}(k))^\beta x^k, \ \beta < 1$

(l) $\displaystyle\sum_{k=0}^{\infty} \frac{\sin(k) - \cos(k)}{|k - 99.5|} x^k$

(m) $\displaystyle\sum_{k=1}^{\infty} \cos(k)\sin(1/k) x^k$

(n) $\displaystyle\sum_{k=0}^{\infty} \frac{-9k^2 - 5}{k^3 + 1} (\sin(k)\cos(k/2) + \cos(k)\sin(k/2)) x^k$

(8) Find the radius of convergence R of the following series from VARI-ATIONS 5.46. Discuss convergence at R and $-R$ if $R < \infty$.

(a) $\displaystyle\sum_{k=1}^{\infty} \frac{k^k}{k!} x^k$

(b) $\displaystyle\sum_{k=1}^{\infty} \frac{(k/4)^k}{k!} x^k$

(c) $\displaystyle\sum_{k=1}^{\infty} \frac{k!x^k}{k(k+1)\cdots(2k-1)}$

(d) $\displaystyle\sum_{k=1}^{\infty} \frac{(xk!)^{2k}}{(k^2)!}$

(e) $\displaystyle\sum_{k=1}^{\infty} \frac{(2k)!x^k}{k^2(k^2-1)\cdots(k^2-k+1)}$

(f) $\displaystyle\sum_{k=5}^{\infty} \left(\frac{1}{\ln(\ln(k))}\right)^{\ln(k)} x^k$

(g) $\displaystyle\sum_{k=5}^{\infty} x^k \left(\frac{1}{\ln(k)}\right)^{\ln(\ln(k))}$

(h) $\displaystyle\sum_{k=2}^{\infty} \frac{k^{\ln(k)}}{(\ln(k))^k} x^k$

(i) $\displaystyle\sum_{k=0}^{\infty} (-x)^k(\operatorname{arccot}(k))^\beta, \ \beta > 1$

(j) $\displaystyle\sum_{k=0}^{\infty} \frac{\sin(k) - \cos(k)}{\ln(|k - 99.5|)} x^k$

(k) $\displaystyle\sum_{k=1}^{\infty} (\sin(k) - \sin(k + (1/k))) x^k$

(l) $\displaystyle\sum_{k=0}^{\infty} \ln(1 + (1/k))\sin(2k + \pi)x^k$

(9) Consider the sequence of functions

$$a_k(x) = \frac{x}{1 + kx^2}.$$

Show that $a_k(x)$, $k = 0, 1, 2, \ldots$, tends to the zero function, $a(x) = 0$, uniformly on the domain $D = \{x\colon -\infty < x < \infty\}$. Discuss the convergence of the sequence of derivatives $a_k'(x)$ to $a'(x) = 0$.

(10) Consider the sequence of functions on $[0, 1]$

$$a_k(x) = \begin{cases} 0 & \text{if } 0 \le x < \dfrac{1}{k+1} \\[2ex] \sin(\pi/x) & \text{if } \dfrac{1}{k+1} \le x \le \dfrac{1}{k} \\[2ex] 0 & \text{if } \dfrac{1}{k} < x \le 1 \end{cases}$$

for $k = 1, 2, \ldots$. Does this sequence converge to a function on $[0, 1]$? Is convergence uniform? Prove or disprove your assertion.

(11) Consider the series

$$\sum_{k=1}^{\infty} a_k(x)$$

where $a_k(x)$ is as in **(10)**. Discuss the convergence of this series on $[0, 1]$. Does this series converge uniformly? Prove or disprove your assertion. Sketch some of the functions $s_n(x)$ in the sequence of partial sums of this series.

(12) Discuss the convergence of the series

$$s(x) = \sum_{k=0}^{\infty} \frac{1}{1 + k^2 x}.$$

For what x does this series converge absolutely? On what intervals does it converge uniformly? Prove or disprove your assertions.

(13) For what range of x can

(a) $\sin(x)$ be replaced by $x - x^3/6 + x^5/120$ with error less than .0001?

(b) $\cos(x)$ be replaced by $1 - x^2/2$ with error less than .001?

(c) e^x be replaced by $1 + x + x^2/2 + x^3/6$ with error less than .0001?

(d) $\ln(1 - x)$ be replaced by $-x - x^2/2 - x^3/3$ with error less than .01?

(14) Find the Maclaurin series for

(a) $\displaystyle\int_0^x \frac{1}{1 + t^3}\, dt$

(b) $\displaystyle\int_0^x \ln(1 + t^5)\, dt$

(c) $\displaystyle\int_0^x \frac{t^2}{(1 - t^2)^2}\, dt$

(15) Compute the Maclaurin series for $f(x) = (1 - x)^{-2}$ in the following three ways:

(a) Directly from the definition by computing derivatives $f^{(k)}(0)$, $k = 0, 1, 2, \ldots$.

(b) By differentiating the series for $(1 - x)^{-1}$ term by term. Justify your computations

(c) By direct multiplication of $(1 + x + x^2 + \ldots)$ times itself. Show that the coefficient of x^n in the series $(1 + x + x^2 + \ldots)^2$ is the number of pairs of nonnegative integers (i, j) with $i + j = n$. What is the corresponding interpretation of the coefficient of x^n in $(1 + x + x^2 + \ldots)^p$, $p > 2$, p an integer?

(16) Using Abel's limit theorem, discussed just after COROLLARY 5.69, show that $\ln(2) = 1 - (1/2) + (1/3) - (1/4) + \ldots$.

(17) By starting with the Maclaurin series

$$f(x) = a_0 + a_1 x + a_2 x^2 + \ldots$$

with the a_k unknown, substitute this series into the differential equation

$$x^2 f'(x) - x f(x) = \sin(x).$$

Solve for the a_k and thus obtain a power series solution to this differential equation. REMARK: If $f(x)$ is a solution so is $f(x) + cx$ for any constant c.

(18) Obtain the Maclaurin series of the following functions in two different ways. The first way is directly, by computing the derivatives $f^{(k)}(0)$, and the second way is by direct multiplication of the Maclaurin series of the two factors.

(a) $e^{2x}\sin(x)$

(b) $\sin(x)\cos(2x)$

(c) $\sinh(x)\cosh(2x)$

(d) $\sin(x)\ln(1 + x)$

(19) Directly and by long division, obtain the first three nonzero terms of the Maclaurin series of the following functions:

(a) $\dfrac{\ln(1 + x)}{\cos(x)}$

(b) $\dfrac{\ln(1 + x)}{e^{\sin(x)}}$

(c) $\dfrac{e^x}{x^2 + x + 1}$

(d) $\dfrac{\sin(x)}{1 + x}$

(20) Directly and by composition, obtain the first four nonzero terms of the Maclaurin series of the following functions.

(a) $\tan(x^3 + 1)$

(b) $\cos(\sin(x))$

(c) $\cos(\ln(1 + x))$

(d) $\sin(1 + x + x^2)$

Appendix 1

MATH TABLES

DERIVATIVES

$$\frac{d\,(au)}{dx} = \frac{a\,du}{dx}.$$

$$\frac{d\,(u+v)}{dx} = \frac{du}{dx} + \frac{dv}{dx}.$$

$$\frac{d\,(uv)}{dx} = v\,\frac{du}{dx} + u\,\frac{dv}{dx}.$$

$$\frac{d\left(\dfrac{u}{v}\right)}{dx} = \frac{v\,\dfrac{du}{dx} - u\,\dfrac{dv}{dx}}{v^2}.$$

$$\frac{d\,f(u)}{dx} = \frac{d\,f(u)}{du} \cdot \frac{du}{dx}.$$

$$\frac{d^2 f(u)}{dx^2} = \frac{df}{du} \cdot \frac{d^2 u}{dx^2} + \frac{d^2 f}{du^2} \cdot \frac{du^2}{dx^2}.$$

$$\frac{dx^n}{dx} = nx^{n-1}.$$

$$\frac{de^x}{dx} = e^x.$$

$$\frac{da^u}{dx} = a^u \cdot \frac{du}{dx} \cdot \log_e a.$$

$$\frac{dx^x}{dx} = x^x(1 + \log_e x).$$

$$\frac{d\,(\log_a x)}{dx} = \frac{1}{x \cdot \log_e a} = \frac{\log_a e}{x}.$$

$$\frac{d\,\sin x}{dx} = \cos x.$$

$$\frac{d\,\cos x}{dx} = -\sin x.$$

$$\frac{d\,\tan x}{dx} = \sec^2 x.$$

$$\frac{d\,\operatorname{ctn} x}{dx} = -\csc^2 x.$$

$$\frac{d\,\sec x}{dx} = \tan x \cdot \sec x.$$

$$\frac{d\,\csc x}{dx} = -\operatorname{ctn} x \cdot \csc x.$$

$$\frac{d\,\sin^{-1} x}{dx} = \frac{1}{\sqrt{1 - x^2}}.$$

$$\frac{d \cos^{-1}x}{dx} = \frac{-1}{\sqrt{1-x^2}}.$$

$$\frac{d \csch x}{dx} = - \csch x \cdot \ctnh x.$$

$$\frac{d \tan^{-1}x}{dx} = \frac{1}{1+x^2}.$$

$$\frac{d \sinh^{-1}x}{dx} = \frac{1}{\sqrt{x^2+1}}.$$

$$\frac{d \ctn^{-1}x}{dx} = -\frac{1}{1+x^2}.$$

$$\frac{d \cosh^{-1}x}{dx} = \frac{1}{\sqrt{x^2-1}}.$$

$$\frac{d \sec^{-1}x}{dx} = \frac{1}{x\sqrt{x^2-1}}.$$

$$\frac{d \tanh^{-1}x}{dx} = \frac{1}{1-x^2}.$$

$$\frac{d \csc^{-1}x}{dx} = -\frac{1}{x\sqrt{x^2-1}}.$$

$$\frac{d \ctnh^{-1}x}{dx} = \frac{1}{1-x^2}.$$

$$\frac{d \sinh x}{dx} = \cosh x.$$

$$\frac{d \sech^{-1}x}{dx} = \frac{-1}{x\sqrt{1-x^2}}.$$

$$\frac{d \cosh x}{dx} = \sinh x.$$

$$\frac{d \csch^{-1}x}{dx} = \frac{-1}{x\sqrt{x^2+1}}.$$

$$\frac{d \tanh x}{dx} = \sech^2 x.$$

$$\frac{d}{db}\int_a^b f(x)\,dx = f(b).$$

$$\frac{d \ctnh x}{dx} = - \csch^2 x.$$

$$\frac{d}{da}\int_a^b f(x)\,dx = - f(a).$$

$$\frac{d \sech x}{dx} = - \sech x \cdot \tanh x.$$

SERIES AND PRODUCTS

[The expression in brackets attached to an infinite series shows values of the variable which lie within the interval of convergence. If a series is convergent for all finite values of x, the expression $[x^2 < \infty]$ is used.]

$$(a+b)^n = a^n + na^{n-1}b$$
$$+ \frac{n(n-1)}{2!}a^{n-2}b^2 + \cdots + \frac{n!\,a^{n-k}b^k}{(n-k)!\,k!} + \cdots. \quad [b^2 < a^2.]$$

$$(a-bx)^{-1} = \frac{1}{a}\left[1 + \frac{bx}{a} + \frac{b^2x^2}{a^2} + \frac{b^3x^3}{a^3} + \cdots\right]. \quad [b^2x^2 < a^2.]$$

$$(1 \pm x)^n = 1 \pm nx + \frac{n(n-1)}{2!}x^2$$
$$\pm \frac{n(n-1)(n-2)x^3}{3!} + \cdots + \frac{(\pm 1)^k n!\,x^k}{(n-k)!\,k!} + \cdots.$$
$$[x^2 < 1.]$$

$$(1 \pm x)^{-n} = 1 \mp nx + \frac{n(n+1)}{2!}x^2$$

$$\mp \frac{n(n+1)(n+2)x^3}{3!} + \cdots (\mp)^k \frac{(n+k-1)!\, x^k}{(n-1)!\, k!} + \cdots$$
$$[x^2 < 1.]$$

$$(1 \pm x)^{\frac{1}{2}} = 1 \pm \tfrac{1}{2}x - \frac{1 \cdot 1}{2 \cdot 4}x^2 \pm \frac{1 \cdot 1 \cdot 3}{2 \cdot 4 \cdot 6}x^3 \cdot$$

$$- \frac{1 \cdot 1 \cdot 3 \cdot 5}{2 \cdot 4 \cdot 6 \cdot 8}x^4 \pm \cdots. \qquad [x^2 < 1.]$$

$$(1 \pm x)^{-\frac{1}{2}} = 1 \mp \tfrac{1}{2}x + \frac{1 \cdot 3}{2 \cdot 4}x^2 \mp \frac{1 \cdot 3 \cdot 5}{2 \cdot 4 \cdot 6}x^3$$

$$+ \frac{1 \cdot 3 \cdot 5 \cdot 7}{2 \cdot 4 \cdot 6 \cdot 8}x^4 \mp \cdots. \qquad [x^2 < 1.]$$

$$(1 \pm x)^{\frac{1}{3}} = 1 \pm \tfrac{1}{3}x - \frac{1 \cdot 2}{3 \cdot 6}x^2 \pm \frac{1 \cdot 2 \cdot 5}{3 \cdot 6 \cdot 9}x^3$$

$$- \frac{1 \cdot 2 \cdot 5 \cdot 8}{3 \cdot 6 \cdot 9 \cdot 12}x^4 \pm \cdots. \qquad [x^2 < 1.]$$

SERIES

$$(1 \pm x)^{-\frac{1}{3}} = 1 \mp \tfrac{1}{3}x + \frac{1 \cdot 4}{3 \cdot 6}x^2 \mp \frac{1 \cdot 4 \cdot 7}{3 \cdot 6 \cdot 9}x^3$$

$$+ \frac{1 \cdot 4 \cdot 7 \cdot 10}{3 \cdot 6 \cdot 9 \cdot 12}x^4 \mp \cdots. \qquad [x^2 < 1.]$$

$$(1 \pm x^2)^{\frac{1}{2}} = 1 \pm \tfrac{1}{2}x^2 - \frac{x^4}{2 \cdot 4} \pm \frac{1 \cdot 3\, x^6}{2 \cdot 4 \cdot 6} - \frac{1 \cdot 3 \cdot 5\, x^8}{2 \cdot 4 \cdot 6 \cdot 8} \pm \cdots.$$
$$[x^2 < 1.]$$

$$(1 \pm x^2)^{-\frac{1}{2}} = 1 \mp \tfrac{1}{2}x^2 + \frac{1 \cdot 3}{2 \cdot 4}x^4 \mp \frac{1 \cdot 3 \cdot 5}{2 \cdot 4 \cdot 6}x^6 + \cdots.$$
$$[x^2 < 1.]$$

$$(1 \pm x)^{-1} = 1 \mp x + x^2 \mp x^3 + x^4 \mp x^5 + \cdots. \qquad [x^2 < 1.]$$

$$(1 \pm x)^{\frac{3}{2}} = 1 \pm \tfrac{3}{2}x + \frac{3 \cdot 1}{2 \cdot 4}x^2 \mp \frac{3 \cdot 1 \cdot 1}{2 \cdot 4 \cdot 6}x^3$$

$$+ \frac{3 \cdot 1 \cdot 1 \cdot 3}{2 \cdot 4 \cdot 6 \cdot 8}x^4 \mp \frac{3 \cdot 1 \cdot 1 \cdot 3 \cdot 5}{2 \cdot 4 \cdot 6 \cdot 8 \cdot 10}x^5 + \cdots. \qquad [x^2 < 1.]$$

$$(1 \pm x)^{-\frac{3}{2}} = 1 \mp \tfrac{3}{2}x + \frac{3 \cdot 5}{2 \cdot 4}x^2 \mp \frac{3 \cdot 5 \cdot 7}{2 \cdot 4 \cdot 6}x^3 + \cdots. \qquad [x^2 < 1.]$$

$$(1 \pm x)^{-2} = 1 \mp 2\,x + 3\,x^2 \mp 4\,x^3 + 5\,x^4 \mp 6\,x^5 + \cdots.$$
$$[x^2 < 1.]$$

$$e^x = 1 + x + \frac{x^2}{2!} + \frac{x^3}{3!} + \cdots. \qquad [x^2 < \infty.]$$

$$a^x = 1 + x \log a + \frac{(x \log a)^2}{2!} + \frac{(x \log a)^3}{3!} + \cdots. \ [x^2 < \infty.]$$

$$\tfrac{1}{2}(e^x + e^{-x}) = 1 + \frac{x^2}{2!} + \frac{x^4}{4!} + \frac{x^6}{6!} + \cdots. \qquad [x^2 < \infty.]$$

$$\tfrac{1}{2}(e^x - e^{-x}) = x + \frac{x^3}{3!} + \frac{x^5}{5!} + \frac{x^7}{7!} + \cdots. \qquad [x^2 < \infty.]$$

$$e^{-x^2} = 1 - x^2 + \frac{x^4}{2!} - \frac{x^6}{3!} + \frac{x^8}{4!} - \cdots. \qquad [x^2 < \infty.]$$

$$\log x = 2\left[\frac{x-1}{x+1} + \tfrac{1}{3}\left(\frac{x-1}{x+1}\right)^3 + \tfrac{1}{5}\left(\frac{x-1}{x+1}\right)^5 + \cdots\right].$$
$$[x > 0.]$$

$$\log(1+x) = x - \tfrac{1}{2}x^2 + \tfrac{1}{3}x^3 - \tfrac{1}{4}x^4 + \cdots. \qquad [x^2 < 1.]$$

$$\log\left(\frac{1+x}{1-x}\right) = 2\left[x + \tfrac{1}{3}x^3 + \tfrac{1}{5}x^5 + \tfrac{1}{7}x^7 + \cdots\right]. \quad [x^2 < 1.]$$

$$\log\left(\frac{x+1}{x-1}\right) = 2\left[\frac{1}{x} + \tfrac{1}{3}\left(\frac{1}{x}\right)^3 + \tfrac{1}{5}\left(\frac{1}{x}\right)^5 + \cdots\right]. \ [x^2 > 1.]$$

$$\log(x + \sqrt{1+x^2}) = x - \frac{1}{6}\frac{x^3}{1} + \frac{1 \cdot 3}{2 \cdot 4 \cdot 5}x^5 - \frac{1 \cdot 3 \cdot 5}{2 \cdot 4 \cdot 6 \cdot 7}x^7 + \cdots.$$
$$[x^2 < 1.]$$

Series for denary and other logarithms can be obtained from the foregoing developments by aid of the equations,

$$\log_a x = \log_e x \cdot \log_a e, \ \log_e x = \log_a x \cdot \log_e a,$$
$$\log_e(-z) = (2\,n + 1)\,\pi i + \log_e z.$$

$$\sin x = x - \frac{x^3}{3!} + \frac{x^5}{5!} - \frac{x^7}{7!} + \cdots. \qquad [x^2 < \infty.]$$

$$\cos x = 1 - \frac{x^2}{2!} + \frac{x^4}{4!} - \frac{x^6}{6!} + \cdots = 1 - \text{versin}\, x. \ [x^2 < \infty.]$$

$$\tanh^{-1} x = x + \frac{x^3}{3} + \frac{x^5}{5} + \frac{x^7}{7} + \cdots. \qquad [x^2 < 1.]$$

$$\operatorname{ctnh}^{-1} x = \frac{1}{x} + \frac{1}{3\,x^3} + \frac{1}{5\,x^5} + \cdots. \qquad [x^2 > 1.]$$

$$\operatorname{csch}^{-1} x = \frac{1}{x} - \frac{1}{2 \cdot 3 \cdot x^3} + \frac{1 \cdot 3}{2 \cdot 4 \cdot 5 \cdot x^5} - \frac{1 \cdot 3 \cdot 5}{2 \cdot 4 \cdot 6 \cdot 7 \cdot x^7} + \cdots.$$
$$[x^2 > 1.]$$

$$\int_0^x e^{-x^2}\, dx = x - \tfrac{1}{3} x^3 + \frac{x^5}{5 \cdot 2!} - \frac{x^7}{7 \cdot 3!} + \cdots. \qquad [x^2 < \infty.]$$

$$\int_0^x \cos\,(x^2)\, dx = x - \frac{x^5}{5 \cdot 2!} + \frac{x^9}{9 \cdot 4!} - \frac{x^{13}}{13 \cdot 6!} + \cdots. \quad [x^2 < \infty.]$$

$$\int_0^1 \frac{x^{a-1}\, dx}{1 + x^b} = \frac{1}{a} - \frac{1}{a + b} + \frac{1}{a + 2\,b} - \frac{1}{a + 3\,b} + \cdots.$$

$$f(x + h) = f(x) + h \cdot f'(x + \theta h).$$

$$f(x + h) = f(x) + h \cdot f'(x) + \frac{h^2}{2!} f''(x)$$
$$+ \cdots + \frac{h^n}{n!} \cdot f^n(x + \theta h).$$

$$f(x + h) = f(x) + h \cdot f'(x) + \frac{h^2}{2!} f''(x)$$
$$+ \cdots + \frac{h^n}{(n - 1)!} \cdot (1 - \theta)^{n-1} \cdot f^n(x + \theta h).$$

$$f(x + h,\, y + k) = f(x,\, y) + hf'_x(x + \theta h,\, y + \theta k)$$
$$+ kf'_y(x + \theta h,\, y + \theta k).$$

$$\sin^{-1} x = x + \frac{x^3}{6} + \frac{1 \cdot 3}{2 \cdot 4} \cdot \frac{x^5}{5} + \frac{1 \cdot 3 \cdot 5}{2 \cdot 4 \cdot 6} \cdot \frac{x^7}{7}$$
$$+ \cdots = \tfrac{1}{2} \pi - \cos^{-1} x. \qquad [x^2 < 1.]$$

$$\tan^{-1} x = x - \tfrac{1}{3} x^3 + \tfrac{1}{5} x^5 - \tfrac{1}{7} x^7 + \cdots = \tfrac{1}{2} \pi - \operatorname{ctn}^{-1} x.$$
$$[x^2 < 1.]$$

$$\tan^{-1} x = \frac{\pi}{2} - \frac{1}{x} + \frac{1}{3\,x^3} - \frac{1}{5\,x^5} + \cdots. \qquad [x^2 > 1.]$$

$$\sec^{-1} x = \frac{\pi}{2} - \frac{1}{x} - \frac{1}{6\,x^3} - \frac{1 \cdot 3}{2 \cdot 4 \cdot 5\,x^5} - \frac{1 \cdot 3 \cdot 5}{2 \cdot 4 \cdot 6 \cdot 7\,x^7} - \cdots$$
$$= \tfrac{1}{2} \pi - \csc^{-1} x. \qquad [x^2 > 1.]$$

$$e^{\sin x} = 1 + x + \frac{x^2}{2!} - \frac{3\,x^4}{4!} - \frac{8\,x^5}{5!} - \frac{3\,x^6}{6!} + \frac{56\,x^7}{7!} + \cdots.$$

$$[x^2 < \infty.]$$

$$e^{\cos x} = e\left(1 - \frac{x^2}{2!} + \frac{4\,x^4}{4!} - \frac{31\,x^6}{6!} + \cdots\right). \qquad [x^2 < \infty.]$$

$$e^{\tan x} = 1 + x + \frac{x^2}{2!} + \frac{3\,x^3}{3!} + \frac{9\,x^4}{4!} + \frac{37\,x^5}{5!} + \cdots. \ [x^2 < \tfrac{1}{4}\pi^2.]$$

$$e^{\sin^{-1}x} = 1 + x + \frac{x^2}{2!} + \frac{2\,x^3}{3!} + \frac{5\,x^4}{4!} + \cdots. \qquad [x^2 < 1.]$$

$$e^{\tan^{-1}x} = 1 + x + \frac{x^2}{2} - \frac{x^3}{6} - \frac{7\,x^4}{24} - \cdots. \qquad [x^2 < 1.]$$

$$\sinh x = x + \frac{x^3}{3!} + \frac{x^5}{5!} + \frac{x^7}{7!} + \cdots. \qquad [x^2 < \infty.]$$

$$\cosh x = 1 + \frac{x^2}{2!} + \frac{x^4}{4!} + \frac{x^6}{6!} + \frac{x^8}{8!} + \cdots. \qquad [x^2 < \infty.]$$

A series of numbers, B_1, B_2, $B_3 \cdots$, of odd and even orders, which appear in the developments of many functions, may be computed by means of the equations,

$$B_{2n} - \frac{2\,n\,(2\,n-1)}{2!}\,B_{2n-2}$$

$$+ \frac{2\,n\,(2\,n-1)\,(2\,n-2)\,(2\,n-3)}{4!}\,B_{2n-4} - \cdots(-1)^n \quad = 0.$$

$$\frac{2^{2n}\,(2^{2n}-1)}{2\,n}\,B_{2n-1} = (2\,n-1)\,B_{2n-2}$$

$$- \frac{(2\,n-1)\,(2\,n-2)\,(2\,n-3)}{3!}\,B_{2n-4} + \cdots(-1)^{n-1} = 0.$$

$$\frac{x}{e^x-1} = 1 - \frac{x}{2} + \frac{B_1 x^2}{2!} - \frac{B_3 x^4}{4!} + \frac{B_5 x^6}{6!} - \frac{B_7 x^8}{8!} + \cdots.$$

$$[x < 2\,\pi.]$$

$$\log x = (x-1) - \tfrac{1}{2}(x-1)^2 + \tfrac{1}{3}(x-1)^3 - \cdots.$$

$$[2 > x > 0.]$$

$$\log x = \frac{x-1}{x} + \tfrac{1}{2}\left(\frac{x-1}{x}\right)^2 + \tfrac{1}{3}\left(\frac{x-1}{x}\right)^3 + \cdots.$$

$$[x > \tfrac{1}{2}.]$$

$$\operatorname{ctnh} x = \frac{1}{x}\left(1 + \Sigma\left[(-1)^{n-1} 2^{2n} B_{2n-1} x^{2n}/(2\,n)!\right]\right).$$

$$[x^2 < \pi^2.]$$

$$\operatorname{sech} x = 1 + \Sigma\left[(-1)^n B_{2n} x^{2n}/(2\,n)!\right]. \qquad [x^2 < \tfrac{1}{4}\pi^2.]$$

$$\operatorname{csch} x = \frac{1}{x} - (2-1)\,2\,B_1\frac{x}{2!} + (2^3-1)\,2\,B_3\frac{x^3}{4!} - \cdots$$

$$= \frac{1}{x}\left(1 + 2\,\Sigma\left[(-1)^n(2^{2n-1}-1)\,B_{2n-1} x^{2n}/(2\,n)!\right]\right).$$

$$[x^2 < \pi^2.]$$

$$\sinh^{-1} x = x - \tfrac{1}{6}x^3 + \frac{1\cdot 3\cdot x^5}{2\cdot 4\cdot 5} - \frac{1\cdot 3\cdot 5\cdot x^7}{2\cdot 4\cdot 6\cdot 7} + \cdots. \;\; [x^2 < 1.]$$

$$\log\sin x = \log x - \tfrac{1}{6}x^2 - \tfrac{1}{180}x^4 - \tfrac{1}{2835}x^6$$

$$- \cdots - \frac{2^{2n-1} B_{2n-1} x^{2n}}{n\,(2\,n)!} - \cdots. \qquad [x^2 < \pi^2.]$$

$$\log\cos x = -\tfrac{1}{2}x^2 - \tfrac{1}{12}x^4 - \tfrac{1}{45}x^6 - \tfrac{17}{2520}x^8$$

$$- \cdots - \frac{2^{2n-1}(2^{2n}-1)\,B_{2n-1} x^{2n}}{n\,(2\,n)!} - \cdots. \quad [x^2 < \tfrac{1}{4}\pi^2.]$$

$$\log\tan x = \log x + \tfrac{1}{3}x^2 + \tfrac{7}{90}x^4 + \tfrac{62}{2835}x^6$$

$$+ \cdots + \frac{(2^{2n-1}-1)\,2^{2n} B_{2n-1} x^{2n}}{n\,(2\,n)!} + \cdots. \quad [x^2 < \tfrac{1}{4}\pi^2.]$$

Whence $B_1 = \tfrac{1}{6}$, $B_2 = 1$, $B_3 = \tfrac{1}{30}$, $B_4 = 5$, $B_5 = \tfrac{1}{42}$, $B_6 = 61$, $B_7 = \tfrac{1}{30}$, $B_8 = 1385$, $B_9 = \tfrac{5}{66}$, $B_{10} = 50521$, $B_{11} = \tfrac{691}{2730}$, $B_{12} = 2702765$, $B_{13} = \tfrac{7}{6}$, etc. The B's of odd orders are called Bernoulli's Numbers; those of even orders, Euler's Numbers. What are here denoted by B_{2n-1} and B_{2n} are sometimes represented by B_n and E_n, respectively,

$$\frac{B_{2n-1}}{(2\,n)!} = \frac{2}{(2^{2n}-1)\,\pi^{2n}}\left[1 + \frac{1}{3^{2n}} + \frac{1}{5^{2n}} + \frac{1}{7^{2n}} + \cdots\right],$$

$$\frac{B_{2n}}{(2\,n)!} = \frac{2^{2n+2}}{\pi^{2n+1}}\left[1 - \frac{1}{3^{2n+1}} + \frac{1}{5^{2n+1}} - \frac{1}{7^{2n+1}} + \cdots\right].$$

$$\tan x = x + \frac{x^3}{3} + \frac{2\,x^5}{15} + \frac{17\,x^7}{315} + \frac{62\,x^9}{2835}$$

$$+ \cdots + \frac{2^{2n}(2^{2n}-1)\,B_{2n-1}\,x^{2n-1}}{(2\,n)!} + \cdots. \qquad [x^2 < \tfrac{1}{4}\,\pi^2.]$$

$$\operatorname{ctn} x = \frac{1}{x} - \frac{x}{3} - \frac{x^3}{45} - \frac{2\,x^5}{945} - \frac{x^7}{4725}$$

$$- \cdots - \frac{B_{2n-1}(2\,x)^{2n}}{x\,(2\,n)!} - \cdots. \qquad [x^2 < \pi^2.]$$

$$\sec x = 1 + \frac{x^2}{2!} + \frac{5\,x^4}{4!} + \frac{61\,x^6}{6!} + \cdots + \frac{B_{2n}x^{2n}}{(2\,n)!} + \cdots \left[x^2 < \frac{\pi^2}{4}.\right]$$

$$\csc x = \frac{1}{x} + \frac{x}{3!} + \frac{7\,x^3}{3\cdot 5!} + \frac{31\,x^5}{3\cdot 7!}$$

$$+ \cdots + \frac{2\,(2^{2n+1}-1)}{(2\,n+2)!}\,B_{2n+1}x^{2n+1} + \cdots. \qquad [x^2 < \pi^2.]$$

$$\tanh x = (2^2 - 1)\,2^2\,B_1\frac{x}{2!} - (2^4 - 1)\,2^4\,B_3\frac{x^3}{4!} + \cdots$$

$$= \Sigma\,[(-1)^{n-1}\,2^{2n}(2^{2n}-1)\,B_{2n-1}\,x^{2n-1}/(2\,n)!].$$
$$[x^2 < \tfrac{1}{4}\,\pi^2.]$$

$$\log \sin \tfrac{1}{2}x = -\log 2 - \cos x - \tfrac{1}{2}\cos 2\,x - \tfrac{1}{3}\cos 3\,x - \cdots.$$
$$[0 < x < \tfrac{1}{2}\,\pi.]$$

$$\log \cos \tfrac{1}{2}x = -\log 2 + \cos x - \tfrac{1}{2}\cos 2\,x + \tfrac{1}{3}\cos 3\,x - \cdots.$$
$$[0 < x < \tfrac{1}{2}\,\pi.]$$

$$f(x) = \tfrac{1}{2}\,b_0 + b_1 \cos \frac{\pi x}{c} + b_2 \cos \frac{2\,\pi x}{c} + \cdots$$

$$+ a_1 \sin \frac{\pi x}{c} + a_2 \sin \frac{2\,\pi x}{c} + \cdots, \; [-c < x < c.]$$

$$\text{where } b_m = \frac{1}{c}\int_{-c}^{+c} f(a) \cos \frac{m\pi a}{c}\,da,$$

$$a_m = \frac{1}{c}\int_{-c}^{+c} f(a) \sin \frac{m\pi a}{c}\,da.$$

TABLE OF INTEGRALS

Fundamental Forms

$$\int a\,dx = ax.$$

$$\int af(x)\,dx = a\int f(x)\,dx.$$

$$\int \frac{dx}{x} = \log x. \quad [\log x = \log(-x) + (2k+1)\pi i.]$$

$$\int x^m dx = \frac{x^{m+1}}{m+1}, \text{ when } m \text{ is different from } -1.$$

$$\int e^x dx = e^x.$$

$$\int a^x \log a\,dx = a^x.$$

$$\int \frac{dx}{1+x^2} = \tan^{-1}x, \text{ or } -\operatorname{ctn}^{-1}x.$$

$$\int \frac{dx}{\sqrt{1-x^2}} = \sin^{-1}x, \text{ or } -\cos^{-1}x$$

$$\int \frac{dx}{x\sqrt{x^2-1}} = \sec^{-1}x, \text{ or } -\csc^{-1}x.$$

$$\int \frac{dx}{\sqrt{2x-x^2}} = \operatorname{versin}^{-1}x, \text{ or } -\operatorname{coversin}^{-1}x.$$

$$\int \cos x\,dx = \sin x, \text{ or } -\operatorname{coversin} x.$$

$$\int \sin x\,dx = -\cos x, \text{ or } \operatorname{versin} x.$$

$$\int \operatorname{ctn} x\,dx = \log \sin x.$$

$$\int \tan x\,dx = -\log \cos x.$$

$$\int \tan x \sec x\,dx = \sec x.$$

$$\int \sec^2 x \, dx = \tan x.$$

$$\int \csc^2 x \, dx = -\ctn x.$$

In the following formulas, u, v, w, and y represent any functions of x:

$$\int (u + v + w + \text{etc.}) \, dx = \int u \, dx + \int v \, dx + \int w \, dx + \text{etc.}$$

$$\int u \, dv = uv - \int v \, du.$$

$$\int u \frac{dv}{dx} \, dx = uv - \int v \frac{du}{dx} \, dx.$$

$$\int f(y) \, dx = \int \frac{f(y) \, dy}{\frac{dy}{dx}}$$

Rational Algebraic Functions

EXPRESSIONS INVOLVING $(a + bx)$.

The substitution of y or z for x, where $y \equiv a + bx$, $z \equiv (a + bx)/x$, gives

$$\int (a + bx)^m \, dx = \frac{1}{b} \int y^m \, dy.$$

$$\int x (a + bx)^m \, dx = \frac{1}{b^2} \int y^m (y - a) \, dy.$$

$$\int x^n (a + bx)^m \, dx = \frac{1}{b^{n+1}} \int y^m (y - a)^n \, dy.$$

$$\int \frac{x^n \, dx}{(a + bx)^m} = \frac{1}{b^{n+1}} \int \frac{(y - a)^n \, dy}{y^m}.$$

$$\int \frac{dx}{x^n (a + bx)^m} = -\frac{1}{a^{m+n-1}} \int \frac{(z - b)^{m+n-2} \, dz}{z^m}.$$

Whence

$$\int \frac{dx}{a + bx} = \frac{1}{b} \log (a + bx).$$

$$\int \frac{dx}{(a+bx)^2} = -\frac{1}{b(a+bx)}.$$

$$\int \frac{dx}{(a+bx)^3} = -\frac{1}{2b(a+bx)^2}.$$

$$\int \frac{x\,dx}{a+bx} = \frac{1}{b^2}[a+bx - a\log(a+bx)].$$

$$\int \frac{x\,dx}{(a+bx)^2} = \frac{1}{b^2}\left[\log(a+bx) + \frac{a}{a+bx}\right].$$

EXPRESSIONS INVOLVING $(a+bx^n)$.

$$\int \frac{dx}{c^2+x^2} = \frac{1}{c}\tan^{-1}\frac{x}{c} = \frac{1}{c}\sin^{-1}\frac{x}{\sqrt{x^2+c^2}}.$$

$$\int \frac{dx}{c^2-x^2} = \frac{1}{2c}\log\frac{c+x}{c-x}, \quad \int \frac{dx}{x^2-c^2} = \frac{1}{2c}\log\frac{x-c}{x+c}.^*$$

$$\int \frac{dx}{a+bx^2} = \frac{1}{\sqrt{ab}}\tan^{-1}\left(x\sqrt{\frac{b}{a}}\right), \text{ or } \frac{1}{\sqrt{-ab}}\cdot\tanh^{-1}\left(x\sqrt{\frac{-b}{a}}\right).$$

$$\int \frac{dx}{a+bx^2} = \frac{1}{2\sqrt{-ab}}\log\frac{\sqrt{a}+x\sqrt{-b}}{\sqrt{a}-x\sqrt{-b}}, \text{ if } a>0,\ b<0.$$

$$\int \frac{dx}{(a+bx^2)^2} = \frac{x}{2a(a+bx^2)} + \frac{1}{2a}\int \frac{dx}{a+bx^2}.$$

$$\int \frac{dx}{(a+bx^2)^{m+1}} = \frac{1}{2ma}\frac{x}{(a+bx^2)^m} + \frac{2m-1}{2ma}\int \frac{dx}{(a+bx^2)^m}.$$

$$\int \frac{x\,dx}{a+bx^2} = \frac{1}{2b}\log\left(x^2+\frac{a}{b}\right).$$

$$\int \frac{x\,dx}{(a+bx^2)^{m+1}} = \frac{1}{2}\int \frac{dz}{(a+bz)^{m+1}}, \text{ where } z=x^2.$$

$$\int \frac{dx}{x(a+bx^2)} = \frac{1}{2a}\log\frac{x^2}{a+bx^2}.$$

$$\int \frac{x^2\,dx}{a+bx^2} = \frac{x}{b} - \frac{a}{b}\int \frac{dx}{a+bx^2}.$$

$$\int \frac{dx}{x^2(a+bx^2)} = -\frac{1}{ax} - \frac{b}{a}\int \frac{dx}{a+bx^2}.$$

$$\int \frac{x^2\,dx}{(a+bx^2)^{m+1}} = \frac{-x}{2\,mb\,(a+bx^2)^m} + \frac{1}{2\,mb}\int \frac{dx}{(a+bx^2)^m}.$$

$$\int \frac{dx}{x^2\,(a+bx^2)^{m+1}} = \frac{1}{a}\int \frac{dx}{x^2(a+bx^2)^m} - \frac{b}{a}\int \frac{dx}{(a+bx^2)^{m+1}}.$$

$$*\int \frac{dx}{c^2-x^2} = \frac{1}{c}\tanh^{-1}\left(\frac{x}{c}\right); \quad \int \frac{dx}{x^2-c^2} = -\frac{1}{c}\operatorname{ctnh}^{-1}\left(\frac{x}{c}\right).$$

Expressions Involving $(a + bx + cx^2)$.

Let $X = a + bx + cx^2$ and $q = 4\,ac - b^2$, then

$$\int \frac{dx}{X} = \frac{2}{\sqrt{q}}\tan^{-1}\frac{2\,cx+b}{\sqrt{q}}, \text{ or } -\frac{2}{\sqrt{-q}}\cdot\tanh^{-1}\frac{2\,cx+b}{\sqrt{-q}}.$$

$$\int \frac{dx}{X} = \frac{1}{\sqrt{-q}}\log\frac{2\,cx+b-\sqrt{-q}}{2\,cx+b+\sqrt{-q}}, \text{ when } q<0.$$

$$\int \frac{dx}{X^2} = \frac{2\,cx+b}{qX} + \frac{2\,c}{q}\int \frac{dx}{X}.$$

$$\int \frac{dx}{X^3} = \frac{2\,cx+b}{q}\left(\frac{1}{2\,X^2} + \frac{3\,c}{qX}\right) + \frac{6\,c^2}{q^2}\int \frac{dx}{X}.$$

$$\int \frac{dx}{X^{n+1}} = \frac{2\,cx+b}{nqX^n} + \frac{2\,(2\,n-1)\,c}{qn}\int \frac{dx}{X^n}.$$

$$\int \frac{x\,dx}{X} = \frac{1}{2\,c}\log X - \frac{b}{2\,c}\int \frac{dx}{X}.$$

$$\int \frac{x\,dx}{X^2} = -\frac{bx+2\,a}{qX} - \frac{b}{q}\int \frac{dx}{X}.$$

$$\int \frac{x\,dx}{X^{n+1}} = -\frac{2\,a+bx}{nqX^n} - \frac{b\,(2\,n-1)}{nq}\int \frac{dx}{X^n}.$$

$$\int \frac{x^2}{X}\,dx = \frac{x}{c} - \frac{b}{2\,c^2}\log X + \frac{b^2-2\,ac}{2\,c^2}\int \frac{dx}{X}.$$

$$\int \frac{x^2}{X^2}\,dx = \frac{(b^2-2\,ac)\,x+ab}{cqX} + \frac{2\,a}{q}\int \frac{dx}{X}.$$

$$\int \frac{x^m\,dx}{X^{n+1}} = -\frac{x^{m-1}}{(2\,n-m+1)\,cX^n} - \frac{n-m+1}{2\,n-m+1}\cdot\frac{b}{c}\int \frac{x^{m-1}\,dx}{X^{n+1}}$$
$$+ \frac{m-1}{2\,n-m+1}\cdot\frac{a}{c}\int \frac{x^{m-2}\,dx}{X^{n+1}}.$$

Irrational Algebraic Functions

EXPRESSIONS INVOLVING $\sqrt{a+bx}$.

The substitution of a new variable of integration, $y = \sqrt{a+bx}$, gives

$$\int \sqrt{a+bx}\, dx = \frac{2}{3\,b}\, \sqrt{(a+bx)^3}.$$

$$\int x\sqrt{a+bx}\, dx = -\frac{2\,(2\,a - 3\,bx)\,\sqrt{(a+bx)^3}}{15\,b^2}.$$

$$\int x^2\sqrt{a+bx}\, dx = \frac{2\,(8\,a^2 - 12\,abx + 15\,b^2x^2)\,\sqrt{(a+bx)^3}}{105\,b^3}$$

$$\int \frac{\sqrt{a+bx}}{x}\, dx = 2\sqrt{a+bx} + a\int \frac{dx}{x\sqrt{a+bx}}.$$

$$\int \frac{dx}{\sqrt{a+bx}} = \frac{2\sqrt{a+bx}}{b}.$$

$$\int \frac{x\,dx}{\sqrt{a+bx}} = -\frac{2\,(2\,a - bx)}{3\,b^2}\,\sqrt{a+bx}.$$

$$\int \frac{x^2\,dx}{\sqrt{a+bx}} = \frac{2\,(8\,a^2 - 4\,abx + 3\,b^2x^2)}{15\,b^3}\,\sqrt{a+bx}.$$

$$\int \frac{dx}{x\sqrt{a+bx}} = \frac{1}{\sqrt{a}}\log\left(\frac{\sqrt{a+bx}-\sqrt{a}}{\sqrt{a+bx}+\sqrt{a}}\right), \text{ for } a>0.$$

$$\int \frac{dx}{x\sqrt{a+bx}} = \frac{2}{\sqrt{-a}}\tan^{-1}\sqrt{\frac{a+bx}{-a}}, \text{ or } \frac{-2}{\sqrt{a}}\cdot\tanh^{-1}\sqrt{\frac{a+bx}{a}}.$$

$$\int \frac{x^m\,dx}{\sqrt{a+bx}} = \frac{2\,x^m\sqrt{a+bx}}{(2\,m+1)\,b} - \frac{2\,ma}{(2\,m+1)\,b}\int \frac{x^{m-1}\,dx}{\sqrt{a+bx}}.$$

$$\int \frac{dx}{x^n\sqrt{a+bx}} = -\frac{\sqrt{a+bx}}{(n-1)\,ax^{n-1}} - \frac{(2\,n-3)\,b}{(2\,n-2)\,a}\int \frac{dx}{x^{n-1}\sqrt{a+bx}}.$$

Expressions Involving $\sqrt{x^2 \pm a^2}$ and $\sqrt{a^2 - x^2}$.

$$\int \sqrt{x^2 \pm a^2}\, dx = \tfrac{1}{2}\left[x\sqrt{x^2 \pm a^2} \pm a^2 \log\left(x + \sqrt{x^2 \pm a^2}\right)\right].^{*}$$

$$\int \sqrt{a^2 - x^2}\, dx = \tfrac{1}{2}\left(x\sqrt{a^2 - x^2} + a^2 \sin^{-1}\frac{x}{a}\right).$$

$$\int \frac{dx}{\sqrt{x^2 \pm a^2}} = \log\left(x + \sqrt{x^2 \pm a^2}\right).^{*}$$

$$\int \frac{dx}{\sqrt{a^2 - x^2}} = \sin^{-1}\frac{x}{a}, \text{ or } -\cos^{-1}\frac{x}{a}.$$

$$\int \frac{dx}{x\sqrt{x^2 - a^2}} = \frac{1}{a}\cos^{-1}\frac{a}{x}, \text{ or } \frac{1}{u}\sec^{-1}\frac{x}{a}.$$

$$\int \frac{dx}{x\sqrt{a^2 \pm x^2}} = -\frac{1}{a}\log\left(\frac{a + \sqrt{a^2 \pm x^2}}{x}\right)$$

$$\int \frac{\sqrt{a^2 \pm x^2}}{x}\, dx = \sqrt{a^2 \pm x^2} - a\log\frac{a + \sqrt{a^2 \pm x^2}}{x}.^{*}$$

$$\int \frac{\sqrt{x^2 - a^2}}{x}\, dx = \sqrt{x^2 - a^2} - a\cos^{-1}\frac{a}{x}.$$

$$\int \frac{x\, dx}{\sqrt{a^2 \pm x^2}} = \pm\sqrt{a^2 \pm x^2}.$$

$$\int \frac{x\, dx}{\sqrt{x^2 - a^2}} = \sqrt{x^2 - a^2}.$$

$$\int x\sqrt{x^2 \pm a^2}\, dx = \tfrac{1}{3}\sqrt{(x^2 \pm a^2)^3}.$$

$$\int x\sqrt{a^2 - x^2}\, dx = -\tfrac{1}{3}\sqrt{(a^2 - x^2)^3}.$$

$${}^{*}\log\left(\frac{x + \sqrt{x^2 + a^2}}{a}\right) = \sinh^{-1}\left(\frac{x}{a}\right); \ \log\left(\frac{x + \sqrt{x^2 - a^2}}{a}\right) = \cosh^{-1}\left(\frac{x}{a}\right);$$
$$\log\left(\frac{a + \sqrt{a^2 - x^2}}{x}\right) = \operatorname{sech}^{-1}\left(\frac{x}{a}\right); \ \log\left(\frac{a + \sqrt{a^2 + x^2}}{x}\right) = \operatorname{csch}^{-1}\left(\frac{x}{a}\right).$$

$$\int \sqrt{(x^2 \pm a^2)^3}\, dx$$

$$= \tfrac{1}{4}\left[x\sqrt{(x^2 \pm a^2)^3} \pm \frac{3\,a^2 x}{2}\sqrt{x^2 \pm a^2} + \frac{3\,a^4}{2}\log\left(x + \sqrt{x^2 \pm a^2}\right)\right].^{*}$$

$$\int \sqrt{(a^2 - x^2)^3}\, dx$$

$$= \tfrac{1}{4}\left[x\sqrt{(a^2 - x^2)^3} + \frac{3\,a^2 x}{2}\sqrt{a^2 - x^2} + \frac{3\,a^4}{2}\sin^{-1}\frac{x}{a}\right].$$

$$\int \frac{dx}{\sqrt{(x^2 \pm a^2)^3}} = \frac{\pm x}{a^2\sqrt{x^2 \pm a^2}}.$$

$$\int \frac{dx}{\sqrt{(a^2 - x^2)^3}} = \frac{x}{a^2\sqrt{a^2 - x^2}}.$$

$$\int \frac{x\, dx}{\sqrt{(x^2 \pm a^2)^3}} = \frac{-1}{\sqrt{x^2 \pm a^2}}.$$

$$\int \frac{x\, dx}{\sqrt{(a^2 - x^2)^3}} = \frac{1}{\sqrt{a^2 - x^2}}.$$

$$\int x\sqrt{(x^2 \pm a^2)^3}\, dx = \tfrac{1}{5}\sqrt{(x^2 \pm a^2)^5}.$$

$$\int x\sqrt{(a^2 - x^2)^3}\, dx = -\tfrac{1}{5}\sqrt{(a^2 - x^2)^5}.$$

$$\int x^2\sqrt{x^2 \pm a^2}\, dx$$

$$= \frac{x}{4}\sqrt{(x^2 \pm a^2)^3} \mp \frac{a^2}{8}\, x\sqrt{x^2 \pm a^2} - \frac{a^4}{8}\log\left(x + \sqrt{x^2 \pm a^2}\right).\text{*}$$

$$\int x^2\sqrt{a^2 - x^2}\, dx$$

$$= -\frac{x}{4}\sqrt{(a^2 - x^2)^3} + \frac{a^2}{8}\left(x\sqrt{a^2 - x^2} + a^2\sin^{-1}\frac{x}{a}\right).$$

$$\log z = \sinh^{-1}\left(\frac{z^2 - 1}{2z}\right) = \cosh^{-1}\left(\frac{z^2 + 1}{2z}\right);\quad \tanh^{-1} z = -i\cdot\tan^{-1}(zi).$$

$$\int \frac{\sqrt{a^2 \pm x^2}\, dx}{x^3} = -\frac{\sqrt{a^2 \pm x^2}}{2x^2} \pm \frac{1}{2}\int \frac{dx}{x\sqrt{a^2 \pm x^2}}.$$

$$\int x^3\sqrt{a^2 \pm x^2}\, dx = \left(\pm\tfrac{1}{5}x^2 - \tfrac{2}{15}a^2\right)\sqrt{(a^2 \pm x^2)^3}.$$

$$\int \frac{dx}{x^3\sqrt{a^2 \pm x^2}} = -\frac{\sqrt{a^2 \pm x^2}}{2\,a^2 x^2} \mp \frac{1}{2\,a^2}\int \frac{dx}{x\sqrt{a^2 \pm x^2}}.$$

$$\int \frac{dx}{x^3\sqrt{x^2-a^2}} = \frac{\sqrt{x^2-a^2}}{2\,a^2 x^2} + \frac{1}{2\,a^3}\sec^{-1}\left(\frac{x}{a}\right).$$

$$\int \frac{x^2\,dx}{\sqrt{x^2\pm a^2}} = \frac{x}{2}\sqrt{x^2\pm a^2} \mp \frac{a^2}{2}\log\left(x+\sqrt{x^2\pm a^2}\right).^*$$

$$\int \frac{x^2\,dx}{\sqrt{a^2-x^2}} = -\frac{x}{2}\sqrt{a^2-x^2} + \frac{a^2}{2}\sin^{-1}\frac{x}{a}.$$

$$\int \frac{dx}{x^2\sqrt{x^2\pm a^2}} = \mp\frac{\sqrt{x^2\pm a^2}}{a^2 x}.$$

$$\int \frac{dx}{x^2\sqrt{a^2-x^2}} = -\frac{\sqrt{a^2-x^2}}{a^2 x}.$$

$$\int \frac{\sqrt{x^2\pm a^2}\,dx}{x^2} = -\frac{\sqrt{x^2\pm a^2}}{x} + \log\left(x+\sqrt{x^2\pm a^2}\right).^*$$

$$\int \frac{\sqrt{a^2-x^2}}{x^2}\,dx = -\frac{\sqrt{a^2-x^2}}{x} - \sin^{-1}\frac{x}{a}.$$

$$\int \frac{x^2\,dx}{\sqrt{(x^2\pm a^2)^3}} = \frac{-x}{\sqrt{x^2\pm a^2}} + \log\left(x+\sqrt{x^2\pm a^2}\right).^*$$

$$\int \frac{x^2\,dx}{\sqrt{(a^2-x^2)^3}} = \frac{x}{\sqrt{a^2-x^2}} - \sin^{-1}\frac{x}{a}.$$

EXPRESSIONS INVOLVING $\sqrt{a+bx+cx^2}$.

Let $X = a + bx + cx^2$, $q = 4ac - b^2$, and $k = \dfrac{4c}{q}$. In order to rationalize the function $f(x, \sqrt{a+bx+cx^2})$ we may put $\sqrt{a+bx+cx^2} = \sqrt{\pm c}\sqrt{A+Bx\pm x^2}$, according as c is positive or negative, and then substitute for x a new variable z, such that

$$z = \sqrt{A+Bx+x^2} \pm x, \text{ if } c > 0.$$

$$z = \frac{\sqrt{A+Bx-x^2}-\sqrt{A}}{x}, \text{ if } c < 0 \text{ and } \frac{a}{-c} > 0.$$

$$z = \sqrt{\frac{x-\beta}{a-x}}, \text{ where } a \text{ and } \beta \text{ are the roots of the equation}$$

$$A+Bx-x^2 = 0, \text{ if } c < 0 \text{ and } \frac{a}{-c} < 0.$$

By rationalization, or by the aid of reduction formulas, may be obtained the values of the following integrals:

$$\int \frac{dx}{\sqrt{X}} = \frac{1}{\sqrt{c}} \log\left(\sqrt{X} + x\sqrt{c} + \frac{b}{2\sqrt{c}}\right), \text{ if } c > 0.$$

$$\int \frac{dx}{\sqrt{X}} = \frac{-1}{\sqrt{-c}} \sin^{-1}\left(\frac{2cx+b}{\sqrt{-q}}\right), \text{ or } \frac{1}{\sqrt{c}} \sinh^{-1}\left(\frac{2cx+b}{\sqrt{q}}\right).$$

$$\int \frac{dx}{x^4 + a^4} = \frac{1}{4a^3\sqrt{2}} \left\{ \log\left(\frac{x^2 + ax\sqrt{2} + a^2}{x^2 - ax\sqrt{2} + a^2}\right) + 2\tan^{-1}\left(\frac{ax\sqrt{2}}{a^2 - x^2}\right) \right\}.$$

$$\int \frac{dx}{x^4 - a^4} = \frac{1}{4a^3} \left\{ \log\left(\frac{x-a}{x+a}\right) - 2\tan^{-1}\left(\frac{x}{a}\right) \right\}.$$

Transcendental Functions

$$\int \sin^2 x\, dx = -\tfrac{1}{2} \cos x \sin x + \tfrac{1}{2} x = \tfrac{1}{2} x - \tfrac{1}{4} \sin 2x.$$

$$\int \sin^3 x\, dx = -\tfrac{1}{3} \cos x (\sin^2 x + 2).$$

$$\int \sin^n x\, dx = -\frac{\sin^{n-1} x \cos x}{n} + \frac{n-1}{n} \int \sin^{n-2} x\, dx.$$

$$\int \cos x\, dx = \sin x.$$

$$\int \cos^2 x\, dx = \tfrac{1}{2} \sin x \cos x + \tfrac{1}{2} x = \tfrac{1}{2} x + \tfrac{1}{4} \sin 2x.$$

$$\int \cos^3 x\, dx = \tfrac{1}{3} \sin x (\cos^2 x + 2).$$

$$\int \cos^n x\, dx = \frac{1}{n} \cos^{n-1} x \sin x + \frac{n-1}{n} \int \cos^{n-2} x\, dx.$$

$$\int \sin x \cos x\, dx = \tfrac{1}{2} \sin^2 x.$$

$$\int \sin^2 x \cos^2 x\, dx = -\tfrac{1}{8}\left(\tfrac{1}{4} \sin 4x - x\right).$$

$$\int \sin x \cos^m x\, dx = -\frac{\cos^{m+1} x}{m+1}.$$

$$\int \sin^m x \, \cos x \, dx = \frac{\sin^{m+1} x}{m+1}.$$

$$\int \cos^m x \, \sin^n x \, dx = \frac{\cos^{m-1} x \, \sin^{n+1} x}{m+n}$$
$$+ \frac{m-1}{m+n} \int \cos^{m-2} x \, \sin^n x \, dx.$$

$$\int \cos^m x \, \sin^n x \, dx = - \frac{\sin^{n-1} x \, \cos^{m+1} x}{m+n}$$
$$+ \frac{n-1}{m+n} \int \cos^m x \, \sin^{n-2} x \, dx.$$

$$\int \frac{dx}{\cos^n x} = \frac{1}{n-1} \cdot \frac{\sin x}{\cos^{n-1} x} + \frac{n-2}{n-1} \int \frac{dx}{\cos^{n-2} x}.$$

$$\int \tan x \, dx = - \log \cos x.$$

$$\int \tan^2 x \, dx = \tan x - x.$$

$$\int \tan^n x \, dx = \frac{\tan^{n-1} x}{n-1} - \int \tan^{n-2} x \, dx.$$

$$\int \operatorname{ctn} x \, dx = \log \sin x.$$

$$\int \operatorname{ctn}^2 x \, dx = - \operatorname{ctn} x - x.$$

$$\int \operatorname{ctn}^n x \, dx = - \frac{\operatorname{ctn}^{n-1} x}{n-1} - \int \operatorname{ctn}^{n-2} x \, dx.$$

$$\int \sec x \, dx = \log \tan \left(\frac{\pi}{4} + \frac{x}{2} \right) = \tfrac{1}{2} \log \frac{1 + \sin x}{1 - \sin x}.$$

$$\int \sec^2 x \, dx = \tan x.$$

$$\int \sec^n x \, dx = \int \frac{dx}{\cos^n x} = \frac{\sin x}{(n-1) \cos^{n-1} x} + \frac{n-2}{n-1} \int \frac{dx}{\cos^{n-2} x}$$
$$= \frac{\sin x}{(n-1) \cos^{n-1} x} + \frac{n-2}{n-1} \int \sec^{n-2} x \, dx.$$

$$\int \csc x \, dx = \log \tan \tfrac{1}{2} x.$$

$$\int \csc^n x \, dx = \int \frac{dx}{\sin^n x}$$

$$= -\frac{\cos x}{(n-1)\sin^{n-1}x} + \frac{n-2}{n-1}\int \frac{dx}{\sin^{n-2}x}$$

$$= -\frac{\cos x}{(n-1)\sin^{n-1}x} + \frac{n-2}{n-1}\int \csc^{n-2}x \, dx.$$

$$\int \frac{dx}{1+\sin x} = -\tan\left(\tfrac{1}{4}\pi - \tfrac{1}{2}x\right). \quad [\text{See } 241.]$$

$$\int \frac{dx}{1-\sin x} = \operatorname{ctn}\left(\tfrac{1}{4}\pi - \tfrac{1}{2}x\right) = \tan\left(\tfrac{1}{4}\pi + \tfrac{1}{2}x\right).$$

$$\int \frac{dx}{1+\cos x} = \tan\tfrac{1}{2}x, \text{ or } \csc x - \operatorname{ctn} x.$$

$$\int \frac{dx}{1-\cos x} = -\operatorname{ctn}\tfrac{1}{2}x, \text{ or } -\operatorname{ctn} x - \csc x.$$

$$\int \frac{dx}{a \pm b \sin x} = \frac{2 \sec \theta}{a} \cdot \tan^{-1}\left(\sec\theta \cdot \tan\tfrac{1}{2}x \pm \tan\theta\right),$$

if $a > b$, and $b = a \sin \theta$.

$$\int \frac{dx}{a \pm b \sin x} = \frac{\pm \sec a}{b} \log \frac{\sin\tfrac{1}{2}(a \pm x)}{\cos\tfrac{1}{2}(x \mp a)},$$

if $b > a$, and $a = b \sin a$. $\quad [\text{See } 241.]$

$$\int \frac{dx}{a + b \cos x} = \frac{-1}{\sqrt{a^2 - b^2}} \cdot \sin^{-1}\left[\frac{b + a\cos x}{a + b\cos x}\right],$$

$$\text{or } \frac{1}{\sqrt{a^2 - b^2}} \sin^{-1}\left[\frac{\sqrt{a^2 - b^2} \cdot \sin x}{a + b\cos x}\right],$$

$$\text{or } \frac{2}{\sqrt{a^2 - b^2}} \tan^{-1}\left[\sqrt{\frac{a-b}{a+b}} \tan\tfrac{1}{2}x\right],$$

$$\text{or } \frac{1}{\sqrt{a^2 - b^2}} \tan^{-1}\left[\frac{\sqrt{a^2 - b^2} \cdot \sin x}{b + a\cos x}\right],$$

$$\int x^m \cos x \, dx = x^m \sin x - m \int x^{m-1} \sin x \, dx.$$

$$\int \frac{\sin x}{x^m} \, dx = -\frac{1}{m-1} \cdot \frac{\sin x}{x^{m-1}} + \frac{1}{m-1} \int \frac{\cos x}{x^{m-1}} \, dx.$$

$$\int \frac{\cos x}{x^m} \, dx = -\frac{1}{m-1} \cdot \frac{\cos x}{x^{m-1}} - \frac{1}{m-1} \int \frac{\sin x}{x^{m-1}} \, dx.$$

$$\int \frac{\sin x}{x} \, dx = x - \frac{x^3}{3 \cdot 3!} + \frac{x^5}{5 \cdot 5!} - \frac{x^7}{7 \cdot 7!} + \frac{x^9}{9 \cdot 9!} \cdots.$$

$$\int \frac{\cos x}{x} \, dx = \log x - \frac{x^2}{2 \cdot 2!} + \frac{x^4}{4 \cdot 4!} - \frac{x^6}{6 \cdot 6!} + \frac{x^8}{8 \cdot 8!} \cdots.$$

$$\int \frac{x \, dx}{\sin x} = x + \frac{x^3}{3 \cdot 3!} + \frac{7 \, x^5}{3 \cdot 5 \cdot 5!} + \frac{31 \, x^7}{3 \cdot 7 \cdot 7!} + \frac{127 \, x^9}{3 \cdot 5 \cdot 9!} + \cdots$$

$$\int \frac{x \, dx}{\cos x} = \frac{x^2}{2} + \frac{x^4}{4 \cdot 2!} + \frac{5 \, x^6}{6 \cdot 4!} + \frac{61 \, x^8}{8 \cdot 6!} + \frac{1385 \, x^{10}}{10 \cdot 8!} + \cdots.$$

$$\int \frac{x \, dx}{\sin^2 x} = -x \operatorname{ctn} x + \log \sin x.$$

$$\int \frac{x \, dx}{\cos^2 x} = x \tan x + \log \cos x.$$

$$n^2 \int x^m \sin^n x \, dx$$

$$= x^{m-1} \sin^{n-1} x \, (m \sin x - nx \cos x)$$

$$+ \, n(n-1) \int x^m \sin^{n-2} x \, dx - m(m-1) \int x^{m-2} \sin^n x \, dx.$$

$$n^2 \int x^m \cos^n x \, dx$$

$$= x^{m-1} \cos^{n-1} x \, (m \cos x + nx \sin x)$$

$$+ \, n(n-1) \int x^m \cos^{n-2} x \, dx - m(m-1) \int x^{m-2} \cos^n x \, dx.$$

$$\int \frac{\sin^n x \, dx}{\cos^m x} = \frac{1}{n-m} \left(-\frac{\sin^{n-1} x}{\cos^{m-1} x} + (n-1) \int \frac{\sin^{n-2} x \, dx}{\cos^m x} \right)$$

$$= \frac{1}{m-1} \left(\frac{\sin^{n+1} x}{\cos^{m-1} x} - (n-m+2) \int \frac{\sin^n x \, dx}{\cos^{m-2} x} \right)$$

$$= \frac{1}{m-1} \left(\frac{\sin^{n-1} x}{\cos^{m-1} x} - (n-1) \int \frac{\sin^{n-2} x \, dx}{\cos^{m-2} x} \right).$$

$$\int \frac{\cos^m x\, dx}{\sin^n x} = -\frac{\cos^{m+1} x}{(n-1)\sin^{n-1} x} - \frac{m-n+2}{n-1}\int \frac{\cos^m x\, dx}{\sin^{n-2} x}$$

$$= \frac{\cos^{m-1} x}{(m-n)\sin^{n-1} x} + \frac{m-1}{m-n}\int \frac{\cos^{m-2} x\, dx}{\sin^n x}$$

$$= -\frac{1}{n-1}\frac{\cos^{m-1} x}{\sin^{n-1} x} - \frac{m-1}{n-1}\int \frac{\cos^{m-2} x\, dx}{\sin^{n-2} x}\cdot$$

$$\int \frac{\sin^m x\, dx}{\cos^n x} = -\int \frac{\cos^m\left(\dfrac{\pi}{2}-x\right) d\left(\dfrac{\pi}{2}-x\right)}{\sin^n\left(\dfrac{\pi}{2}-x\right)}\cdot$$

$$\int \frac{dx}{\sin x \cos x} = \log \tan x.$$

$$\int \frac{dx}{\cos x \sin^2 x} = \log \tan\left(\frac{\pi}{4}+\frac{x}{2}\right) - \csc x.$$

$$\int \frac{dx}{\sin^m x \cos^n x}$$

$$= \frac{1}{n-1}\cdot\frac{1}{\sin^{m-1} x \cdot \cos^{n-1} x} + \frac{m+n-2}{n-1}\int \frac{dx}{\sin^m x \cdot \cos^{n-2} x}$$

$$= -\frac{1}{m-1}\cdot\frac{1}{\sin^{m-1} x \cdot \cos^{n-1} x} + \frac{m+n-2}{m-1}\int \frac{dx}{\sin^{m-2} x \cdot \cos^n x}\cdot$$

$$\int \frac{dx}{\sin^m x} = -\frac{1}{m-1}\cdot\frac{\cos x}{\sin^{m-1} x} + \frac{m-2}{m-1}\int \frac{dx}{\sin^{m-2} x}\cdot$$

$$\int \frac{x\, dx}{1+\sin x} = -x\tan\tfrac{1}{2}\left(\tfrac{1}{2}\pi-x\right) + 2\log\cos\tfrac{1}{2}\left(\tfrac{1}{2}\pi-x\right).$$

$$\int \frac{x\, dx}{1-\sin x} = x\operatorname{ctn}\tfrac{1}{2}\left(\tfrac{1}{2}\pi-x\right) + 2\log\sin\tfrac{1}{2}\left(\tfrac{1}{2}\pi-x\right).$$

$$\int \frac{x\, dx}{1+\cos x} = x\tan\tfrac{1}{2}x + 2\log\cos\tfrac{1}{2}x.$$

$$\int \frac{x\, dx}{1-\cos x} = -x\operatorname{ctn}\tfrac{1}{2}x + 2\log\sin\tfrac{1}{2}x.$$

$$\int \frac{\tan x\, dx}{\sqrt{a+b\tan^2 x}} = \frac{1}{\sqrt{b-a}}\cos^{-1}\left(\frac{\sqrt{b-a}}{\sqrt{b}}\cdot\cos x\right)\cdot$$

$$\int \frac{dx}{a+b\tan^2 x} = \frac{1}{a-b}\left[x - \sqrt{\frac{b}{a}}\cdot\tan^{-1}\left(\sqrt{\frac{b}{a}}\cdot\tan x\right)\right].$$

$$\int \frac{\tan x \, dx}{a + b \tan x}$$
$$= \frac{1}{a^2 + b^2} \left\{ bx - a \log (a + b \tan x) + a \log \sec x \right\}.$$

$$\int x \sin x \, dx = \sin x - x \cos x.$$

$$\int x^2 \sin x \, dx = 2 x \sin x - (x^2 - 2) \cos x.$$

$$\int x^3 \sin x \, dx = (3 x^2 - 6) \sin x - (x^3 - 6 x) \cos x.$$

$$\int x^m \sin x \, dx = - x^m \cos x + m \int x^{m-1} \cos x \, dx.$$

$$\int x \cos x \, dx = \cos x + x \sin x.$$

$$\int x^2 \cos x \, dx = 2 x \cos x + (x^2 - 2) \sin x.$$

$$\int x^3 \cos x \, dx = (3 x^2 - 6) \cos x + (x^3 - 6 x) \sin x.$$

$$\int \frac{x^m \, dx}{\sin^n x}$$
$$= \frac{1}{(n - 1)(n - 2)} \left[- \frac{x^{m-1}(m \sin x + (n - 2) x \cos x)}{\sin^{n-1} x} \right.$$
$$\left. + (n - 2)^2 \int \frac{x^m \, dx}{\sin^{n-2} x} + m (m - 1) \int \frac{x^{m-2} \, dx}{\sin^{n-2} x} \right].$$

$$\int \frac{x^m \, dx}{\cos^n x}$$
$$= \frac{1}{(n - 1)(n - 2)} \left[- \frac{x^{m-1}(m \cos x - (n - 2) x \sin x)}{\cos^{n-1} x} \right.$$
$$\left. + (n - 2)^2 \int \frac{x^m \, dx}{\cos^{n-2} x} + m (m - 1) \int \frac{x^{m-2} \, dx}{\cos^{n-2} x} \right].$$

$$\int \frac{\sin^n x \, dx}{x^m}$$
$$= \frac{1}{(m - 1)(m - 2)} \left[- \frac{\sin^{n-1} x ((m - 2) \sin x + nx \cos x)}{x^{m-1}} \right.$$
$$\left. - n^2 \int \frac{\sin^n x \, dx}{x^{m-2}} + n (n - 1) \int \frac{\sin^{n-2} x \, dx}{x^{m-2}} \right].$$

$$\int \frac{\cos^n x \, dx}{x^m}$$

$$= \frac{1}{(m-1)(m-2)} \left[\frac{\cos^{n-1} x \, (nx \cos x - (m-2) \cos x)}{x^{m-1}} \right.$$

$$\left. - n^2 \int \frac{\cos^n x \, dx}{x^{m-2}} + n(n-1) \int \frac{\cos^{n-2} x \, dx}{x^{m-2}} \right].$$

$$\int x^p \sin^m x \, \cos^n x \, dx$$

$$= \frac{1}{(m+n)^2} \left[x^{p-1} \sin^m x \, \cos^{n-1} x \, (p \cos x + (m+n) x \sin x) \right.$$

$$+ (n-1)(m+n) \int x^p \sin^m x \, \cos^{n-2} x \, dx$$

$$- mp \int x^{p-1} \sin^{m-1} x \, \cos^{n-1} x \, dx$$

$$\left. - p(p-1) \int x^{p-2} \sin^m x \, \cos^n x \, dx \right].$$

$$= \frac{1}{(m+n)^2} \left[x^{p-1} \sin^{m-1} x \, \cos^n x (p \sin x - (m+n) x \cos x) \right.$$

$$+ (m-1)(m+n) \int x^p \sin^{m-2} x \, \cos^n x \, dx$$

$$+ np \int x^{p-1} \sin^{m-1} x \, \cos^{n-1} x \, dx$$

$$\left. - p(p-1) \int x^{p-2} \sin^m x \, \cos^n x \, dx \right].$$

In this book, we use $\sin^{-1} x$ to denote $\dfrac{1}{\sin x}$. For this part of the table we use the classical notation $\sin^{-1} x = arcsin(x)$, $\cos^{-1} x = arccos(x)$, etc.

$$\int \sin^{-1} x \, dx = x \sin^{-1} x + \sqrt{1 - x^2}.$$

$$\int \cos^{-1} x \, dx = x \cos^{-1} x - \sqrt{1 - x^2}.$$

$$\int \tan^{-1} x \, dx = x \tan^{-1} x - \tfrac{1}{2} \log(1 + x^2).$$

$$\int \mathrm{ctn}^{-1} x \, dx = x \, \mathrm{ctn}^{-1} x + \tfrac{1}{2} \log(1 + x^2).$$

$$\int \sin mx \sin nx\, dx = \frac{\sin (m - n)\, x}{2\, (m - n)} - \frac{\sin (m + n)\, x}{2\, (m + n)}.$$

$$\int \sin mx \cos nx\, dx = - \frac{\cos (m - n)\, x}{2\, (m - n)} - \frac{\cos (m + n)\, x}{2\, (m + n)}.$$

$$\int \cos mx \cos nx\, dx = \frac{\sin (m - n)\, x}{2\, (m - n)} + \frac{\sin (m + n)\, x}{2\, (m + n)}.$$

$$\int \sin^2 mx\, dx = \frac{1}{2\, m}\, (mx - \sin mx \cos mx).$$

$$\int \cos^2 mx\, dx = \frac{1}{2\, m}\, (mx + \sin mx \cos mx).$$

$$\int \sin mx \cos mx\, dx = - \frac{1}{4\, m} \cos 2\, mx.$$

$$\int \sin nx \sin^m x\, dx = \frac{1}{m + n} \left[- \cos nx \sin^m x \right.$$
$$\left. + m \int \cos (n - 1)\, x \cdot \sin^{m-1} x\, dx \right]$$

$$\int \sec^{-1} x\, dx = x \sec^{-1} x - \log (x + \sqrt{x^2 - 1}).$$

$$\int \csc^{-1} x\, dx = x \csc^{-1} x + \log (x + \sqrt{x^2 - 1}).$$

$$\int \mathrm{versin}^{-1} x\, dx = (x - 1)\, \mathrm{versin}^{-1} x + \sqrt{2\, x - x^2}.$$

$$\int (\sin^{-1} x)^2\, dx = x\, (\sin^{-1} x)^2 - 2\, x + 2 \sqrt{1 - x^2}\, \sin^{-1} x.$$

$$\int (\cos^{-1} x)^2\, dx = x\, (\cos^{-1} x)^2 - 2\, x - 2 \sqrt{1 - x^2}\, \cos^{-1} x.$$

$$\int x \sin^{-1} x\, dx = \tfrac{1}{4} [(2\, x^2 - 1) \sin^{-1} x + x \sqrt{1 - x^2}].$$

$$\int x \cos^{-1} x\, dx = \tfrac{1}{4} [(2\, x^2 - 1) \cos^{-1} x - x \sqrt{1 - x^2}].$$

$$\int x \tan^{-1} x\, dx = \tfrac{1}{2} [(x^2 + 1) \tan^{-1} x - x].$$

$$\int x \operatorname{ctn}^{-1}x\, dx = \tfrac{1}{2}\big[(x^2+1)\operatorname{ctn}^{-1}x + x\big].$$

$$\int x \sec^{-1}x\, dx = \tfrac{1}{2}\big[x^2 \sec^{-1}x - \sqrt{x^2-1}\big].$$

$$\int x \csc^{-1}x\, dx = \tfrac{1}{2}\big[x^2 \csc^{-1}x + \sqrt{x^2-1}\big].$$

$$\int x^n \sin^{-1}x\, dx = \frac{1}{n+1}\left(x^{n+1}\sin^{-1}x - \int \frac{x^{n+1}\,dx}{\sqrt{1-x^2}}\right).$$

$$\int x^n \cos^{-1}x\, dx = \frac{1}{n+1}\left(x^{n+1}\cos^{-1}x + \int \frac{x^{n+1}\,dx}{\sqrt{1-x^2}}\right).$$

$$\int x^n \tan^{-1}x\, dx = \frac{1}{n+1}\left(x^{n+1}\tan^{-1}x - \int \frac{x^{n+1}\,dx}{1+x^2}\right).$$

$$\int x^n \operatorname{ctn}^{-1}x\, dx = \frac{1}{n+1}\left(x^{n+1}\operatorname{ctn}^{-1}x + \int \frac{x^{n+1}\,dx}{1+x^2}\right).$$

$$\int \frac{\sin^{-1}x\,dx}{x^2} = \log\left(\frac{1-\sqrt{1-x^2}}{x}\right) - \frac{\sin^{-1}x}{x}.$$

$$\int \frac{\tan^{-1}x\,dx}{x^2} = \log x - \tfrac{1}{2}\log(1+x^2) - \frac{\tan^{-1}x}{x}.$$

$$\int e^{ax}\, dx = \frac{e^{ax}}{a}. \qquad \int f(e^{ax})\, dx = \int \frac{f(y)\,dy}{ay},\quad y = e^{ax}.$$

$$\int x\, e^{ax}\, dx = \frac{e^{ax}}{a^2}(ax-1).$$

$$\int x^m\, e^{ax}\, dx = \frac{x^m e^{ax}}{a} - \frac{m}{a}\int x^{m-1}e^{ax}\, dx.$$

$$\int \frac{e^{ax}}{x^m}\, dx = \frac{1}{m-1}\left[-\frac{e^{ax}}{x^{m-1}} + a\int \frac{e^{ax}\,dx}{x^{m-1}}\right].$$

$$\int a^{bx}\, dx = \frac{a^{bx}}{b\log a}. \qquad \int f(a^{bx})\, dx = \int \frac{f(y)\,dy}{b\cdot\log a\cdot y},\quad y = a^{bx}.$$

$$\int x^n a^x\, dx = \frac{a^x x^n}{\log a} - \frac{na^x x^{n-1}}{(\log a)^2} + \frac{n(n-1)a^x x^{n-2}}{(\log a)^3}\cdots$$
$$\pm \frac{n(n-1)(n-2)\cdots 2.1\, a^x}{(\log a)^{n+1}}.$$

$$\int \frac{a^x\, dx}{x^n} = \frac{1}{n-1}\left[-\frac{a^x}{x^{n-1}} - \frac{a^x \cdot \log a}{(n-2)\, x^{n-2}} \right.$$
$$\left. -\frac{a^x \cdot (\log a)^2}{(n-2)\,(n-3)\, x^{n-3}} - \cdots + \frac{(\log a)^{n-1}}{(n-2)\,(n-3)\cdots 2.1} \int \frac{a^x\, dx}{x}\right].$$

$$\int \frac{a^x\, dx}{x} = \log x + x\log a + \frac{(x\log a)^2}{2\cdot 2!} + \frac{(x\log a)^3}{3\cdot 3!} + \cdots.$$

$$\int \frac{\log x\, dx}{(a+bx)^m}$$
$$= \frac{1}{b\,(m-1)}\left[-\frac{\log x}{(a+bx)^{m-1}} + \int \frac{dx}{x\,(a+bx)^{m-1}}\right].$$

$$\int \frac{\log x\, dx}{a+bx} = \frac{1}{b}\log x \cdot \log(a+bx) - \frac{1}{b}\int \frac{\log(a+bx)\, dx}{x}.$$

$$\int (a+bx)\log x\, dx = \frac{(a+bx)^2}{2\,b}\log x - \frac{a^2\log x}{2\,b} - ax - \tfrac{1}{4}bx^2.$$

$$\int \frac{\log x\, dx}{\sqrt{a+bx}}$$
$$= \frac{2}{b}\left[(\log x - 2)\sqrt{a+bx} + \sqrt{a}\,\log(\sqrt{a+bx} + \sqrt{a}) \right.$$
$$\left. - \sqrt{a}\,\log(\sqrt{a+bx} - \sqrt{a})\right],\ \text{if } a > 0$$
$$= \frac{2}{b}\left[(\log x - 2)\sqrt{a+bx} + 2\sqrt{-a}\,\tan^{-1}\sqrt{\frac{a+bx}{-a}}\right],\ \text{if } a < 0.$$

$$\int \sin\log x\, dx = \tfrac{1}{2}x\,[\sin\log x - \cos\log x].$$

$$\int \cos\log x\, dx = \tfrac{1}{2}x\,[\sin\log x + \cos\log x].$$

$$\int \frac{(\log x)^n\, dx}{x} = \frac{(\log x)^{n+1}}{n+1}.$$

$$\int \frac{dx}{\log x} = \log(\log x) + \log x + \frac{(\log x)^2}{2\cdot 2!} + \frac{(\log x)^3}{3\cdot 3!} + \cdots.$$

$$\int \frac{dx}{(\log x)^n} = -\frac{x}{(n-1)\,(\log x)^{n-1}} + \frac{1}{n-1}\int \frac{dx}{(\log x)^{n-1}}.$$

$$\int \frac{x^m\, dx}{(\log x)^n} = -\frac{x^{m+1}}{(n-1)\,(\log x)^{n-1}} + \frac{m+1}{n-1}\int \frac{x^m\, dx}{(\log x)^{n-1}}.$$

$$\int \log x \, dx = x \log x - x.$$

$$\int x^m \log x \, dx = x^{m+1} \left[\frac{\log x}{m+1} - \frac{1}{(m+1)^2} \right].$$

$$\int (\log x)^n \, dx = x (\log x)^n - n \int (\log x)^{n-1} \, dx.$$

$$\int x^m (\log x)^n \, dx = \frac{x^{m+1} (\log x)^n}{m+1} - \frac{n}{m+1} \int x^m (\log x)^{n-1} \, dx.$$

$$\int \frac{dx}{x \log x} = \log (\log x), \text{ and } \int \frac{(n-1) \, dx}{x (\log x)^n} = \frac{-1}{(\log x)^{n-1}}.$$

$$\int \log (a^2 + x^2) \, dx = x \cdot \log (a^2 + x^2) - 2x + 2a \cdot \tan^{-1} \left(\frac{x}{a} \right).$$

$$\int \frac{dx}{1 + e^x} = \log \frac{e^x}{1 + e^x}.$$

$$\int \frac{dx}{a + be^{mx}} = \frac{1}{am} [mx - \log (a + be^{mx})].$$

$$\int \frac{dx}{ae^{mx} + be^{-mx}} = \frac{1}{m \sqrt{ab}} \tan^{-1} \left(e^{mx} \sqrt{\frac{a}{b}} \right).$$

$$\int \frac{dx}{\sqrt{a + be^{mx}}} = \frac{1}{m \sqrt{a}} \{ \log (\sqrt{a + be^{mx}} - \sqrt{a})$$

$$- \log (\sqrt{a + be^{mx}} + \sqrt{a}) \}, \text{ or } \frac{2}{m \sqrt{-a}} \tan^{-1} \frac{\sqrt{a + be^{mx}}}{\sqrt{-a}}.$$

$$\int \frac{xe^x \, dx}{(1 + x)^2} = \frac{e^x}{1 + x}, \quad \int x^n \cdot e^{ax^{n+1}} \, dx = \frac{e^{ax^{n+1}}}{a (n+1)}.$$

$$\int e^{ax} \sin px \, dx = \frac{e^{ax} (a \sin px - p \cos px)}{a^2 + p^2}.$$

$$\int e^{ax} \cos px \, dx = \frac{e^{ax} (a \cos px + p \sin px)}{a^2 + p^2}.$$

$$\int e^{ax} \log x \, dx = \frac{e^{ax} \log x}{a} - \frac{1}{a} \int \frac{e^{ax} \, dx}{x}.$$

$$\int e^{ax} \sin^2 x \, dx = \frac{e^{ax}}{4 + a^2} \left(\sin x (a \sin x - 2 \cos x) + \frac{2}{a} \right).$$

$$\int e^{ax} \cos^2 x \, dx = \frac{e^{ax}}{4 + a^2} \left(\cos x \, (2 \sin x + a \, \cos x) + \frac{2}{a} \right).$$

$$\int e^{ax} \sin^n bx \, dx = \frac{1}{a^2 + n^2 b^2} \bigg((a \sin bx$$

$$- nb \, \cos bx) \, e^{ax} \sin^{n-1} bx + n \, (n-1) \, b^2 \int e^{ax} \sin^{n-2} bx \cdot dx \bigg).$$

$$\int e^{ax} \cos^n bx \, dx = \frac{1}{a^2 + n^2 b^2} \bigg((a \cos bx$$

$$+ nb \, \sin bx) \, e^{ax} \cos^{n-1} bx + n \, (n-1) \, b^2 \int e^{ax} \cos^{n-2} bx \, dx \bigg).$$

$$\int e^{ax} \tan^n x \, dx$$

$$= \frac{e^{ax} \tan^{n-1} x}{n-1} - \frac{a}{n-1} \int e^{ax} \tan^{n-1} x \, dx - \int e^{ax} \tan^{n-2} x \, dx.$$

$$\int e^{ax} \operatorname{ctn}^n x \, dx$$

$$= - \frac{e^{ax} \operatorname{ctn}^{n-1} x}{n-1} + \frac{a}{n-1} \int e^{ax} \operatorname{ctn}^{n-1} x \, dx - \int e^{ax} \operatorname{ctn}^{n-2} x \, dx.$$

$$\int \frac{e^{ax} \, dx}{\sin^n x} = - e^{ax} \frac{a \sin x + (n-2) \cos x}{(n-1)(n-2) \sin^{n-1} x}$$

$$+ \frac{a^2 + (n-2)^2}{(n-1)(n-2)} \int \frac{e^{ax} \, dx}{\sin^{n-2} x}.$$

$$\int \frac{e^{ax} \, dx}{\cos^n x} = - e^{ax} \frac{a \cos x - (n-2) \sin x}{(n-1)(n-2) \cos^{n-1} x}$$

$$+ \frac{a^2 + (n-2)^2}{(n-1)(n-2)} \int \frac{e^{ax} \, dx}{\cos^{n-2} x}.$$

$$\int e^{ax} \sin^m x \, \cos^n x \, dx$$

$$= \frac{1}{(m+n)^2 + a^2} \bigg\{ e^{ax} \sin^{mx} x \, \cos^{n-1} x \, (a \cos x + (m+n) \sin x)$$

$$- ma \int e^{ax} \sin^{m-1} x \, \cos^{m-1} x \, dx$$

$$+ (n-1)(m+n) \int e^{ax} \sin^m x \, \cos^{n-2} x \, dx \bigg\}$$

$$\int \frac{x^m\, dx}{\log x} = \int \frac{e^{-y}}{y}\, dy, \text{ where } y = -(m+1)\log x.$$

MISCELLANEOUS DEFINITE INTEGRALS

$$\int_0^\infty \frac{a\, dx}{a^2 + x^2} = \frac{\pi}{2}, \text{ if } a > 0;\ 0, \text{ if } a = 0;\ -\frac{\pi}{2}, \text{ if } a < 0.$$

$$\int_0^\infty x^{n-1} e^{-x}\, dx = \int_0^1 \left[\log \frac{1}{x}\right]^{n-1} dx \equiv \Gamma(n).$$

$$\Gamma(z+1) = z \cdot \Gamma(z), \text{ if } z > 0.$$

$$\Gamma(y) \cdot \Gamma(1-y) = \frac{\pi}{\sin \pi y}, \text{ if } 1 > y > 0. \quad \Gamma(2) = \Gamma(1) = 1.$$

$$\Gamma(n+1) = n!, \text{ if } n \text{ is an integer.} \qquad \Gamma(z) = \Pi(z-1).$$

$$\Gamma(\tfrac{1}{2}) = \sqrt{\pi}. \qquad Z(y) = D_y[\log \Gamma(y)]. \quad Z(1) = -0.577216.$$

$$\int_0^1 x^{m-1}(1-x)^{n-1}\, dx = \int_0^\infty \frac{x^{m-1}\, dx}{(1+x)^{m+n}} = \frac{\Gamma(m)\,\Gamma(n)}{\Gamma(m+n)}.$$

$$\int_0^{\frac{\pi}{2}} \sin^n x\, dx = \int_0^{\frac{\pi}{2}} \cos^n x\, dx$$

$$= \frac{1 \cdot 3 \cdot 5 \cdots (n-1)}{2 \cdot 4 \cdot 6 \cdots (n)} \cdot \frac{\pi}{2}, \text{ if } n \text{ is an even integer,}$$

$$= \frac{2 \cdot 4 \cdot 6 \cdots (n-1)}{1 \cdot 3 \cdot 5 \cdot 7 \cdots n}, \text{ if } n \text{ is an odd integer,}$$

$$= \tfrac{1}{2}\sqrt{\pi}\,\frac{\Gamma\left(\dfrac{n+1}{2}\right)}{\Gamma\left(\dfrac{n}{2}+1\right)}, \text{ for any value of } n \text{ greater than } -1.$$

$$\int_0^\infty \frac{\sin mx\, dx}{x} = \frac{\pi}{2}, \text{ if } m > 0;\ 0, \text{ if } m = 0;\ -\frac{\pi}{2}, \text{ if } m < 0.$$

$$\int_0^1 \log\left(\frac{1+x}{1-x}\right) \cdot \frac{dx}{x} = \frac{\pi^2}{4}.$$

$$\int_0^1 \frac{\log x\, dx}{\sqrt{1-x^2}} = -\frac{\pi}{2}\log 2.$$

$$\int_0^1 \frac{(x^p - x^q)\, dx}{\log x} = \log \frac{p+1}{q+1}, \text{ if } p+1 > 0,\, q+1 > 0.$$

$$\int_0^1 (\log x)^n \, dx = (-1)^n \cdot n!.$$

$$\int_0^1 \left(\log \frac{1}{x}\right)^{\frac{1}{2}} dx = \frac{\sqrt{\pi}}{2}.$$

$$\int_0^1 \left(\log \frac{1}{x}\right)^n dx = n!.$$

$$\int_0^1 \frac{dx}{\sqrt{\log\left(\dfrac{1}{x}\right)}} = \sqrt{\pi}.$$

$$\int_0^1 x^m \log\left(\frac{1}{x}\right)^n dx = \frac{\Gamma(n+1)}{(m+1)^{n+1}}, \text{ if } m+1 > 0,\, n+1 > 0.$$

$$\int_0^\infty \log\left(\frac{e^x + 1}{e^x - 1}\right) dx = \frac{\pi^2}{4}.$$

$$\int_0^{\frac{\pi}{2}} \log \sin x \, dx = \int_0^{\frac{\pi}{2}} \log \cos x \, dx = -\frac{\pi}{2} \cdot \log 2.$$

$$\int_0^\pi x \cdot \log \sin x \, dx = -\frac{\pi^2}{2} \log 2.$$

$$\int_0^\pi \log(a \pm b \cos x) \, dx = \pi \log\left(\frac{a + \sqrt{a^2 - b^2}}{2}\right). \qquad a \geq b.$$

$$\int_0^\infty \frac{dx}{e^{nx} + e^{-nx}} = \frac{\pi}{4\,n}.$$

$$\int_0^\infty \frac{x \, dx}{e^{nx} - e^{-nx}} = \frac{\pi^2}{8\,n^2}.$$

$$\int_0^{\pi i} \sinh(mx) \cdot \sinh(nx)\, dx = \int_0^{\pi i} \cosh(mx) \cdot \cosh(nx)\, dx$$
$$= 0, \text{ if } m \text{ is different from } n.$$

$$\int_0^{\pi i} \cosh^2(mx)\, dx = -\int_0^{\pi i} \sinh^2(mx)\, dx = \frac{\pi i}{2}.$$

$$\int_{-\pi i}^{+\pi i} \sinh(mx)\, dx = 0.$$

$$\int_0^{\pi i} \cosh{(mx)}\, dx = 0.$$

$$\int_{-\pi i}^{\pi i} \sinh{(mx)} \cosh{(nx)}\, dx = 0.$$

$$\int_0^{\pi i} \sinh{(mx)} \cosh{(mx)}\, dx = 0.$$

$$\int_0^{\infty} e^{-ax} \cos{mx}\, dx = \frac{a}{a^2 + m^2}, \text{ if } a > 0.$$

$$\int_0^{\infty} e^{-ax} \sin{mx}\, dx = \frac{m}{a^2 + m^2}, \text{ if } a > 0.$$

$$\int_0^{\infty} e^{-a^2 x^2} \cos{bx}\, dx = \frac{\sqrt{\pi} \cdot e^{-\frac{b^2}{4a^2}}}{2a}. \qquad\qquad a > 0.$$

$$\int_0^1 \frac{\log x}{1 - x}\, dx = -\frac{\pi^2}{6}.$$

$$\int_0^1 \frac{\log x}{1 + x}\, dx = -\frac{\pi^2}{12}.$$

$$\int_0^1 \frac{\log x}{1 - x^2}\, dx = -\frac{\pi^2}{8}.$$

$$\int_0^{\infty} \frac{\sin x \cdot \cos{mx}\, dx}{x} = 0, \text{ if } m < -1 \text{ or } m > 1;$$

$$\frac{\pi}{4}, \text{ if } m = -1 \text{ or } m = 1; \ \frac{\pi}{2}, \text{ if } -1 < m < 1.$$

$$\int_0^{\infty} \frac{\sin^2 x\, dx}{x^2} = \frac{\pi}{2}.$$

$$\int_0^{\infty} \cos{(x^2)}\, dx = \int_0^{\infty} \sin{(x^2)}\, dx = \tfrac{1}{2}\sqrt{\frac{\pi}{2}}.$$

$$\int_0^{\pi} \sin{kx} \cdot \sin{mx}\, dx = \int_0^{\pi} \cos{kx} \cdot \cos{mx}\, dx = 0,$$

if k is different from m.

$$\int_0^{\pi} \sin^2{mx}\, dx = \int_0^{\pi} \cos^2{mx}\, dx = \frac{\pi}{2}.$$

$$\int_0^{\infty} \frac{\cos{mx}\, dx}{1 + x^2} = \frac{\pi}{2} \cdot e^{-m}. \qquad\qquad m > 0.$$

$$\int_0^{\infty} \frac{\cos x\, dx}{\sqrt{x}} = \int_0^{\infty} \frac{\sin x\, dx}{\sqrt{x}} = \sqrt{\frac{\pi}{2}}.$$

$$\int_0^\infty e^{-a^2x^2}\,dx = \frac{1}{2\,a}\,\sqrt{\pi}\cdot = \frac{1}{2\,a}\,\Gamma\left(\tfrac{1}{2}\right).$$

$$\int_0^\infty x^n e^{-ax}\,dx = \frac{\Gamma\,(n+1)}{a^{n+1}} = \frac{n\,!}{a^{n+1}}\cdot$$

$$\int_0^\infty x^{2\,n} e^{-ax^2}\,dx = \frac{1\cdot3\cdot5\cdots(2\,n-1)}{2^{n+1}\,a^n}\sqrt{\frac{\pi}{a}}\cdot$$

$$\int_0^\infty e^{-x^2-\frac{a^2}{x^2}}\,dx = \frac{e^{-2\,a}\,\sqrt{\pi}}{2}\cdot \qquad\qquad a>0.$$

$$\int_0^\infty e^{-nx}\sqrt{x}\,dx = \frac{1}{2\,n}\sqrt{\frac{\pi}{n}}\cdot$$

$$\int_0^\infty \frac{e^{-nx}}{\sqrt{x}}\,dx = \sqrt{\frac{\pi}{n}}\cdot \qquad\qquad a>0.$$

TRIGONOMETRIC FUNCTIONS

	0°.	30°.	45°.	60°.	90°.	120°.	135°.	150°.	180°.
sin	0	$\frac{1}{2}$	$\frac{1}{2}\sqrt{2}$	$\frac{1}{2}\sqrt{3}$	1	$\frac{1}{2}\sqrt{3}$	$\frac{1}{2}\sqrt{2}$	$\frac{1}{2}$	0
cos	1	$\frac{1}{2}\sqrt{3}$	$\frac{1}{2}\sqrt{2}$	$\frac{1}{2}$	0	$-\frac{1}{2}$	$-\frac{1}{2}\sqrt{2}$	$-\frac{1}{2}\sqrt{3}$	-1
tan	0	$\frac{1}{\sqrt{3}}$	1	$\sqrt{3}$	∞	$-\sqrt{3}$	-1	$-\frac{1}{\sqrt{3}}$	0
ctn	∞	$\sqrt{3}$	1	$\frac{1}{\sqrt{3}}$	0	$-\frac{1}{\sqrt{3}}$	-1	$-\sqrt{3}$	∞
sec	1	$\frac{2}{\sqrt{3}}$	$\sqrt{2}$	2	∞	-2	$-\sqrt{2}$	$-\frac{2}{\sqrt{3}}$	-1
csc	∞	2	$\sqrt{2}$	$\frac{2}{\sqrt{3}}$	1	$\frac{2}{\sqrt{3}}$	$\sqrt{2}$	2	∞

$$\sin\tfrac{1}{2}\,a = \sqrt{\tfrac{1}{2}\,(1-\cos a)}.$$

$$\cos\tfrac{1}{2}\,a = \sqrt{\tfrac{1}{2}\,(1+\cos a)}.$$

$$\tan\tfrac{1}{2}\,a = \sqrt{\frac{1-\cos a}{1+\cos a}} = \frac{1-\cos a}{\sin a} = \frac{\sin a}{1+\cos a}\cdot$$

$$\sin 2\,a = 2\,\sin a\,\cos a.$$

$$\sin 3\,a = 3\,\sin a - 4\,\sin^3 a.$$

$$\sin 4\,a = 8\,\cos^3 a\cdot\sin a - 4\,\cos a\,\sin a.$$

$\sin 5\,a = 5 \sin a - 20 \sin^3 a + 16 \sin^5 a.$

$\sin 6\,a = 32 \cos^5 a \sin a - 32 \cos^3 a \sin a + 6 \cos a \sin a$

$\cos 2\,a = \cos^2 a - \sin^2 a = 1 - 2 \sin^2 a = 2 \cos^2 a - 1.$

$\cos 3\,a = 4 \cos^3 a - 3 \cos a.$

$\cos 4\,a = 8 \cos^4 a - 8 \cos^2 a + 1.$

$\cos 5\,a = 16 \cos^5 a - 20 \cos^3 a + 5 \cos a.$

$\cos 6\,a = 32 \cos^6 a - 48 \cos^4 a + 18 \cos^2 a - 1.$

$\tan 2\,a = \dfrac{2 \tan a}{1 - \tan^2 a}.$

$\operatorname{ctn} 2\,a = \dfrac{\operatorname{ctn}^2 a - 1}{2 \operatorname{ctn} a}.$

$\sin (a \pm \beta) = \sin a \cdot \cos \beta \pm \cos a \cdot \sin \beta.$

$\cos (a \pm \beta) = \cos a \cdot \cos \beta \mp \sin a \cdot \sin \beta.$

$\tan (a \pm \beta) = \dfrac{\tan a \pm \tan \beta}{1 \mp \tan a \cdot \tan \beta}.$

$\operatorname{ctn} (a \pm \beta) = \dfrac{\operatorname{ctn} a \cdot \operatorname{ctn} \beta \mp 1}{\operatorname{ctn} a \pm \operatorname{ctn} \beta}.$

$\sin a \pm \sin \beta = 2 \sin \tfrac{1}{2} (a \pm \beta) \cdot \cos \tfrac{1}{2} (a \mp \beta).$

$\cos a + \cos \beta = 2 \cos \tfrac{1}{2} (a + \beta) \cdot \cos \tfrac{1}{2} (a - \beta).$

$\cos a - \cos \beta = - 2 \sin \tfrac{1}{2} (a + \beta) \cdot \sin \tfrac{1}{2} (a - \beta).$

$\tan a \pm \tan \beta = \dfrac{\sin (a \pm \beta)}{\cos a \cdot \cos \beta}.$

$\operatorname{ctn} a \pm \operatorname{ctn} \beta = \pm \dfrac{\sin (a \pm \beta)}{\sin a \cdot \sin \beta}.$

$\sin \alpha \cos \beta = \tfrac{1}{2} [\sin (\alpha + \beta) + \sin (\alpha - \beta)]$

$\cos \alpha \cos \beta = \tfrac{1}{2} [\cos (\alpha + \beta) + \cos (\alpha - \beta)]$

$\sin \alpha \sin \beta = \tfrac{1}{2} [\cos (\alpha - \beta) - \cos (\alpha + \beta)]$

HYPERBOLIC FUNCTIONS

$$\sinh x = \tfrac{1}{2}(e^x - e^{-x}) = -\sinh(-x) = -i\sin(ix)$$
$$= (\operatorname{csch} x)^{-1} = 2\tanh \tfrac{1}{2}x \div (1 - \tanh^2 \tfrac{1}{2}x).$$

$$\cosh x = \tfrac{1}{2}(e^x + e^{-x}) = \cosh(-x) = \cos(ix) = (\operatorname{sech} x)^{-1}$$
$$= (1 + \tanh^2 \tfrac{1}{2}x) \div (1 - \tanh^2 \tfrac{1}{2}x).$$

$$\tanh x = (e^x - e^{-x}) \div (e^x + e^{-x}) = -\tanh(-x)$$
$$= -i\tan(ix) = (\operatorname{ctnh} x)^{-1} = \sinh x \div \cosh x.$$

$$\cosh xi = \cos x.$$

$$\sinh xi = i\sin x.$$

$$\cosh^2 x - \sinh^2 x = 1.$$

$$1 - \tanh^2 x = \operatorname{sech}^2 x.$$

$$1 - \operatorname{ctnh}^2 x = -\operatorname{csch}^2 x.$$

$$\sinh(x \pm y) = \sinh x \cdot \cosh y \pm \cosh x \cdot \sinh y.$$

$$\cosh(x \pm y) = \cosh x \cdot \cosh y \pm \sinh x \cdot \sinh y.$$

$$\tanh(x \pm y) = (\tanh x \pm \tanh y) \div (1 \pm \tanh x \cdot \tanh y).$$

$$\sinh(2x) = 2\sinh x \cosh x.$$

$$\cosh(2x) = \cosh^2 x + \sinh^2 x = 2\cosh^2 x - 1 = 1 + 2\sinh^2 x.$$

$$\tanh(2x) = 2\tanh x \div (1 + \tanh^2 x).$$

$$\sinh(\tfrac{1}{2}x) = \sqrt{\tfrac{1}{2}(\cosh x - 1)}.$$

$$\cosh(\tfrac{1}{2}x) = \sqrt{\tfrac{1}{2}(\cosh x + 1)}.$$

$$\tanh(\tfrac{1}{2}x) = (\cosh x - 1) \div \sinh x = \sinh x \div (\cosh x + 1).$$

$$\sinh x + \sinh y = 2\sinh \tfrac{1}{2}(x+y) \cdot \cosh \tfrac{1}{2}(x-y).$$

$$\sinh x - \sinh y = 2\cosh \tfrac{1}{2}(x+y) \cdot \sinh \tfrac{1}{2}(x-y).$$

$$\frac{\sin a \pm \sin \beta}{\cos a + \cos \beta} = \tan \tfrac{1}{2}(a \pm \beta).$$

$$\frac{\sin a \pm \sin \beta}{\cos a - \cos \beta} \Rightarrow -\operatorname{ctn} \tfrac{1}{2}(a \mp \beta).$$

$$\frac{\sin a + \sin \beta}{\sin a - \sin \beta} = \frac{\tan \tfrac{1}{2}(a + \beta)}{\tan \tfrac{1}{2}(a - \beta)}.$$

$$\sin^2 a - \sin^2 \beta = \sin(a + \beta) \cdot \sin(a - \beta).$$

$$\cos^2 a - \cos^2 \beta = -\sin(a + \beta) \cdot \sin(a - \beta).$$

$$\cos^2 a - \sin^2 \beta = \cos(a + \beta) \cdot \cos(a - \beta).$$

$$\sin xi = \tfrac{1}{2} i (e^x - e^{-x}) = i \sinh x.$$

$$\cos xi = \tfrac{1}{2} (e^x + e^{-x}) = \cosh x.$$

$$\tan xi = \frac{i(e^x - e^{-x})}{e^x + e^{-x}} = i \tanh x.$$

$$e^{x + yi} = e^x \cos y + i e^x \sin y.$$

$$a^{x + yi} = a^x \cos(y \cdot \log a) + i a^x \sin(y \cdot \log a).$$

$$(\cos \theta \pm i \cdot \sin \theta)^n = \cos n\theta \pm i \cdot \sin n\theta.$$

$$\sin x = -\tfrac{1}{2} i (e^{xi} - e^{-xi}).$$

$$\cos x = \tfrac{1}{2} (e^{xi} + e^{-xi}).$$

$$\tan x = -i \frac{e^{2xi} - 1}{e^{2xi} + 1}.$$

$$\sin(x \pm yi) = \sin x \cos yi \pm \cos x \sin yi$$
$$= \sin x \cosh y \pm i \cos x \sinh y.$$

$$\cos(x \pm yi) = \cos x \cos yi \mp \sin x \sin yi$$
$$= \cos x \cosh y \mp i \sin x \sinh y.$$

$$\frac{a}{\sin A} = \frac{b}{\sin B} = \frac{c}{\sin C}.$$

$$a^2 = b^2 + c^2 - 2bc \cos A.$$

$$\frac{a + b}{a - b} = \frac{\sin A + \sin B}{\sin A - \sin B} = \frac{\tan \tfrac{1}{2}(A + B)}{\tan \tfrac{1}{2}(A - B)} = \frac{\operatorname{ctn} \tfrac{1}{2} C}{\tan \tfrac{1}{2}(A - B)}.$$

$$\sin \tfrac{1}{2} A = \sqrt{\frac{(s - b)(s - c)}{bc}}, \quad \text{where } 2s = a + b + c.$$

$$\cos \tfrac{1}{2} A = \sqrt{\frac{s(s - a)}{bc}}.$$

$$\tan \tfrac{1}{2} A = \sqrt{\frac{(s - b)(s - c)}{s(s - a)}}.$$

$$\text{Area} = \tfrac{1}{2} bc \sin A = \sqrt{s(s - a)(s - b)(s - c)}.$$

$$\tan^{-1}x = \sin^{-1}\frac{x}{\sqrt{1+x^2}} = \cos^{-1}\frac{1}{\sqrt{1+x^2}} = \tfrac{1}{2}\sin^{-1}\frac{2x}{1+x^2}$$

$$= \operatorname{ctn}^{-1}\frac{1}{x} = \tfrac{1}{2}\pi - \operatorname{ctn}^{-1}x = \sec^{-1}\sqrt{1+x^2}$$

$$= \tfrac{1}{2}\pi - \tan^{-1}\frac{1}{x}$$

$$= \csc^{-1}\frac{\sqrt{1+x^2}}{x} = \tfrac{1}{2}\cos^{-1}\left[\frac{1-x^2}{1+x^2}\right]$$

$$= 2\cos^{-1}\left[\frac{1+\sqrt{1+x^2}}{2\sqrt{1+x^2}}\right]^{\frac{1}{2}} = 2\sin^{-1}\left[\frac{\sqrt{1+x^2}-1}{2\sqrt{1+x^2}}\right]^{\frac{1}{2}}$$

$$= \tfrac{1}{2}\tan^{-1}\frac{2x}{1-x^2} = 2\tan^{-1}\left[\frac{\sqrt{1+x^2}-1}{x}\right]$$

$$= -\tan^{-1}c + \tan^{-1}\left[\frac{x+c}{1-cx}\right] = -\tan^{-1}(-x)$$

$$= \tfrac{1}{2}i\log\frac{1-xi}{1+xi} = \tfrac{1}{2}i\log\frac{i+x}{i-x}$$

$$= -\tfrac{1}{2}i\log\frac{1+xi}{1-xi}.$$

Appendix 2

ADDITIONAL READING AND EXERCISES

We have used L'Hopital's Rule on many occasions. Here is a good opportunity for you to learn the proof of L'Hopital's Rule and explore the mathematics journals in your library. Read the article "L'Hopital's Rule" by A. E. Taylor, *American Mathematical Monthly* 59 (1952): 20–24.

As an excellent general reference and source of extra problems see the following books from the SCHAUM'S OUTLINE SERIES (McGraw-Hill):

Beginning Calculus, Mendelson, E.

Calculus, Ayres, F.

If you need to review algebra or analytic geometry the following books from SCHAUM'S OUTLINE SERIES are recommended:

College Algebra, Spiegel, M.

Analytic Geometry, Kindle, J.

We now relate the material in this book to certain standard textbooks. This gives you the opportunity for extra practice and some perspective on different treatments of calculus.

We refer to the following standard calculus texts:

[1] Ellis, R., and Gulick, D. *Calculus with Analytic Geometry*. 3d ed. New York: Harcourt, Brace, Jovanovich, 1986.

[2] Faires, J. D., and Faires, B. T. *Calculus and Analytic Geometry*. Boston: Prindle, Weber, Schmidt, 1983.

[3]   Goldstein, L. J., Lay, D. C., and Schneider, D. I. *Calculus and Its Applications*, 3d ed. Englewood Cliffs, N.J.: Prentice-Hall, 1984.

[4]   Shenk, A. *Calculus and Analytic Geometry*, 3d ed. Glenview, Ill.: Scott, Foresman & Co., 1984.

[5]   Swokowski, E. W. *Calculus with Analytic Geometry*, alternate ed. Boston: Prindle, Weber, Schmidt, 1983.

[6]   Thomas, G. B., and Finney, R. L. *Calculus and Analytic Geometry*, 6th ed. Reading, Mass.: Addison-Wesley.

For each exercise in *Top-Down Calculus* we inform the reader where in the above-listed books he or she can find additional reading and exercise material.

EXERCISE 1.6

[1]   Read Section 3.2.

[2]   Read Section 3.1.

[3]   Read Section 1.3.

[4]   Read Section 3.1.

[5]   Read Section 3.1 and work Exercise 2.

[6]   Read Section 1.7.

EXERCISE 1.16 AND ITS VARIATIONS

[1]   See Section 1.6 on Combining Functions and work Exercises 1–45. To practice graphing a function see Section 1.4 and do some of Exercises 1–40. For more problems like 1.16(5) see Sections 3.2–3.6 and work Exercises 3.2, 1–39, 3.3, 1–7, 10–17, 25–30, 3.4, 1–12, 3.5, 1–10, 35–38, 43–45, 59, 64, 3.6, 1–10, and 39–41.

[2]   See Section 1.4 and work Exercises 1–8 ((e) and (f) parts only), 9–30, and 32–36. For Exercise 1.16.5 see Sections 3.3, and 3.5–3.7 and work Exercises 3.3, 1–41, 3.5, 1–16, 33, 34, 37–54, 3.6, 13–30, 39–41, 3.7, 1–14, and 27–32.

[3]   See Section 0.3, Examples 5, 6, 7, Practice Problem 3, and work Exercises 25–36 in the same section. For Problem 1.16.5 see Section 1.3 and Exercise 3, 1–48; Section 1.6 and Exercise 6, 1–48; and Section 1.7 and Exercise 7, 1–34. Even more exercises of this type can be found in Supplementary Exercises for Chapter 1. See also Section 3.1, Exercises 1–18, and Section 3.2, Exercises 1–30.

[4] See Section 3.2, Study Problem 3.2, and work Exercise 3.2, 1–20; read Section 3.3, Study Problem 3.3, and work Exercise 3.3, 1–37; read part of Section 3.4 about normal lines and do "study exercise" 6 and Exercises 25 and 26; read part of Section 3.6 about higher order derivatives and work an example by yourself; read Section 3.7, do the study problems and Exercise 3.7, 1–28, 37–48.

[5] See Section 1.5 and work Exercise 1.5, 7–20; read Section 3.2 and do Exercises 1–20 (you may use your own method for finding a derivative); read 3.3 except quotient rule and do Exercise 3.3, 1–32; read 3.6 and do Exercises 1–12, 33–35, 39–42, 61–66; read 3.9 and do some of Exercise 3.9 (skip the problems involving derivatives of trigonometric functions).

[6] Read part of Section 1.5 about composition of functions and work Exercises 48–53. Read Sections 2.1 and 2.2 and work exercises following Section 2.2; read part of Section 2.6 about the Chain Rule and do Exercises 1–15.

EXERCISE 2.23 AND ITS VARIATIONS

[1] You may now work all exercises in Sections 3.4 and 3.5. For the derivative of the function e^x, see Section 6.3 and work Exercise 6.3, 1–16, 19–22; the derivatives of power and logarithmic functions are in Section 6.4 and related exercises are in 6.4, 1–22.

[2] For differentiation of trigonometric functions see Section 3.4 and exercises that follow it. Also, you may now do the next of the exercises in Section 3.5. Exponential functions are to be found in Section 7.3 (skip Examples 5 and 6 and the second part of the corollary), work Exercises 1–22. See Section 7.4 on logarithmic functions and work Exercises 1–10.

[3] Now you can do the rest of the exercises following Sections 3.1 and 3.2. Exponential and logarithmic functions and their differentiation are discussed in Chapter 4. You may work any of the exercises in the chapter.

[4] See again Section 3.4 and do all of Exercise 3.4. Read Sections 3.8 and 3.9 and do Exercise 3.9. Read parts of Section 7.3 about the exponential function and its derivative; do Exercises 21–35. Read Section 7.4 (skip integration part) and do Exercises 1, 3–10, 14–16.

[5] See Section 3.4 and do all of Exercise 3.4. Work Exercise 3.6, 23–38. Read Section 7.2 and work Exercises 1–28. Read Section 7.3 and work Exercises 1–30. Read Section 7.5 and work Exercises 1–28.

[6]   See part of Section 2.3 about the quotient rule and do Exercises 18–33.
Read Sections 2.7 and 2.8 and work Exercises 1–36. There are also
some nice review exercises following Chapter 2. Read Section 6.4, work
Exercises 1–22; read 6.5, work Exercises 11–21 and 23–31; read 6.6,
work Exercises 21–44; read 6.7, work Exercises 1–15; read 6.8, work
Exercises 3–15.

EXERCISE 3.20 AND ITS VARIATIONS

[1]   See Section 2.3 and work Exercises 1–52. Read Section 4.1 and work
Exercises 1–30. Read Section 4.4 and do Exercises 1–42. See Section
6.8 and work Exercises 1–50.

[2]   See Section 2.3 and work Exercises 31–46. Look at examples in Sections
2.4, 2.5, and 2.6 and work some of the review exercises. See Section
4.5 and work Exercises 1–40. See Section 3.8 and work Exercises 1–
62.

[3]   See Section 1.4, Practice Problems 4, and work Exercises 1–28. See
Sections 2.1, 2.2, 2.3, 2.4, and corresponding practice exercises. Work
Exercise 2.4.

[4]   See Section 2.4 and work study problems and Exercise 2.4. See Section
10.2 and work study problems and Exercises 1–22. See Sections 4.1,
4.2, 4.3, 4.4 and do study problems in all of these plus Exercise 4.4.

[5]   See Sections 2.2 and 2.3. Work Exercises 1–36. See Section 2.5 and
work Exercises 1–30. See Section 10.1 and work Exercises 1–54. See
Section 4.1 and work Exercises 1–26.

[6]   See Section 1.9 and work Exercises 1–87. See Section 3.9 and work
problems 1–25. See Sections 3.1 and 3.2 and work Exercises 15–26.

EXERCISE 3.25 AND ITS VARIATIONS

[1]   See Section 4.5 and work Exercise 4.5.

[2]   See Sections 4.1, 4.2, and 4.3 and from each one choose some exercises
to work.

[3]   See Sections 2.5 and 2.6 and work some exercises from each one.

[4]   See Sections 4.6 and 4.7 and work some exercises from each one.

[5]   See Sections 4.6 and 4.7 and work some exercises from each one.

[6]   See Sections 3.4, 3.5, and 3.6 and do some exercises from each one.

EXERCISE 4.5 AND ITS VARIATIONS

[1]   See Sections 5.5 and 5.6 and work Exercises 5.6, 1–22, 25–28, and 30–33.

[2]   See Sections 5.5 and 5.6 and work Exercises 1–47.

[3]   See Section 9.1 and do practice problems plus Exercises 1–40.

[4]   See Sections 5.5 and 5.6 and work Exercises   1–20, and 7.3, 46–54. See Chapter 8.1 and do Exercises 1–62.

[5]   See Section 5.6 and work Exercises 1–34.

[6]   See Section 4.8 (skip definite integrals) and work Exercises 1–30 (skip definite integral ones).

EXERCISE 4.15 AND ITS VARIATIONS

[1]   Look at Sections 5.1, 5.2, 5.3, and 5.4 and do exercises in 5.4, 37–46, and 66; Section 5.6, Problems 45–50; Section 5.8, Problems 1–18.

[2]   Look at Sections 5.1, 5.2, 5.3, and 5.4 and do Exercises 5.4, 27–36; see Section 6.1 and work Exercises 1–38.

[3]   See Section 9.3 up to Example 4 and work Exercises 1–20.

[4]   See Sections 5.2, 5.3, 5.4, and 5.5 and work Exercises 5.4, 10–23; 5.5, 32–39.

[5]   See Sections 5.2, 5.3, 5.5, and 6.1 and do Exercises 5.5, 1–32, and 6.1, 33–36.

[6]   See Sections 4.5 and 4.7 and work exercises in Section 4.7, 1–17.

EXERCISE 4.26 AND ITS VARIATIONS

[1]   See Section 7.1 and work Exercises 1–42.

[2]   See Section 8.1 and work Exercises 1–46.

[3]   See Section 9.2 and work all of the exercises and practice problems.

[4]   See Section 8.2 and work Exercises 1–36.

[5]   See Section 9.1 and work Exercises 1–38.

[6]   See Section 7.2 and work Exercises 1–34.

EXERCISE 4.27 AND ITS VARIATIONS

[1] See Section 7.2 and work Exercises 1–58.

[2] See Section 8.2 and work Exercises 1–40.

[3] No assignment.

[4] See Section 8.3 and work Exercises 1–18.

[5] See Section 9.2 and work Exercises 1–30.

[6] See Sections 7.3 and 7.4 and work Exercises in 7.3, 1–49.

EXERCISE 4.28 AND ITS VARIATIONS

[1] See Section 7.3 and work Exercises 1–42.

[2] See Section 8.3 and work Exercises 1–32.

[3] No assignment.

[4] See Section 8.4 and work Exercises 1–23.

[5] See Section 9.3 and work Exercises 1–22.

[6] See Section 7.5 and work Exercises 1–50.

EXERCISE 4.42 AND ITS VARIATIONS

[1] See Section 7.4 and work Exercises 1–29.

[2] See Section 8.4 and work Exercises 1–32.

[3] No assignment.

[4] See Section 8.5 and work Exercises 1–49.

[5] See Section 9.4 and work Exercises 1–37.

[6] See Section 7.7 and work Exercises 1–50.

EXERCISE 4.48 AND ITS VARIATIONS

[1] See Sections 8.1, 8.3, 8.4, 8.5, 8.10, and 14.2. Select some exercises in each section and look at review exercises for Chapter 8.

[2] See Sections 6.1, 6.2, 6.4, 6.5, and 16.3. Select some exercises in each section and look at review exercises for Chapter 6.

[3] The only reference is to double integrals in Section 7.7.

[4] See Sections 6.1, 6.2, 6.3, 6.4, 6.5, 6.6, and 17.3. Select some exercises in each section and look at miscellaneous exercises in Chapter 6.

[5] See Sections 6.1, 6.2, 6.3, 6.4, 6.7, and 17.5. Select some exercises in each section.

[6] See Sections 5.2, 5.3, 5.4, 5.5, 5.6, 5.7, and 16.5. Select some exercises in each section and look at miscellaneous problems at the end of Chapter 5.

As for CHAPTER 5, INFINITE SERIES, our approach is a bit different from any of these texts, so you are invited to explore the appropriate chapters on your own!

Appendix 3

SOLUTIONS, PARTIAL SOLUTIONS, AND HINTS TO EXERCISES

This appendix contains solutions, partial solutions, and hints to a variety of exercises. It is *very important* that you attempt to solve a problem yourself before looking up its solution. Everybody develops his or her own style of solving problems. After some practice in calculus, you will find it easier to work quickly through the problems yourself than to fool around looking up solutions. Where these solutions will be most helpful is in picking up tricks and ways of thinking about problems that you may have missed, and in checking your answers. Don't forget the MATH TABLES and, in particular, the TABLE OF INTEGRALS. It is both a realistic and valuable experience for you to use the TABLE OF INTEGRALS. If you have access to computer software that computes indefinite integrals, learn to use it. You will quickly learn that using tables and computer software will help, not hurt, your progress in learning the basic classical techniques of integration.

1.20 VARIATIONS ON EXERCISE 1.16

(1)

    (a) In the expression of $f(x) = 3x^4 + 4x^2 + 1$, replace x by $g(x) = x + 2$, we get

$$h(x) = 3(x+2)^4 + 4(x+2)^2 + 1 \quad .$$

    (b) In the expression of $f(x) = \dfrac{1}{x-2}$, replace x by $g(x) = x - 2$, we get

$$h(x) = \frac{1}{(x-2) - 2} = \frac{1}{x-4} \quad .$$

    (c) In the expression of $\gamma(s) = \sqrt{s} - 4s+1$, replace s by $s(x)=2x$, we get

$$(s(x)) = \sqrt{2x} - 4(2x) + 1 = \sqrt{2x} - 8x+1 \quad .$$

    (d) $m(n) = (n+4)^3 - 1$ and $n(x) = 14x - 37$ so replace n by $14x - 37$, we get

$$m(n(x)) = ((14x-37)+4)^3 - 1 = (14x-33)^3 - 1 \quad .$$

    (e) $f(x) = \dfrac{(x+1)^2-1}{(x+1)^2+1}$, $g(x) = 2x^2 + 3$.

        Replace x by $2x^2 + 3$ in the expression for $f(x)$ we get

$$\frac{((2x^2+3)+1)^2-1}{((2x^2+3)+1)^2+1} = \frac{(2x^2+4)^2-1}{(2x^2+4)^2+1} \quad .$$

    (f) $f(g)) = 2/(g-3)$, $g(u) = \sqrt{u+2}$,

$$h(u) = f(g(u)) = \frac{2}{\sqrt{u+2}-3}.$$

        In order that $h(u)$ is well defined, we must have $u + 2 \geq 0$ and $\sqrt{u+2} - 3 \neq 0$, i.e., $u \geq -2$ and $u \neq 7$.

    (g) $f(x) = (x+2)^{-2}$, $g(x) = x^3 - 1$

$$f(g(x)) = (g(x) + 2)^{-2} = (x^3 - 1 + 2)^{-2} = (x^3 + 1)^{-2}$$

$$g(f(x)) = (g(x))^3 - 1 = [(x+2)^{-2}]^3 - 1 = (x+2)^{-6} - 1.$$

For $x^3 + 1 \neq 0$, i.e., $x \neq -1$ $f(g(x))$ is defined

and for $x + 2 \neq 0$, i.e., $x \neq -2$, $g(f(x))$ is defined.

(h) $f(x) = \sqrt{x}, \; g(x) = x^2 + x$.

$$f(g(x)) = \sqrt{g(x)} = \sqrt{x^2+x}$$

$$g(f(x)) = (f(x))^2 + f(x) = x + \sqrt{x},$$

for $x^2 + x \geq 0$, i.e., $x \geq 0$, or $x \leq -1$, $f(g(x))$ is defined

for $x \geq 0$, $g(f(x))$ is defined.

(2)

(a) The natural choice of $f(g) = \sqrt{g}$, and $g(x) = x^2 + 4x$.

(b) Choose $f(g) = g^3 - 3g^2 + 4$, and $g(x) = x + 1$.

(c) $h(x) = (x-2)^2 + 2(x^2-4x+5) = (x-2)^2 + 2((x-2)^2+1)$.

Choose $f(g) = g^2 + 2(g+1)$ and $g(x) = (x-2)^2$.

(d) For $g(x) = x^2 - 2x + 3$, we have

$$h(x) = \frac{(x^2-2x+3)}{(x^2-2x+3)+2} = \frac{g(x)}{g(x)+2} \; , \; \therefore \; f(g) = \frac{g}{g+2} \qquad .$$

for $g(x) = x-1$, we have $h(x) = \dfrac{(x-1)^2+2}{(x-1)^2+4} = \dfrac{g(x)^2+2}{g(x)^2+4}$ so $f(g) = \dfrac{g^2+2}{g^2+4}$ $\square$

(3)

(a) $h(x) = \left(\dfrac{1}{x^{1/3}} - \dfrac{1}{x^{2/3}} \right)^3,$

for $g(x) = \dfrac{1}{x^{2/3}} - \dfrac{1}{x^{2/3}}$, we have $h(x) = (g(x))^3$ so $f(g) = g^3$.

For $g(x) = x^{1/3}$, we have $h(x) = \left(\dfrac{1}{g(x)} - \dfrac{1}{g^2(x)} \right)^3$, so $f(g) = \left(\dfrac{1}{g} - \dfrac{1}{g^2} \right)^3$.

(b) $h(x) = \dfrac{1}{\sqrt{x^2+2}}$,

for $g(x) = \sqrt{x^2+2}$, we have $h(x) = \dfrac{1}{g(x)}$, so $f(g) = \dfrac{1}{g}$.

For $g(x) = \sqrt{x}$, we have $x = g^2(x)$.

$h(x) = \dfrac{1}{\sqrt{g^4(x)+2}}$, $f(g) = \dfrac{1}{\sqrt{g^4+2}}$.

(c) $h(x) = (1-4x^2)^{1/2}$,

for $g(x) = 1 - 4x^2$, we have $h(x) = (g(x))^{1/2}$, so $f(g) = g^{1/2}$.

for $g(x) = 2x$, we have $h(x) = (1 - (2x)^2)^{1/2} = (1-(g(x))^2)^{1/2}$ so $f(g) = (1-g^2)^{1/2}$.

(d) $h(x) = x^2 + 2x + 5$,

for $g(x) = \sqrt{x}$, we have $x = g^2(x))$ so $h(x) = g^4(x) + 2g^2(x)) + 5$, $f(g) = g^4 + 2g^2 + 5$,

For $g(x) = x + 1$, we have

$h(x) = (x+1)^2 + 4$, $\therefore f(g) = g^2 + 4$.

(e) $h(x) = x^{2/3} - x^{4/3} + 1$

for $g(x) = x^{2/3} - x^{4/3}$, we have $h(x) = g(x) + 1$,

$f(g) = g + 1$.

For $g(x) = x^{1/3}$, we have $g^2(x) = x^{2/3}$, $g^4(x) = x^{4/3}$,

$\therefore h(x) = g^2(x) - g^4(x) + 1$, $\therefore f(g) = g^2 - g^4 + 1$.

(f) $h(x) = \sqrt{x^{1/2}+x^2}$ for $g(x) = x^{1/2}$, we have $x^2 = g^4(x)$,

$\therefore h(x) = \sqrt{g(x)+g^4(x)}$, $\therefore f(g) = \sqrt{g+g^4}$.

For $g(x) = x^2$, we have $x^{1/2} = g(x)^{1/4}$,

$\therefore h(x) = \sqrt{g(x)^{1/4} + g(x)}$, $\therefore f(g) = \sqrt{g^{1/4} + g}$.

(g) $h(x) = \dfrac{x^2 + x + 1}{x^2 + x + 2}$ for $g(x) = x^2 + x$, we have $h(x) = \dfrac{g(x) + 1}{g(x) + 2}$,

$\therefore f(g) = \dfrac{g + 1}{g + 2}$.

For $g(x) = 2x$, we have $x = \dfrac{g(x)}{2}$,

$\therefore h(x) = \dfrac{\left(\dfrac{g(x)}{2}\right)^2 + \dfrac{g(x)}{2} + 1}{\left(\dfrac{g(x)}{2}\right)^2 + \dfrac{g(x)}{2} + 2} = \dfrac{g(x)^2 + 2g(x) + 4}{g(x)^2 + 2g(x) + 8}$

$\therefore f(g) = \dfrac{g^2 + 2g + 4}{g^2 + 2g + 8}$. $\qquad\qquad\qquad\square$

(4)

(a) $f(g) = -g + 2$, $g(x) = \sin x + 1$, $\therefore f(g(x)) = -(\sin x + 1) + 2 = -\sin x + 1$.

(b) $f(g) = \sqrt{g}$, $g(x) = 3 \sin x$, $\therefore f(g(x)) = \sqrt{3 \sin x} = \sqrt{3} \cdot \sqrt{\sin x}$.

(c) $f(g) = |g|$, $g(x) = \cos x$, $\therefore f(g(x)) = |\cos x|$, even function.

(5)

(a) $\dfrac{d}{dt}(t) = 1$.

(b) $\dfrac{d}{dx}\left(x + \dfrac{2}{(x-1)^{2/3}}\right) = \dfrac{d}{dx}(x) + \dfrac{d}{dx}\left(\dfrac{2}{(x-1)^{2/3}}\right)$

$\qquad = 1 + 2\dfrac{d}{dx}\left((x-1)^{-2/3}\right) = 1 + 2 \cdot \left(-\dfrac{2}{3}\right)(x-1)^{-2/3-1} \cdot \dfrac{d}{dx}(x-1)$

$\qquad = 1 - \dfrac{4}{3}(x-1)^{-5/3}$.

(c) $\dfrac{d}{dx}(x^2+x+1)^5 = 5(x^2+x+1)^{5-1} \cdot \dfrac{d}{dx}(x^2+x+1)$ (chain rule)

$$5(x^2+x+1)^4\left[\dfrac{d}{dx}(x^2) + \dfrac{d}{dx}(x)) + \dfrac{d}{dx}(1)\right]$$

$$= 5(x^2+x+1)^4\,[2x+1].$$

(d) $\dfrac{d}{dx}\left(\left(\dfrac{2}{x^2+1}\right)^8\right) = 8\left(\dfrac{2}{x^3+1}\right)^7 \cdot \dfrac{d}{dx}\left(\dfrac{2}{x^2+1}\right)$ (chain rule)

$$= 8\left(\dfrac{2}{x^2+1}\right)^7 \cdot 2\,\dfrac{d}{dx}((x^2+1)^{-1}) = 16\left(\dfrac{2}{x^2+1}\right)^7 \cdot (-1)(x^2+1)^{-2}\,\dfrac{d}{dx}(x^2+1)$$

$$= -16\left(\dfrac{2}{x^2+1}\right)^7 \cdot (x^2+1)^{-2} \cdot 2x = \dfrac{-32\cdot 2^7}{(x^2+1)^9}\,x = -\dfrac{2^{12}}{(x^2+1)^9}\,x,$$

or $\dfrac{d}{dx}\left(\left(\dfrac{2}{x^2+1}\right)^8\right) = \dfrac{d}{dx}(2^8(x^2+1)^{-8}) = 2^8\,\dfrac{d}{dx}((x^2+1)^{-8})$

$$= 2^8(-8)(x^2+1)^{-9}\,\dfrac{d}{dx}(x^2+1) = -2^{11}(x^2+=1)^{-9}\,2x = -\dfrac{2^{12}}{(x^2+1)^9}\,x.$$

(e) $g(t) = \sqrt{3t^2-4} = (3t^2-4)^{1/2}.$

$$\therefore \; g'(t) = \dfrac{1}{2}(3t^2-4)^{1/2-1}\cdot(3t^2-4)' = \dfrac{1}{2}(3t^2-4)^{-1/2}\cdot 6t.$$

$$= \dfrac{3t}{\sqrt{3t^2-4}}.$$

(f) $f(x) = \dfrac{1}{(3x^2-5x)^2} = (3x^2-5x)^{-2}.$

$$\therefore \; \dfrac{df}{dx} = -2(3x^2-5x)^{-3}\cdot\dfrac{d}{dx}(3x^2-5x)$$

$$= -2(3x^2-5x)^{-3}\cdot(3\,\dfrac{d}{dx}x^2 - 5\,\dfrac{d}{dx}x)$$

$$= -2(3x^2-5x)^{-3}\cdot(6x-5).$$

(g) $\dfrac{d}{dy}(y^4-(y+1)^3) = 4y^3 - 3(y+1)^2.$

(h) $c(x) = (A(x)-3B(x))^4$. $A(-2) = B(-2) = 1$, $A'(-2) = 0$, $B'(-2) = -2$.

$$\therefore\ C'(x) = 4(A(x)-3B(x))^3 \cdot \frac{d}{dx}(A(x)-3B(x))$$

$$= 4(A(x)-3B(x))^3 (A'(x)-3B'(x))$$

$$\Rightarrow\ C'(-2) = 4(A(-2)-3B(-2))^3(A'(-2)-3B'(-2))$$

$$= 4(1-3)^3(0-3(-2)) = 4(-2)^3 \cdot 6 = -192.$$

(i) $y = \sqrt{x^2-1} = (x^2-1)^{1/2}$,

$$\therefore\ y'(x) = \frac{1}{2}(x^2-1)^{-1/2}(x^2-1)' = \frac{1}{2}(x^2-1)^{-1/2} \cdot 2x = (x^2-1)^{-1/2} \cdot x$$

$$\Rightarrow\ y'(2) = 3^{-1/2} \cdot 2 = \frac{2}{\sqrt{3}}.$$

The slope of the tangential line at $x = 2$ is $\frac{2}{\sqrt{3}}$,

and $\therefore\ y(2) = \sqrt{3}$.

$\therefore$ the equation of the tangential line is

$$y - \sqrt{3} = \frac{2}{\sqrt{3}}(x-2),\ \text{ or }\ 2x - \sqrt{3}\,y-1 = 0.$$

(j) $y = x^4+2$, $\therefore\ y'(x) = 4x^3$, $y'(-1) = -4$,

i.e., the slope of the tangential line at $x = -1$ is -4. Therefore, the slope of the normal line at $x = -1$ is $\frac{-1}{-4} = \frac{1}{4}$,

and $\therefore\ y(-1) = 3$, $\therefore$ the equation of the normal line is $y - 3 = \frac{1}{4}(x+1)$,
i.e., $x - 4y + 13 = 0$.

(k) $f(x) = \frac{1}{\sqrt{x+1}} = (x+1)^{-1/2}$.

$$\therefore\ f'(x) = -\frac{1}{2}(x+1)^{-1/2-1} \cdot (x+1)' = -\frac{1}{2}(x+1)^{-3/2}.$$

$$f''(x) = \frac{d}{dx} \cdot \left(-\frac{1}{2}(x+1)^{-3/2} \right) = -\frac{1}{2}\frac{d}{dx}\left((x+1)^{-3/2}\right)$$

$$= -\frac{1}{2}\left(-\frac{3}{2} \right)(x+1)^{-3/2-1} \cdot \frac{d}{dx}(x+1)$$

$$= \frac{3}{4}(x+1)^{-5/2}.$$

(l) $g(t) = t^3 + 2t^2 + t + 4.$

$\therefore$ $g'(t) = 3t^2 + 4t + 1,$ $g''(t) = 6t + 4.$ $\dfrac{d^3 g(t)}{dt^3} = 6.$

1.21 PARTIAL SOLUTIONS OR HINTS

(1)

(a) $h(x) = f(g(x)) = f(x^3+2x) = (x^3+2x+1)^2 + 1.$
(Note: $f(x) = (x+1)^2 + 1 = (x-1)^2 + 4x + 1.$)

(c) $h(x) = s(t(x))$
$= s(x^2+2x-1) = (x^2+2x-1)^4 - (x^2+2x-1)^3 + 2(x^2+2x-1) + 1.$

(e) $h(x) = f(g(x)) = f(x^{1/3}) = \dfrac{x^{1/3}}{x^{1/3}-1}.$ $x \neq 1.$

(g) $f(g(x)) = f(x-1) = (x-1)^2 + x - 1 = x(x-1),$
$g(f(x)) = g(x^2+x) = (x^2+x) - 1.$

(2)

(a) $g(x) = x^2 + x - 1$ $f(g) = g^3 + \sqrt{g} + 4.$

(b) $g(x) = x^3 - 1,$ $f(g) = g^2 + g - 1.$

(c) $g(x) = (x-2)^2,$ $f(g) = \dfrac{g-1}{g+1}.$

(d) Note $x^2 + 8x + 3 = (x-4)^2 + 16x - 13 = (x-4)^2 + 16(x-4) + 51$ so choose
$g(x) = (x-4)$ then $f(g) = (g^2+16g+51)/(g^2-1).$

(3)

    (a) $f(g) = g^2 - g + 5.$

    (c) For $g(x) = 2x^2 - 3,$ $f(g) = \sqrt{g}.$
         For $g(x) = 2x^2,$ $f(g) = \sqrt{g-3}.$

    (e) For $g(x) = \sqrt{x},$ $f(g) = (3+g^2+2g)^{5/2}.$
         For $g(x) = 1 + x + 2\sqrt{x},$ $f(g) = (2+g)^{5/2}.$

(4)

    (a) $f(g(x)) = |x|^2 + 1 = x^2 + 1.$
    (c) $f(g(x)) = \sqrt{|x|}$ even function

(5)

    (a) $\dfrac{d}{dx}(3) = 0.$

    (c) $\dfrac{d}{dx}(x^4+3x^3+2x^2+x-1) = 4x^3 + 9x^2 + 4x + 1.$

    (e) $g'(x) = \dfrac{1}{2}\left(\sqrt{x}+1\right)^{-1/2} \cdot \dfrac{1}{2} x^{-1/2} = \dfrac{1}{4\sqrt{x(\sqrt{x}+1)}}.$
    (g) $g'(x) = -2/x^3,$ $g''(x) = 6/x^4.$

    (i) $y'(x) = \dfrac{1}{2} x^{-1/2},$ $y'(4) = \dfrac{1}{2} \cdot 4^{-1/2} = \dfrac{1}{4}$
        $\therefore$ the equation is

$$y - 2 = \frac{1}{4}(x-4) \quad \text{or} \quad \frac{1}{4}x - y + 1 = 0$$

    (k) $g^{(3)}(x) = 24x - 6$
        $\Rightarrow g^{(3)}(0) = -6$

1.22 VARIATIONS ON EXERCISE 1.16

(5) Computing derivitives:

    (a) $\dfrac{d}{dx}((1+x^2)^{-1}) = (-1)(1+x^2)^{-2} \cdot \dfrac{d}{dx}(1+x^2) = -(1+x^2)^{-2} \cdot 2x.$

(b) $\dfrac{d}{dz} (2z^2-4/z)^2 = 2(2z^2-4/z) \dfrac{d}{dz} (2z^2-4/z)$

$= 2(2z^2-4/z) \cdot (4z+4/z^2)$

(c) $g(x) = \sqrt{2x^3+1} = (2x^3+1)^{1/2}.$

$\therefore\ g'\ (x) = \dfrac{1}{2}\ (2x^3+1)^{-1/2}(2x^3+1)' = \dfrac{1}{2}\ (2x^3+1)^{-1/2} \cdot 6x^2$

$\therefore\ g'\ (1) = \dfrac{1}{2}\ (2+1)^{-1/2} \cdot 6 = \dfrac{3}{\sqrt{3}} = \sqrt{3}.$

(d) $\dfrac{d}{dd}\ (2) = 0.$

(e) $\dfrac{d}{dx} \left(\dfrac{2}{1+ |\ x\ |} \right) = 2 \dfrac{d}{dx} ((1+ |\ x\ |\ )^{-1}) = 2 \cdot (-1)(1+ |\ x\ |\ )^{-2} \cdot \dfrac{d}{dx} (1\ \ |\ x\ |\ )$

$= -2(1+ |\ x\ |\ )^{-2} \cdot \begin{cases} 1 & x > 0 \\ -1 & x < 0 \\ \text{does not exist if}\ \ x = 0 \end{cases}$

$= \begin{cases} -2(1+x)^{-2} & \text{for}\ \ x > 0 \\ 2(1-x)^{-2} & \text{for}\ \ x < 0. \\ \text{does not exist if for}\ \ x = 0 \end{cases}$

(f) $f(x)) = (x-1)^{10}.$

$\therefore\ f'\ (x) = 10(x-1)^9\ ,\ f''\ (x) = 90(x-1)^8.$

(g) $\therefore\ (1+t^{-1})^{-1} = \dfrac{1}{1 + \dfrac{1}{t}} = \dfrac{t}{1+t} = 1 - \dfrac{1}{1+t}.$

$\therefore\ \dfrac{d}{dt} (1+t^{-1})^{-1})^{-1} = -\dfrac{d}{dt} (1+t)^{-1} = -(-1)(1+t)^{-2} = (1+t)^{-2}$

$\therefore\ \dfrac{d^2}{dt^2} (1+t^{-1} = \dfrac{d}{dt} (1+t)^{-2} = (-2)(1+t)^{-3}.$

(h) $y = 1/x.\ \therefore\ y'\ (x) = -1/x^2.\ \Rightarrow\ y'\ (-1) = -1.$

i.e., the slope of the tangential line at $x = -1$ is -1, $y(-1) = -1$, therefore the equation of the tangential line is

$$y+1 = -(x+1), \quad \text{or} \quad x+y+2 = 0 \quad .$$

(i) $y = \sqrt{x-1} = (x-1)^{1/2}$, $\therefore y'(x) = \dfrac{1}{2}(x-1)^{-1/2}$

$\Rightarrow y'(2) = \dfrac{1}{2}$, i.e., the slope of the tangential line at $x = 2$ is $1/2$, so the slope of the normal line at $x = 2$ is -2, and $y(2) = 1$, so the equation of the normal line is

$$y-1 = -2(x-2), \quad \text{or} \quad 2x+y-5 = 0 \quad .$$

(j) $f(x) = \sqrt{(g(x))^2+1} = (g(x)^2+1)^{1/2}$.

$\therefore f'(x) = \dfrac{1}{2}(g(x)^2+1)^{-1/2}(g(x)^2+1)' = \dfrac{1}{2}(g(x)^2+1)^{-1/2} \cdot 2g(x) \cdot g'(x)$

$= g(x)g'(x)(g(x)^2+1)^{-1/2}$.

$g(0) = 1$, $g'(0) = 2$, $\therefore f'(0) = g(0)g'(0) \cdot (g(0)^2+1)^{-1/2} = 2(2)^{-1/2} = \sqrt{2}$.

2.26 PARTIAL ANSWERS OR HINTS

(1)

(a) $\dfrac{d}{dx}\left(\dfrac{x^2-4}{x-2}\right) = \dfrac{d}{dx}(x+2) = 1$.

(c) $\dfrac{d}{dx}\left(\dfrac{x^{2/3}-1}{x^{1/3}-1}\right) = \dfrac{d}{dx}(x^{1/3}+1) = \dfrac{1}{3}x^{-2/3}$.

(e) $\dfrac{d}{dx}\sin\left(\dfrac{x^3+6}{x^2+2x+3}\right) = \cos\left(\dfrac{x^3+6}{x^2+2x+3}\right)\dfrac{(x^2+2x+3)\cdot 3x^2-(x^3+6)(2x+2)}{(x^2+2x+3)^2}$.

(2)

(a) $\dfrac{d}{dt}\left(\dfrac{1-\cos^2 t}{\sin^2 t}\right) = \dfrac{d}{dt}(1) = 0$.

(c) $\dfrac{d}{dx} (\sin x \cdot e^x) = \sin x \cdot e^x + \cos x \cdot e^x = e^x (\sin x + \cos x).$

(e) $\left(\log_a |x^{1/3}|\right)^{1/3} = \left(\dfrac{1}{3} \log_a |x|\right)^{1/3} = 3^{-1/3} \left(\dfrac{\ln|x|}{\ln a}\right)^{1/3}.$

$\Rightarrow \dfrac{d}{dx} \left(\log_a |x^{1/3}|\right)^{1/3} = (3 \ln a)^{-1/3} \cdot \dfrac{1}{x} \cdot \dfrac{1}{3} (\ln |x|)^{-2/3}.$

(g) Note $\dfrac{\ln x^2+1}{\ln x^2+2} = 1 - \dfrac{1}{\ln x^2+2} = 1 - \dfrac{1}{2 \ln |x| +2}$

$= 1 - \dfrac{1}{2} (\ln |x| +1)^{-1}$

$\therefore \dfrac{d}{dx} \left(\dfrac{\ln x^2+1}{\ln x^2+2}\right)^3 = 3 \left(\dfrac{\ln x^2+1}{\ln x^2+2}\right)^2 \cdot \left(\dfrac{-1}{2}\right) (-1) (\ln |x| +1)^{-2} \cdot \dfrac{1}{x}$

$= \dfrac{3}{2} \left(\dfrac{\ln x^2+1}{\ln x^2+2}\right) \dfrac{1}{x} \cdot \dfrac{1}{(\ln |x| +1)^2}.$

(3)

(b) $\dfrac{d}{dx} \csc x = - \cot (x) \csc x$

$\therefore \dfrac{d^2}{dx^2} \csc x = - (\cot x \cdot (- \cot x \csc x) - \csc^2 x \cdot \csc x)$

$= \csc x (\cot^2 x + \csc^2 x).$

(d) $\dfrac{d}{dx} \ln(|\tan x|)\Big|_{x=\pi/8} = \dfrac{1}{\tan x} \cdot \tan x \sec x \Big|_{x=\pi/8} = \dfrac{1}{\cos \dfrac{\pi}{8}}.$

(f) $f'(x) = \dfrac{1}{e^{ex}} e^{ex} e = e.$

(h) $\dfrac{d}{dx} \cot^2 x = 2 \cot x \cdot \sec^2 x.$

$\Rightarrow \dfrac{d^2x}{dx^2} \cot^2 x = 2[(-\csc^2 x)\sec^2 x + \cot x \cdot 2 \sec^2 x \cdot \tan x]$

$$= 2(2 \sec^2 x - \sec^2 x \cdot \csc^2 x) = 2 \sec^2 x \, (2 - \csc^2 x).$$

(4)

    (b) $(\ln x)^{1/4}$ defined for $x > 1$.

      and $\dfrac{d}{dx} (\ln x)^{1/4} = \dfrac{1}{4} (\ln x)^{-3/4} \cdot \dfrac{1}{x}.$

    (d) Note $(x+1)^{x+1} = e^{(x+1)\ln(x+1)}$

      $\therefore \dfrac{d}{dx} (x+1)^{x+1} = e^{(x+1)\ln(x+1)}(\ln(x+1)+1) = (x+1)^{x+1}(\ln(x+1)+1).$

    (f) Note $\log_2(e^x) = x \log_2 e$

      $\therefore \dfrac{d}{dx} \log_2(e^x) = \log_2 e.$

    (g) Note $\log_{10}(\log_2 x) = \log_{10} \left(\dfrac{\ln x}{\ln 2} \right) = \log_{10}(\ln x) - \log_{10}(\ln(2))$

      $\therefore \dfrac{d}{dx} \log_{10}(\log_2 x) = \dfrac{d}{dx} \log_{10}(\ln x) = \dfrac{d}{dx} \dfrac{\ln(\ln x)}{\ln (10)}$

      $= \dfrac{1}{\ln 10} \cdot \dfrac{1}{\ln x} \cdot \dfrac{1}{x}.$

2.27 PARTIAL ANSWERS

(1)

    (a) $\dfrac{d}{dx} \left(\dfrac{x^4 - 25}{x^2 + 5} \right) = \dfrac{d}{dx} (x^2 - 5) = 2x.$

    (c) $\dfrac{d}{dx} \left(\dfrac{x^{1/2} - 1}{x^{1/4} + 1} \right) = \dfrac{d}{dx} (x^{1/4} - 1) = \dfrac{1}{4} x^{-3/4}.$

    (e) $\dfrac{d}{dx} \cos\left(\dfrac{x^{1/2} + 1}{2x^2} \right) = -\sin\left(\dfrac{x^{1/2} + 1}{2x^2} \right) \cdot \dfrac{d}{dx} \left(\dfrac{1}{2} x^{-3/2} + \dfrac{1}{2} x^{-2} \right)$

      $= -\sin \left(\dfrac{x^{1/2} - 1}{2x^2} \right) \cdot \left(-\dfrac{3}{4} x^{-5/2} - x^{-3} \right)$

$$= \left(\frac{3}{4}\, x^{-5/2}+x^{-3}\right) \sin\left(\frac{x^{1/2}+1}{2x^2}\right).$$

(2)

(a) $\dfrac{d}{dt}\,(4\cos^2(2t) + 4\sin^2(2t)) = \dfrac{d}{dt}\,(4) = 0.$

(c) $\dfrac{d}{dx}\,(x^2+3)e^{x\,\tan x} = 2xe^{x\,\tan x} + (x^2+3)e^{x\,\tan x}(x\tan x)$

$\qquad = e^{x\,\tan x}(2x+(x^2+3)(\tan x+x\sec^2 x)).$

(e) $\dfrac{d}{dx}\,\log_a(\,|\,x^3+1\,|\,) = \dfrac{1}{\ln a}\cdot\dfrac{d}{dx}\,\ln(\,|\,x^3+1\,|\,)$

$\qquad = \dfrac{1}{\ln a}\cdot\dfrac{1}{x^3+1}\cdot 3x^2$

(g) $\dfrac{d}{dx}\left(\dfrac{\ln(\sin x)+3}{\ln(\cos x)+1}\right)^3 = 3\left(\dfrac{\ln(\sin x)+3}{\ln(\cos x)+1}\right)^2 \cdot \left(\dfrac{\ln(\sin x)+3}{\ln(\cos x)+1}\right)'$

$\qquad = 3\left(\dfrac{\ln(\sin x)+3}{\ln(\cos x)+1}\right)^2 \cdot \dfrac{(\ln(\cos x)+1)\cot x+(\ln(\sin x)+3)\tan x}{(\ln(\cos x)+1)^2}.$

(3)

(a) $\dfrac{d}{dx}\,\csc(x+2) = -\cot(x+2)\csc(x+2)$

$\qquad \dfrac{d^2}{dx^2}\,\csc(x+2) = -(-\csc^3(x+2) - \cot^2(x+2)\csc(x+2))$

$\qquad = \csc(x+2)(\csc^2(x+2) + \cot^2(x+2))$

$\qquad \therefore\ \dfrac{d^2}{dx^2}\,\csc(x+2)\,\bigg|_{x=\frac{\pi}{4}-2} = \csc\dfrac{\pi}{4}\left(\csc^2\dfrac{\pi}{4} + \cot^2\dfrac{\pi}{4}\right)$

$\qquad = \sqrt{2}\,(2+1) = 3\sqrt{2}.$

(c)   Note $(5\cos(2x)\sin(2x))^{1/2} = \left(\dfrac{5}{2}\right)^{1/2}(\sin(4x))^{1/2}.$

$\qquad \therefore\ \dfrac{d}{dx}\,(5\cos(2x)\sin(2x))^{1/2} = \left(\dfrac{5}{2}\right)^{1/2}\cdot\dfrac{1}{2}\,(\sin 4x)^{-1/2}\,4\cos 4x$

$$= 2\left(\frac{5}{2}\right)^{1/2} (\sin 4x)^{-1/2} \cdot \cos 4x.$$

(e)   Note $\log_{10}(x^2+1) > 0$,

$$\therefore f(x) = 10^{\log_{10}(x^2+1)} = x^2+1$$

$$\Rightarrow f'(x) = 2x.$$

(g) $\dfrac{d}{dx} (\ln x^{3/2} + \ln x^{1/2} - \ln x^{5/2}) = \dfrac{d}{dx} \left(\dfrac{3}{2} \ln x + \dfrac{1}{2} \ln x - \dfrac{5}{2} \ln x\right)$

$$= \frac{d}{dx}\left(-\frac{1}{2}\ln x\right) = -\frac{1}{2} x^{-1}$$

$$\therefore \frac{d^5}{dx^5}(\ln x^{3/2} + \ln x^{1/2} - \ln x^{5/2}) = \frac{d^4}{dx^4}\left(-\frac{1}{2}x^{-1}\right)$$

$$= -\frac{1}{2}\frac{d^4}{dx^4}(x^{-1}) = -\frac{1}{2}(-1)(-2)(-3)(-4)x^{-5} = -12\, x^5.$$

(4)

(a) $\dfrac{d}{dx} \ln(\ln(\ln(\ln(3x^2+2))))$

$$= \frac{1}{\ln(\ln(\ln(3x^2+2)))} \cdot \frac{1}{\ln(\ln(3x^2+2))} \frac{1}{\ln(3x^2+2)} \cdot \frac{6x}{3x^2+2}.$$

(c) $\dfrac{d}{dx} (x^{1/2}+1)^{\ln x} = \dfrac{d}{dx} e^{(\ln x)\ln(x^{1/2}+1)}$

$$= e^{(\ln x)\ln(x^{1/2}+1)} \left(\frac{\ln(x^{1/2}+1)}{x} + \frac{1}{x^{1/2}+1} \cdot \frac{1}{2} x^{-1/2} \ln x\right)$$

$$= e^{(\ln x)\ln(x^{1/2}+1)} \left(\frac{\ln(x^{1/2}+1)}{x} + \frac{\ln x}{2(x+x^{1/2})}\right).$$

(e) $\dfrac{d}{dx} (x+1)^{1/x^2} = \dfrac{d}{dx} e^{\frac{1}{x^2}\ln(x+1)}$

$$= e^{\frac{1}{x^2}\ln(x+1)} \left(\frac{1}{x^2(x+1)} - \frac{2}{x^3}\ln(x+1)\right).$$

(g) $\quad \dfrac{d}{dx} \ln(\log_7 e^{x^2}) = \dfrac{d}{dx} \ln(x^2 \log_7 e) = \dfrac{d}{dx} \ln x^2$

$\qquad = \dfrac{d}{dx}\,(2 \ln |\,x\,|\,) = \dfrac{2}{x}.\ \square$

2.28 VARIATIONS ON EXERCISE 2.23

(1)

(a) $\quad \dfrac{d}{dx}\left(\dfrac{x^2-16}{x+4}\right) = \dfrac{d}{dx}\,(x-4) = 1.$

(c) $\quad \dfrac{d}{dx}\left(\dfrac{x^3-1}{x^{3/2}-1}\right) = \dfrac{d}{dx}\,(x^{3/2}+1) = \dfrac{3}{2}\,x^{1/2}.$

(e) $\quad \dfrac{d}{dx}\cos\left(\dfrac{4x^2+2}{16x^4+8x+2}\right) = -\sin\left(\dfrac{4x^2+2}{16x^4+8x+2}\right)\dfrac{8x(16x^4+8x+2)-4x^2(64x^3+8)}{(16x^4+8x+2)^2}$

(2)

(a) $\quad \dfrac{d}{dt}\left(\dfrac{\cos^2 t}{1-\sin^2 t}\right) = \dfrac{d}{dt}\,(1) = 0.$

(c) $\quad \dfrac{d}{dx}\,(x^3 e^{\cos x}) = 3x^2 e^{\cos x} + x^3 e^{\cos x}(-\sin x).$

(e) $\quad \dfrac{d}{dx}\,(\log_a(\,|\,x^2\,|\,))^{1/2} = \dfrac{d}{dx}\,(2\log_a |\,x\,|\,)^{1/2} = \left(\dfrac{2}{\ln a}\right)^{\frac{1}{2}}\dfrac{d}{dx}\,(\ln |\,x\,|\,)^{1/2}$

$\qquad = \left(\dfrac{2}{\ln a}\right)^{\frac{1}{2}} \cdot \dfrac{1}{2}\,(\ln |\,x\,|\,)^{-1/2}\,\dfrac{1}{x}.$

(g) $\quad \dfrac{d}{dx}\left(\dfrac{\ln x^{1/2}+1}{\ln x^{1/5}+2}\right)^{2/7} = \dfrac{d}{dx}\left(\dfrac{\frac{1}{2}\ln x+1}{\frac{1}{5}\ln x+2}\right)^{2/7}$

$\qquad = \dfrac{2}{7}\left(\dfrac{\frac{1}{2}\ln x+1}{\frac{1}{5}\ln x+2}\right)^{-5/7}\dfrac{\frac{1}{2x}\left(\frac{1}{5}\ln x+2\right) - \frac{1}{5x}\left(\frac{1}{2}\ln x+1\right)}{\left(\frac{1}{5}\ln x+2\right)^2}$

$$= \frac{2}{7} \left(\frac{\frac{1}{2} \ln x + 1}{\frac{1}{5} \ln x + 2} \right)^{-5/7} \cdot \frac{\frac{4}{5x}}{\left(\frac{1}{5} \ln x + 2 \right)^2} \cdot$$

(3)

(b) $\dfrac{d}{dx} \csc(2x+1) = -\cot(2x+1) \csc(2x+1) \cdot 2$

$\therefore \dfrac{d^2}{dx^2} \csc(2x+1) = -2[-2\csc^3(2x+1) - \cot^2(2x+1) \cdot \csc(2x+1)]$

$= 4\csc(2x+1)[\csc^2(2x+1) + \cot^2(2x+1)].$

(d) $\dfrac{d}{dx} \ln(|\cot 3x|) \Big|_{x=\frac{\pi}{21}} = \dfrac{-1}{\cot 3x} \cdot \csc^2 3x \cdot 3 \Big|_{x=\frac{\pi}{21}}$

$= -\dfrac{3\csc^2 \dfrac{\pi}{7}}{\cot \dfrac{\pi}{7}} \cdot$

(f) $f(x) = \log_5 5^{(x^2+1)} = x^2+1, \therefore f(x) = 2x.$

(h) $\dfrac{d}{dx} \tan^2(x^3) = 2\tan(x^3) \cdot \sec^2(x^3) \cdot 3x^2$

$= 6x^2 \tan(x^3)(\tan^2(x^3)+1) = 6x^2(\tan^3(x^3) + \tan(x^3))$

$\dfrac{d^2}{dx^2} \tan^2(x^3) = 6x^2(3\tan^2(x^3)\sec^2(x^3) 3x^2 + \sec^2(x^3) \cdot 3x^2))$

$+ 12x(\tan^3(x^3) + \tan(x^3))$

$= 18x^4(3\tan^4(x^3) + 4\tan^2(x^3)+1) + 12x(\tan^3(x^3) + \tan(x^3))$

$\dfrac{d^2}{dx^2} \tan^2(x^3) = 18x^4(12\tan^3(x^3) \cdot \sec^2(x^3) 3x^2 + 8\tan(x^3)\sec^2(x^3) 3x^2)$

$+ 72x^3(3\tan^4(x^3) + 4\tan^2(x^3) + 1)$

$$+ 12x(3 \tan^2 (x^3) \sec^2 (x^3) \, 3x^2 + \sec^2 (x^3) \, 3x^2$$

$$+ 12(\tan^3 (x^3) + \tan (x^3)).$$

(4)

(a) $\dfrac{d}{dx} \, 2^{3^{4^x}} = \ln 2 \cdot 2^{3^{4^x}} \cdot \dfrac{d}{dx} (3^{4^x}) = \ln 2 \cdot 2^{3^{4^x}} \cdot \ln 3 \cdot 3^{4^x} \cdot \dfrac{d}{dx} (4^x)$

$\qquad = (\ln 2) \cdot (\ln 3)(\ln 4) \cdot 2^{3^{4^x}} \cdot 3^{4^x} \cdot 4^x.$

(c) $\dfrac{d}{dx} \, x^{\ln^2 x} = \dfrac{d}{dx} \, e^{\ln^3 x} = e^{\ln^3 x} \cdot (3 \, \ln^2 x) \cdot \dfrac{1}{x}.$

(e) $\dfrac{d}{dx} \, (x^{1/4x}) = \dfrac{d}{dx} \, e^{\frac{\ln x}{4x}} = e^{\frac{\ln x}{4x}} \cdot \dfrac{1 - \ln x}{4x^2}$

(g) $\dfrac{d}{dx} \, \ln (\log_{10}(x)) = \dfrac{d}{dx} \, \ln \left(\dfrac{\ln x}{\ln 10} \right) = \dfrac{d}{dx} \, [\ln(\ln x) - \ln(\ln 10)] = \dfrac{1}{\ln x} \cdot \dfrac{1}{x}.$

2.29 VARIATIONS ON EXERCISE 2.23

(1)

(b) $\cos \left(\dfrac{1-x^2}{1+x^2} \right) = \cos \left(\dfrac{2}{1+x^2} - 1 \right)$

$\therefore \, \dfrac{d}{dx} \, \cos \left(\dfrac{1-x^2}{1+x^2} \right) = - \sin \left(\dfrac{1-x^2}{1+x^2} \right) \cdot 2 \cdot (-1)(1+x^2)^{-2} \cdot 2x$

$\qquad = 4x(1+x^2)^{-2} \sin \left(\dfrac{1-x^2}{1+x^2} \right)$

(d) $\dfrac{d}{dx} \, \ln (1+2\sqrt{x}) = \dfrac{1}{1+2\sqrt{x}} \cdot x^{-1/2} = \dfrac{1}{2x+\sqrt{x}}.$

(f) $\dfrac{d}{dx} \left(\dfrac{2}{3x^2+4x+1} \right) = 2 \cdot (-1)(3x^2 \, 4x+1)^{-2}(6x+4)$

$\qquad = - 4(3x+2)(3x^2+4x+1)^{-2}.$

(2)

(b) $\dfrac{d}{dx} \left(e^{\sin x} \cdot e^{\cos x}\right) = \dfrac{d}{dx} \left(e^{\sin x + \cos x}\right)$

$= e^{\sin x + \cos x} (\cos x - \sin x).$

(d) $\dfrac{d}{dx} (\ln x^2)^{1/2} = \dfrac{d}{dx} (2\ln |x|)^{1/2} = \sqrt{2} \cdot \dfrac{1}{2} (\ln |x|)^{-1/2} \cdot \dfrac{1}{x}$

$= \dfrac{1}{x\sqrt{\ln x^2}}.$ Or, $\dfrac{d}{dx} (\ln x^2)^{\frac{1}{2}} = \dfrac{1}{2} (\ln x^2)^{-\frac{1}{2}} \cdot \dfrac{2x}{x^2} = \dfrac{1}{x\sqrt{\ln x^2}}$

(f) $\dfrac{d}{dx} (e^{1+\ln x}) = e^{1+\ln x} \cdot \dfrac{1}{x} = e.$ (Note: $e^{1+\ln x} = e^1 e^{\ln x} = ex$).

(3)

(b) $\dfrac{d}{dx} \sec(3x) = 3 \tan(3x) \cdot \sec(3x)$

$\therefore \dfrac{d^2}{dx^2} \sec(3x) = 3[3 \sec^2(3x)\sec(3x) + 3 \tan^2(3x) \cdot \sec(3x)]$

$= 3 \sec(3x)[\sec^2(3x) + \tan^2(3x)].$

(d) $\dfrac{d}{dx} \ln(e^{|x|}) = \dfrac{d}{dx} |x| = \begin{cases} 1 & \text{for } x > 0 \\ -1 & \text{for } x < 0. \\ \text{not exist} & \text{for } x = 0 \end{cases}$

(f) $\dfrac{d}{dx} \left(e^{x^2} \cdot e^{3x} e^{\sin x}\right) = \dfrac{d}{dx} e^{x^2 + 3x + \sin x}$

$= e^{x^2 + 3x + \sin x} (2x + 3 + \cos x).$

(h) $\dfrac{d}{dx} (5 \cos x \cdot \sin x)^3 = \dfrac{d}{dx} \left(\dfrac{5}{2} \sin 2x\right)^3 = \left(\dfrac{5}{2}\right)^3 \dfrac{d}{dx} (\sin 2x)^3$

$= \left(\dfrac{5}{2}\right)^3 3 \sin^2 2x \cdot \cos 2x \cdot 2 = 6 \left(\dfrac{5}{2}\right)^3 \sin^2 2x \cdot \cos 2x.$

(4)

(a) $\dfrac{d}{dx} \left(\dfrac{1}{x}\right)^x = \dfrac{d}{dx} e^{x \ln \frac{1}{x}} = \dfrac{d}{dx} e^{-x \ln x} = e^{-x \ln x}(-1 - \ln x) - \left(\dfrac{1}{x}\right)^x (1 + \ln x).$

(c) $\dfrac{d}{dx} (\ln x)^x = \dfrac{d}{dx} e^{x \ln(\ln x)} = e^{x \ln(\ln x)}(\ln(\ln x) + \dfrac{1}{\ln x}) = (1{-}x)^x \left(\ln(\ln x) + \dfrac{1}{\ln x}\right).$

(e) Let $f(x) = \left(\dfrac{x^2+x+1}{2x+4}\right)^3 \left(\dfrac{x^2+1}{2-3x}\right)^{-4}.$

then $\ln f(x) = 3[\ln (x^2+x+1) - \ln (2x+4)] - 4[\ln(x^2+1) - \ln (2-3x)],$

differentiate on both sides to get

$$\frac{f'\ (x)}{f(x)} = 3\left[\frac{2x+1}{x^2+x+1} - \frac{1}{x+2}\right] - 4\left[\frac{2x}{x^2+1} - \frac{-3}{2-3x}\right]$$

therefore

$$f'\ (x) = \left(\frac{x^2+x+1}{2x+4}\right)^3 \left(\frac{x^2+1}{2-3x}\right)^{-4} \cdot \left[\frac{3(x^2+4x+1)}{(x^2+x+1)(x+2)} + \frac{4(3x^2-4x-3)}{(x^2+1)(2-3x)}\right] \qquad \cdot \qquad \square$$

2.30 VARIATIONS ON EXERCISE 2.23

(1)

(a) $\dfrac{d}{dx} [(3x+1)^2 (x^2+4x+1)^3].$ Use product rule.

(b) $\dfrac{d}{dx} \sin^2 (x^3{-}1) = 2 \sin (x^3{-}1) \dfrac{d}{dx} \sin (x^3{-}1)$ (chain rule).

(c) $\dfrac{d}{dx} (\sqrt{\sin x}\ (x^3+1/x)) = \sqrt{\sin x} \cdot \dfrac{d}{dx} (x^3+1/x) + (x^3+1/x) \dfrac{d}{dx} \sqrt{\sin x}$

$= \sqrt{\sin x} \left(3x^2{-}\dfrac{1}{x^2}\right) + \left(x^3{+}\dfrac{1}{x}\right) \cdot \dfrac{1}{2} (\sin x)^{-1/2} \dfrac{d}{dx} \sin x$ (chain rule).

(d) $\dfrac{d}{dx} \left(\dfrac{x^3+1}{x+1}\right) = \dfrac{d}{dx} (1{-}x+x^2) = -1 + 2x.$

(e) $\ln\left(\dfrac{x^2+\sin x}{2+\cos 3x}\right) = \ln(x^2+\sin x) - \ln (2+\cos 3x)$

$\therefore \quad \dfrac{d}{dx} \ln \left(\dfrac{x^2+\sin x}{2+\cos 3x}\right) = \dfrac{d}{dx} \ln (x^2+\sin x) - \dfrac{d}{dx} \ln (2+\cos x))$

(f) $\dfrac{d}{dx} \left(\dfrac{2}{1+x} \right)^{3/2} = 2^{3/2} \cdot \dfrac{d}{dx} (1+x)^{-3/2} = 2^{3/2} \cdot \left(-\dfrac{3}{2} \right) (1+x)^{-3/2-1} \cdot \dfrac{d}{dx} (x+1)$

$\qquad = 2^{3/2} \cdot \left(-\dfrac{3}{2} \right) (1+x)^{-5/2} = -3 \sqrt{2} \cdot (1+x)^{-5/2}.$ $\qquad\qquad$ □

(2)

(a) $\dfrac{d}{dx} e^{(\tan x - \sec x)} = e^{(\tan x - \sec x)} \cdot \dfrac{d}{dx} (\tan x - \sec x)$ $\quad$ (chain rule)

$\qquad = e^{(\tan x - \sec x)} (\sec^2 x - \tan x \cdot \sec x)$

(b) $\dfrac{d}{dx} \left((\ln x)e^x \right) = (\ln x) \dfrac{d}{dx} e^x + e^x \dfrac{d}{dx} \ln x$ $\quad$ (product rule)

$\qquad = (\ln x)e^x + e^x \cdot \dfrac{1}{x} = e^x \left(\dfrac{1}{x} + \ln x \right).$

(e) $\dfrac{d}{dx} (\ln \sqrt{x}) = \dfrac{1}{\sqrt{x}} \cdot \dfrac{d}{dx} (\sqrt{x})$ $\quad$ (chain rule).

(d) $\dfrac{d}{dx} (\sin (\ln x^{3/2})) = \dfrac{d}{dx} \left(\sin \left(\dfrac{3}{2} \cdot \ln x \right) \right).$ $\quad$ Use chain rule.

(e) $\dfrac{d}{dx} \left(\dfrac{\tan x + 4\sin x + 1}{\cos(2x) - 3\cot x + x} \right)^4 = 4 \left(\dfrac{\tan x + 4\sin x + 1}{\cos(2x) - 3\cot x + x} \right)^3 \cdot \dfrac{d}{dx} \left(\dfrac{\tan x + 4\sin x + 1}{\cos(2x) - 3\cot x + x} \right).$

(f) $\dfrac{d}{dx} \left[\left(\dfrac{2x+1}{x^3-4} \right)^2 \sin 3x \right] = \left(\dfrac{2x+1}{x^3-4} \right)^2 \cdot \dfrac{d}{dx} \sin 3x + \sin 3x \cdot \dfrac{d}{dx} \left(\dfrac{2x+1}{x^3-4} \right)^2.$

(g) $\dfrac{d}{dx} (\sin e^x) = (\cos e^x) \cdot \dfrac{d}{dx} e^x = e^x \cos e^x.$

(h) $\dfrac{d}{dx} \left[\tan^2 \left(\dfrac{-3x+4}{2x+1} \right) \right] = 2\tan \left(\dfrac{-3x+4}{2x+1} \right) \cdot \dfrac{d}{dx} \tan \left(\dfrac{-3x+4}{2x+1} \right)$ $\qquad$ □

(3)

(a) $\dfrac{d}{dx} (\sec^2 x + \tan x) \Big|_{x = \frac{\pi}{3}}$

$$= (2\sec x \cdot \frac{d}{dx}\sec x + \sec^2 x)\Big|_{x=\frac{\pi}{3}} = (2\sec^2 x \cdot \tan x + \sec^2 x)\Big|_{x=\frac{\pi}{3}}$$

$$= (2 \cdot 2^2 \cdot \sqrt{3} + 2^2) = 4(2\sqrt{3}+1).$$

(b) $\quad \dfrac{d}{dx} \ln (3+4x^2)\Big|_{x=1} = \dfrac{1}{3+4x^2} \cdot \dfrac{d}{dx} (3+4x^2)\Big|_{x+1}$

$$= \frac{1}{3+4^2}(8x)\Big|_{x=1} = \frac{8}{7}.$$

(c) $\quad \dfrac{d}{dx} \ln |x| = \dfrac{1}{|x|} \cdot \dfrac{d}{dx} |x|$

$$= \begin{cases} \dfrac{1}{|x|} \cdot 1 & x > 0 \\ \dfrac{1}{|x|} \cdot (-1) & x < 0 \\ \text{not exist} & x = 0 \end{cases} = \begin{cases} \dfrac{1}{x} & x > 0 \\ \dfrac{1}{x} & x < 0 \\ \text{not exist} & x = 0 \end{cases} = \begin{cases} \dfrac{1}{x} & \text{for } x \neq 0 \\ \text{not exist} & \text{for } x = 0 \end{cases}$$

$$\therefore \frac{d^2}{dx^2} \ln |x| \equiv \begin{cases} \dfrac{d}{dx}\left(\dfrac{1}{x}\right) \\ \text{does not exist for } x = 0 \end{cases} = \begin{cases} -\dfrac{1}{x^2} & \text{for } x \neq 0 \\ \text{does not exist} & \text{for } x = 0 \end{cases}$$

(d) $\quad \dfrac{d}{dx}(x \cdot e^{\cos x})\Big|_{x=\pi} = (x\dfrac{d}{dx}e^{\cos x} + e^{\cos x}\dfrac{d}{dx}x)\Big|_{x=\pi}$

$$= \pi \cdot e^{\cos \pi} \cdot \frac{d}{dx}(\cos x)\Big|_{x=\pi} + e^{\cos \pi} = \pi e^{-1}(-\sin \pi) + e^{-1} = e^{-1}.$$

(e) Let $y(x) = \sec x + \tan x$, then by chain rule and (c), we get

$$\frac{d}{dx} \ln |\sec x + \tan x| = \frac{d}{dx} \ln |y(x)| = \left(\frac{d}{dy} \ln |y|\right) \cdot \frac{dy}{dx}$$

$$= \frac{1}{y}(\sec x \tan x + \sec^2 x) = \frac{1}{y}\sec x(\tan x + \sec x) = \sec x \quad .$$

(for $\tan x + \sec x \neq 0$.)

(f) $\quad \dfrac{d}{dx} \sin^5(\cos x) = 5\sin^4(\cos x) \cdot \dfrac{d}{dx}(\cos x) = 5\sin^4(\cos x) \cdot (-\sin x).$

(g) $\dfrac{d}{dx} \left(\dfrac{x^2\sin x - 3x + 1}{\ln x + e^x} \right)^2 = 2\left(\dfrac{x^2\sin x - 3x + 1}{\ln x + e^x} \right) \dfrac{d}{dx} \left(\dfrac{x^3\sin x - 3x + 1}{\ln x + e^x} \right)$

(4)

(a) $\dfrac{d}{dx} \left(\dfrac{1}{x} \right)^{2x} = \dfrac{d}{dx} \ e^{2x\ln \frac{1}{x}} = \dfrac{d}{dx} \ e^{-2x\ln x}$

(b) $\dfrac{d}{dx}(\ln e^{3x/2}) = \dfrac{d}{dx} \left(\dfrac{3x}{2} \right) = \dfrac{3}{2}.$

(c) $\dfrac{d}{dx} \ (e^x x^x) = e^x \cdot \dfrac{d}{dx} \ x^x + x^x \cdot \dfrac{d}{dx} \ e^x$

(d) Let $y(x) = \ln \ |x|$

then $\ln \dfrac{1}{|y(x)|} = - \ln \ |y(x)|$

by chain rule and (3)(c), we get

$\dfrac{d}{dx} \ \ln \left| \dfrac{1}{\ln |x|} \right| = - \dfrac{d}{dx} \ \ln \ |y(x)| = - \dfrac{d}{dy} \ \ln |y| \cdot \dfrac{dy}{dx}$

$= - \dfrac{1}{y} \cdot \dfrac{d \ln |x|}{dx} = - \dfrac{1}{y} \cdot \dfrac{1}{x} = - \dfrac{1}{x \ln |x|}$ for $x \neq \pm 1.$

(e) $\dfrac{d}{dx} \ \ln \ (\sin(e^x)) = (\cot e^x) \cdot e^x.$

(f) $\dfrac{d}{dx} \ \tan(\log_5 x^2) = \sec^2(\log_5 x^2) \cdot \dfrac{d}{dx} \ \log_5 x^2$

$= \sec^2(\log_5 x^2) \cdot \dfrac{d}{dx} \ (2 \log_5 x) = 2\sec^2(\log_5 x^2) \cdot \dfrac{d}{dx} \left(\dfrac{\ln x}{\ln 5} \right)$

$= \dfrac{2}{\ln 5} \ \sec^2(\log_5 x^2) \ \dfrac{1}{x}.$

(g) $\dfrac{d}{dx} \ (x(\ln x)^{1/2}) = x \ \dfrac{d}{dx} \ (\ln x)^{1/2} + (\ln x)^{1/2} \cdot \dfrac{dx}{dx}$

$= x \cdot \dfrac{1}{2} \ (\ln x)^{-1/2} \cdot \dfrac{d}{dx} \ \ln x + (\ln x)^{1/2}$

$= \dfrac{1}{2} \ (\ln x)^{-1/2} + (\ln x)^{1/2}.$

2.33 PRACTICE EXAM 1

(1)

    (a) $f(t) = 2t^4 + 3$, $g(x) = (-x-1)^{-2}$ so $h(x)z = (f(g(x)) =$
$f((-x-1)^{-2}) = 2(-x-1)^{-8} + 3 = 2(x+1)^{-8} + 3$.

    (b) $f(g) = ln\ g$, $g(z) = \sin z$

        $\therefore\ h(z) = f(g(z)) = f(\sin z) = ln(\sin z)$.

        $h(z)$ is defined for $\sin z > 0$, i.e., $2k\pi < z < (2k+1)\pi$

        where k is any integer.

(2)

    (a) $h(x) = (x-2)^3 + x^2 - 4x-1$, $g(x) = x-2$.

        Rewrite $h(x) = (x-2)^3 + (x-2)^2 - 5 = g^3 + g^2 - 5$

        $\therefore\ f(g) = g^3 + g^2 - 5$.

    (b) $h(x) = \dfrac{x^2-6x}{(x-3)^3}$, $g(x) = x - 3$.

        Rewrite $h(x) = \dfrac{[x-3)^2-9]}{(x-3)^3} = \dfrac{(g^2-9)}{g^3}$.

        $\therefore\ f(g) = \dfrac{g^2-9}{g^3}$.

(3)

    (a) $D(x) = (A(x) + B(x) + C(x))^2$, $A(2) = B(2) = 1$, $C(2) = 3$. $A'\ (2) = B'\ (2) = 1$, $C'\ (2) = 2$.

        $\therefore\ D'\ (x) = 2(A(x) + B(x) + C(x))\ (A'\ (x) + B'\ (x) + C'\ (x))$.

        $\Rightarrow\ D'\ (2) = 2(2+3)\ (2+2) = 40$.

(b) $y = 2x^3 + x^2 + 5.$ ∴ $y'(x) = 6x^2 + 2x.$

⟹ $y(1) = 8.$ $y'(1) = 89.$

so the equation of the tangential line at $x = 1$ is $y - 8 = 8(x-1)$, or $8x - y = 0.$

(4)

(a) $\dfrac{d}{dx}\left(\dfrac{x-1}{x^{1/2}+1}\right)^3 = \dfrac{d}{dx}(x^{1/2} - 1)^3 = 3(x^{1/2} - 1)^2 \cdot \dfrac{1}{2}\, x^{-1/2}.$

(b) $\dfrac{d}{dx}\sin\left(\dfrac{x^2+1}{x^3+1}\right) = \cos\left(\dfrac{x^2+1}{x^3+1}\right) = \dfrac{2x(x^3+1)-3x^2(x^2+1)}{(x^3+1)^2}.$

(5)

(a) $\dfrac{d}{dx}\, t^{\sin^2 t + \cos^2 t} = \dfrac{d}{dx} \cdot T = 1.$

(b) $\dfrac{d}{dx}\, x^2\, e^{\cos x} = 2x\, e^{\cos x} + x^2 \cdot e^{\cos x}(-\sin x)$

$$= e^{\cos x}(2x - x^2\sin x).$$

(c) $\left[\dfrac{d^2}{dx^2}\sin 2x\right]_{x=\frac{\pi}{4}} = -4\sin 2x\,\Big|_{x=\frac{\pi}{4}} = -4.$

(6)

(a) $\dfrac{d}{dx}\, \ln(\,|\cos x\,|\,) = \dfrac{1}{\cos x}(-\sin x) = -\tan x.$

(b) $f'(x) = \dfrac{d}{dx}\, \ln e^{sx} = \dfrac{d}{dx}(5x) = 5.$

(c) $\dfrac{d}{dx}(\ln x^3 - \ln x^7)^3 = \dfrac{d}{dx}(-4\ln x)^3 = -4^3 \cdot 3(\ln x)^2 \cdot \dfrac{1}{x}.$

(7)

(a) $\dfrac{d}{dx}(2x+3)^{\ln x} = \dfrac{d}{dx}e^{(\ln x)\ln (2x+3)}$

$$= e^{(\ln (x))\ln(2x+3)} \cdot \left(\dfrac{\ln (2x+3)}{x} + \dfrac{2\ln x}{2x+3}\right).$$

(b) $\dfrac{d}{dx} (2x+3)^x = \dfrac{d}{dx} e^{x \ln (2x+3)} = e^{x \ln (2x+3)} \left(\ln (2x+3) + \dfrac{2x}{2x+3} \right).$

□

2.43 ANSWERS

(1)

(a) $\arcsin x = \dfrac{\pi}{2} - \arccos x$

$\therefore (\arcsin x)' = -(\arccos x)' = \dfrac{1}{(1-x^2)^{1/2}}.$

(b) $(\arctan x)' = \dfrac{1}{\tan' (\arctan x)} = \dfrac{1}{\sec^2(\arctan x)}.$

Let $\arctan x = \theta,$ then $\tan \theta = x.$

$\Rightarrow \sec^2(\arctan x) = \sec^2\theta = 1 + \tan^2\theta = 1 + x^2.$

$\therefore (\arctan x)' = \dfrac{1}{1+x^2}.$

(c) Note $\text{arccot } x = \dfrac{\pi}{2} - \arctan x.$

$\therefore (\text{arccot } x)' = -(\arctan x)' = \dfrac{-1}{1+x^2}.$

(d) $(\text{arcsec } (x))' = \dfrac{1}{\sec' (\text{arcsec } x)} = \dfrac{1}{\tan(\text{arcsec} x)\sec(\text{arcsec} x)}$

Note

$\sec(\text{arcsec } x) = x$

$\tan(\text{arcsec } x) = \sqrt{\sec^2(\text{arcsec } x)-1} = \sqrt{x^2-1},$

$\therefore (\text{arcsec } x)' = \dfrac{1}{x\sqrt{x^2-1}}.$

(e) $(\text{arccsc } x)' = \left(\dfrac{\pi}{2} - \text{arcsec } x \right)' = -(\text{arcsec } x)' = \dfrac{-1}{x\sqrt{x^2-1}}.$ □

3.20 ANSWERS

(1)

(a) $\lim\limits_{x\to-2} \dfrac{x^2+5x+6}{x+2} = \lim\limits_{x\to-2} \dfrac{(x+2)(x+3)}{x+2} = \lim\limits_{x\to-2} (x+3) = 1.$

(b) Note $\dfrac{x^2-81}{2(x-9)} = \dfrac{(x+9)(x-9)}{2(x-9)} = \dfrac{x+9}{2}$

$\therefore \lim\limits_{x\to9} \dfrac{x^2-81}{2(x-9)} = \lim\limits_{x\to9} \dfrac{x+9}{2} = 9.$

(2)

(a) Using L'HOPITAL's RULE, we get

$$\lim\limits_{x\to0} \frac{\sin x}{x} = \lim\limits_{x\to0} \frac{(\sin x)'}{(x)'} = \lim\limits_{x\to0} \frac{\cos x}{1} = \cos 0 = 1 \ .$$

(b) $\lim\limits_{x\to0} \dfrac{\sin 2x}{\sin x} = \lim\limits_{x\to0} \dfrac{(\sin 2x)'}{(\sin x)'} = \lim\limits_{x\to0} \dfrac{2\cos 2x}{\cos x}$

$= \dfrac{2\cos(2\cdot0)}{\cos 0} = \dfrac{2}{1} = 2.$

(3)

(a) $\lim\limits_{x\to\infty} \dfrac{x+\sin x}{2x} = \lim\limits_{x\to\infty} \left(\dfrac{1}{2} + \dfrac{\sin x}{2x}\right) = \dfrac{1}{2} + \lim\limits_{x\to\infty} \dfrac{\sin x}{2x}$

$\left|\dfrac{\sin x}{2x}\right| \le \dfrac{1}{|2x|} \quad \text{and} \quad \dfrac{1}{|2x|} \to 0 \ \text{as} \ x \to \infty$

therefore $\lim\limits_{x\to\infty} \dfrac{\sin x}{2x} = 0.$

Hence $\lim\limits_{x\to\infty} \dfrac{x+\sin x}{2x} = \dfrac{1}{2}.$

(b) $\lim\limits_{x\to\infty} \dfrac{\ln x}{x^a} = \lim\limits_{x\to\infty} \dfrac{(\ln x)'}{(x^a)'} = \lim\limits_{x\to\infty} \dfrac{1/x}{a\,x^{a-1}}$

$= \lim\limits_{x\to\infty} \dfrac{1}{a\,x^a} = \dfrac{1}{a} \lim\limits_{x\to\infty} \dfrac{1}{x^a} = \dfrac{1}{a}\cdot 0 = 0$

(for $a > 0$)

(4)

(a) $y = -2x^2 + 3x + 5$, $y'(x) = -4x + 3$, $y'(x) = 0 \Rightarrow x = \dfrac{3}{4}$.

It is the maxima.

(b) $y = 12(1-x)/x^2 = 12(x^{-2} - x^{-1})$

$\therefore y'(x) = 12(-2x^{-3} + x^{-2})$

$\qquad y'(x) = 0 \Rightarrow -2x^{-3} + x^{-2} = 0 \Rightarrow x = 2$

is the minima,

$y''(x) = 6x^{-4} - 2x^{-3}$; $x = 3$ is the inflection point.

(5)

(a) $f_\alpha(x) = 2\alpha x(x-3\alpha) \qquad \alpha > 0$.

$\therefore f_\alpha'(x) = 2\alpha(2x-3\alpha) = 4\alpha\left(x - \dfrac{3}{2}\alpha\right)$

$f_\alpha'(x) < 0$ for $x < \dfrac{3}{2}\alpha$.

$f_\alpha'(x) = 0$ for $x = \dfrac{3}{2}\alpha$ (the minima).

$f_\alpha'(x) > 0$ for $x > \dfrac{3}{2}\alpha$.

(b) $f_\alpha(x) = \dfrac{1}{x^2} - \dfrac{\alpha}{x}$

$\therefore f_\alpha'(x) = -\dfrac{2}{x^3} + \dfrac{\alpha}{x^2} = x^{-3}(\alpha x - 2) = \alpha x^{-3}\left(x - \dfrac{2}{\alpha}\right)$.

$\therefore f_\alpha'(x) < 0$ for $0 < x < \dfrac{2}{\alpha}$, $f_\alpha'(x) = 0$ for $x = \dfrac{2}{\alpha}$ and $f_\alpha'(x) > 0$ for $x < 0$ or $x > \dfrac{2}{\alpha}$. $f_\alpha''(x) = 6x^{-4} - 2\alpha x^{-3}$.

$\therefore x = \dfrac{3}{\alpha}$ is the inflection point.

3.22 PARTIAL SOLUTIONS

(1)

$$\text{(a)} \quad \lim_{x \to 3} \frac{3x^3 + 4x^2 - x + 1}{2x^2 + 4x - 5} = \frac{3 \times 3^3 + 4 \times 3^2 - 3 + 1}{2 \times 3^2 + 4 \times 3 - 5}$$

(2)

(a) $\therefore \lim_{x \to 0} \tan x = 0, \quad \lim_{x \to 0} \ln x = -\infty$

$\therefore \lim_{x \to 0} \ln (\tan x) = -\infty$ and note $\lim_{x \to 0} \dfrac{1}{x^3} = \infty$. Therefore

$\lim_{x \to 0} \dfrac{\ln (\tan x)}{x^3} = -\infty.$

(b) $\lim_{x \to 0} \dfrac{\sin x}{2 \tan \sqrt{x}} = \lim_{x \to 0} \dfrac{\cos x}{2 \sec^2 \sqrt{x} \cdot \dfrac{1}{2\sqrt{x}}}$

$= \lim_{x \to 0} \dfrac{\sqrt{x} \cos x}{\sec^2 \sqrt{x}} = \dfrac{0 \cdot \cos 0}{\sec^2(0)} = \dfrac{0}{1} = 0.$

(3)

(a) $\lim_{x \to 0} \dfrac{ex - e \sqrt{x}}{\sqrt{x}} = \lim_{x \to 0} (e \sqrt{x} - e) = -e.$

(4)

(a) $y = x \ln |x| \quad y = 0 \Rightarrow x = \pm 1$

$y'(x) = \ln |x| + 1. \quad y'(x) = 0 \Rightarrow x = \pm e^{-1}$

$y''(x) = \dfrac{1}{x}.$

So $x = -e^{-1}$ is the maxima,

$x = e^{-1}$ is the minima.

(5)

(a) $f_\alpha(x) = x - \dfrac{\alpha}{x}, \quad \alpha > 0.$

$$f_\alpha(x) = 0 \Rightarrow x = \pm \sqrt{\alpha}$$

$$f'_\alpha(x) = 1 + \frac{\alpha}{x^2} > 1.$$

$$f''_\alpha(x) = -\frac{2\alpha}{x^3}.$$

No maxima, minima and inflection points.

3.31 VARIATIONS ON EXERCISE 3.25

(1) Let x be the radius of the base, then the height of the can should be

$$h = \frac{(s - 2 \cdot \pi x^2)}{2\pi x}$$

and hence the volume of the can is

$$V = \pi x^2 h = \frac{x(s - 2\pi x^2)}{2}$$

$$V'(x) = \frac{(s - 6\pi x^2)}{2} \quad V'(x) = 0 \Rightarrow x = \left(\frac{s}{6\pi}\right)^{1/2}$$

and $h = \frac{4\pi x^2}{2\pi x} = 2x = \left(\frac{s}{6\pi}\right)^{1/2}$ therefore, when the radius of the base is $\left(\frac{s}{6\pi}\right)^{1/2}$ and the depth of the can is $2\left(\frac{s}{6\pi}\right)^{1/2}$ the can has the maximum volume.

(2) Suppose he runs x miles first then swims to B. Then it takes time T :

$$T = \frac{x}{\rho} + \frac{\sqrt{\omega^2 + (\beta - x)^2}}{\sigma}$$

$$\therefore T'(x) = \frac{1}{\rho} + \frac{x - \beta}{\sigma\sqrt{\omega^2 + (\beta - x)^2}}$$

$T'(x) = 0 \Rightarrow \rho^2(x-\beta)^2 = \sigma^2(\omega^2 + (x-\beta)^2)$ i.e., $(\rho^2 - \sigma^2)(x-\beta)^2 = \sigma^2\omega^2$. If $\rho > \sigma$ then $x - \beta = -\frac{\sigma\omega}{\sqrt{\rho^2 - \sigma^2}}$, or $x = \beta - \frac{\sigma\omega}{\sqrt{\rho^2 - \sigma^2}}$.

(3) Let l be the length of the board

$$l = \frac{\beta}{\cos t} + \frac{\lambda}{\sin t} \qquad 0 \le t \le \frac{\pi}{2}$$

$\therefore \ l'(t) = \beta \tan t \sec t - \lambda \cot t \csc t$

$$= \beta \frac{\sin t}{\cos^2 t} - \frac{\lambda \cos t}{\sin^2 t}.$$

$l'(t) = 0 \Rightarrow \beta \sin^3 t = \lambda \cos^3 t, \text{ i.e., } \tan^3 t = \frac{\lambda}{\beta}$

or $\tan t = \left(\frac{\lambda}{\beta}\right)^{1/3}$

$\therefore \ \sec t = \sqrt{1 + \left(\frac{\lambda}{\beta}\right)^{2/3}} \quad \csc t = \sqrt{1 + \left(\frac{\beta}{\lambda}\right)^{2/3}}$

and $l_{\min} = \beta \sqrt{1 + \left(\frac{\lambda}{\beta}\right)^{2/3}} + \lambda \sqrt{1 + \left(\frac{\beta}{\lambda}\right)^{2/3}}$

(4) Suppose the length of the shadow is s feet when the person is x feet away from the lamppost. Clearly $\frac{s}{s+x} = \frac{\tau}{\lambda}$.

or $\frac{s}{x} = \frac{\tau}{\lambda - \tau} \quad \therefore \ s = \frac{\tau}{\lambda - \tau} \cdot x.$

$\therefore \ s'(t) = \frac{\tau}{\lambda - \tau} x'(t) = \frac{\tau}{\lambda - \tau} \cdot \rho.$

i.e., the length of his shadow changes at the rate of $\frac{\tau}{\lambda - \tau} \rho$ feet per second.

(5) We have $\omega(r) = R(1 + 0.00025r)^{-2}$

$\therefore \ \frac{d\omega(r)}{dt} = \frac{d\omega}{dr} \cdot \frac{dr}{dt} = -2R(1+0.00025r)^{-3} \cdot 0.00025 \cdot v$

$\therefore \ \frac{d\omega}{dt}\bigg|_{r=M} = -0.0005R \cdot v \cdot (1+0.00025M)^{-3}.$

(6) Let g denote the acceleration caused by gravity, then $h(t) = -\frac{1}{2} gt^2 + v_0 t$

$$h'(t) = 0 \Rightarrow -gt + v_0 = 0, \ \Rightarrow t = \frac{v_0}{g}$$

$\therefore \ h_{\max} = -\frac{1}{2} g\left(\frac{v_0}{g}\right)^2 + v_0 \frac{v_0}{g} = \frac{v_0^2}{2g}. \ h(t) = 0 \Rightarrow t = \frac{2v_0}{g}$

$$\Rightarrow h'(t) = -gt + v_0 = -v_0, \quad \text{when} \quad t = \frac{2v_0}{g}.$$

(7) We have $\tan \phi(t) = \dfrac{dy}{dx}$ or

$$\phi(t) = \arctan\left(\frac{dy}{dx}\right).$$

$$x = a + b \cos t$$
$$\therefore y = c + d \sin t$$
$$x'(t) = -b \sin t$$

$$y'(t) = d \cos t$$

$$\therefore \frac{dy}{dx} = \frac{dy}{dt} \cdot \frac{dt}{dx} = \frac{y'(t)}{x'(t)} = \frac{d \cos t}{-b \sin t} = -\frac{d}{b} \cot(t)$$

therefore $\phi(t) = \arctan\left(-\dfrac{d}{b} \cot(t)\right)$

$$\phi'(t) = \frac{1}{1 + \left(\dfrac{d}{b} \cot(t)\right)^2} \left(-\frac{d}{b}\right)(-\csc^2(t))$$

$$\Rightarrow$$

$$= \frac{b \cdot d \csc^2 t}{b^2 + d^2 \cot^2(t)}$$

(8) We have $\dfrac{dl}{dt} = \sqrt{x'(t)^2 + y'(t)^2} = \sqrt{b^2 \sin^2 t + d^2 \cos^2 t}.$

4.5 SOLUTIONS

(1) $\displaystyle \int x \sqrt{x^2+2} \, dx = \int \sqrt{x^2+2} \cdot \frac{1}{2} \, dx^2 = \frac{1}{2} \int (x^2+2)^{1/2} \, d(x^2+2))$

$$= \frac{1}{2} \cdot \frac{1}{1 + \dfrac{1}{2}} (x^2+2)^{1+1/2} = \frac{1}{3} (x^2+2)^{3/2} + C$$

(2) $\displaystyle \int \frac{x^{1/4}}{5+x^{5/4}} \, dx = \int \frac{\dfrac{4}{5} \, dx^{5/4}}{5+x^{5/4}} = \frac{4}{5} \int (5+x^{5/4})^{-1} \, d(5+x^{5/4})$

$$= \frac{4}{5} \ln (5+x^{5/4}) + C$$

(3) $\int \frac{\sqrt{x}-1}{\sqrt{x}} \, dx = \int \left(1 - \frac{1}{\sqrt{x}}\right) dx = \int dx - \int x^{-1/2} \, dx = x - 2\sqrt{x} + C$

(4) $\int \frac{8x}{1+e^2x} \, dx \quad \int \left(\frac{8e^{-2}(1+e^2x)-8e^{-2}}{1+e^2x}\right) dx = 8e^{-2} \int dx - 8e^{-2} \int (1+e^2x)^{-1} dx$

$$= 8e^{-2}x - 8e^{-4} \, ln(1+e^2x) + C$$

(5) $\int \frac{x \cos \sqrt{5x^2+1}}{\sqrt{5x^2+1}} \, dx = \int \cos \sqrt{5x^2+1} \cdot \frac{1}{5} \, d\sqrt{5x^2+1} = \frac{1}{5} \sin \sqrt{5x^2+1} + C$

(6) $\int \sin^{3/4}(2x)\cos(2x)dx = \int \sin^{3/4}(2x) \cdot \frac{1}{2} \, d\sin(2x) = \frac{1}{2} \cdot \frac{4}{7} \sin^{7/4}(2x) + C$

(7) $\int \tan^2(3t)\sec^2(3t)dt = \int \tan^2(3t) \cdot \frac{1}{3} \, d\tan(3t) = \frac{1}{9} \tan^3(3t) + C$

(8) $\int e^x\sec^2(e^x)dx = \int \sec^2(e^x)de^x = \tan(e^x) + C$

(9) $\int \frac{2 \, ln^3x}{x} \, dx = 2 \int ln^3x \, d \, ln \, x = \frac{1}{2} \, ln^4x + C$

(10) $\int \frac{6x}{(x^2+9)^3} \, dx = \int \frac{3 \, dx^2}{(x^2+9)^3} = 3 \int (x^2+9)^{-3} \, d(x^2+9) = -\frac{3}{2} (x^2+9)^{-2} + C$

(11)

$\int \frac{e^{log_2x}}{x} \, dx = \int \frac{e^{(log_2(e)ln(x))}}{x} \, dx = \int \frac{x^{log_2e}}{x} \, dx$

$\qquad = \int x^{log_2e-1} \, dx = \frac{1}{log_2e} x^{log_2e} + C \qquad$ or

$\int \frac{e^{log_2x}}{x} \, dx = ln(2) \int e^g dg = ln(2)e^g = ln(2) \, e^{log_2(x)} = ln(2) \, x^{log_2(e)}$

(12) $\int x \sec x^2\tan x^2 \, dx = \int \sec x^2 \tan x^2 \cdot \frac{1}{2} \, dx^2 = \frac{1}{2} \sec x^2 + C$

(13)

$\int \frac{u}{(u^2+29)^{40}} \, du = \int \frac{\frac{1}{2} \, d(u^2+29)}{(u^2+29)^{40}} = \frac{1}{2} \int (u^2+29)^{-40}d(u^2+29)$

$\qquad = -\frac{1}{2\times39} (u^2+29)^{-39} + C$

$\square$

(14) $\int \dfrac{\csc^2 x}{\sqrt{1+5\cot x}}\, dx = \int \dfrac{-d\cot x}{\sqrt{1+5\cot x}} = -\dfrac{1}{5}\int (1+5\cot x))^{-1/2} d(1+5\cot x)$

$\qquad\qquad\qquad\qquad\qquad = -\dfrac{2}{5}(1+5\cot x)^{1/2} + C$

(15) $\int \dfrac{(\arctan x)^{3/2}}{1+x^2}\, dx = \int (\arctan x)^{3/2}\, d\,(\arctan x)$

$\qquad\qquad\qquad\qquad\qquad = \dfrac{2}{5}(\arctan x)^{5/2} + C$

4.9 PARTIAL SOLUTIONS

(8) $\int \dfrac{\cos\sqrt{x}}{2\sqrt{x}}\, dx = \int \cos\sqrt{x}\, d\sqrt{x} = \sin\sqrt{x} + C$

(9) $\int 5x^2(4x^3+1)^{4/5}dx = \dfrac{5}{3}\int (4x^3+1)^{4/5}dx^3$

$\qquad\qquad\qquad = \dfrac{5}{12}\int (4x^3+1)^{4/5}d(4x^3+1) = \dfrac{5}{12}\cdot\dfrac{5}{9}(4x^3+1)^{9/5} + C$

(10) $\int \dfrac{\ln^6\sqrt{2x+1}}{2x+1}\, dx = \dfrac{1}{2}\int \ln^6\sqrt{2x+1}\; d\,\ln(2x+1)$

$\qquad = \dfrac{1}{2}\cdot\left(\dfrac{1}{2}\right)^6\int \ln^6(2x+1)d\,\ln(2x+1) = \dfrac{1}{7}\left(\dfrac{1}{2}\right)^7 \ln^7(2x+1) + C$

(11) $\int \sqrt{\sin(e\theta)}\,\cos(e\theta)d\theta = \dfrac{1}{e}\int (\sin\theta)^{1/2}\, d\sin e\theta$

$\qquad = \dfrac{2}{3e}(\sin(e\theta))^{3/2} + C$

(12) $\int (10\sin t)^7 \cos t\, dt = \dfrac{1}{10}\int (10\sin t)^7 d(10\sin t)$

$\qquad = \dfrac{1}{10}\cdot\dfrac{1}{8}(10\sin t)^8 + C$

(13) $\int (\sin x + x\cos x)e^{x\sin x}dx = \int e^{x\sin x}d(x\sin x)$

$\qquad = e^{x\sin x} + C$

(14) $\int \sin(\sin \theta)\cos \theta \ d\theta = \int \sin(\sin \theta)d\sin \theta$

$$= - \cos(\sin \theta) + C$$

(15) $\int \dfrac{\sin \theta}{(5-\cos \theta)^{10}} \ d\theta = \int (5-\cos \theta)^{-10}d(5-\cos \theta)$

$$= - \dfrac{1}{9} (5-\cos \theta)^{-9} + C.$$

4.21 VARIATIONS ON EXERCISE 4.15

(1) $A = \int\limits_0^1 3^x \ dx = \int\limits_0^1 e^{x \ln 3} \ dx = \dfrac{1}{\ln 3} e^{x \ln 3} \Big|_0^1 = \dfrac{2}{\ln 3}.$

(2) $y = 3^x, \quad x = \log_3 y$

$$\therefore A = \int\limits_{3/2}^3 \log_3 y \ dy = \dfrac{1}{\ln 3} \int\limits_{3/2}^3 \ln y \ dy$$

$$= \dfrac{1}{\ln 3} y(\ln y - 1) \Big|_{3/2}^3 = \dfrac{3}{\ln 3} (\ln 3 - 1) - \dfrac{3}{2 \ln 3} (\ln 3/2 - 1)$$

(3) $y = - x^2 + 3. \ y = 3^x.$ We have $3^a = - a^2 + 3, 3^b = - b^2 + 3.$

$$A = \int\limits_a^b (-x^2 + 3)dx - \int\limits_a^b 3^x \ dx$$

$$= \left[\dfrac{-x^3}{3} + 3x \right]_a^b - \dfrac{1}{\ln 3} 3^x \Big|_a^b$$

$$= - \dfrac{1}{3} (b^3 - a^3) + 3(b-a) - \dfrac{1}{\ln 3} [(-b^2 + 3) - (-a^2 + 3)]$$

$$= - \dfrac{1}{3} (b^3 - a^3) + 3(b-a) + \dfrac{1}{\ln 3} (b^2 - a^2),$$

where $b = 0.7885, \ a = - 1.686.$

$$\therefore A \approx 0.1277.$$

(5) $A = \int\limits_0^{\pi/2} \frac{1}{2} r^2(\phi)d\phi = \int\limits_0^{\pi/2} \frac{1}{2} (1+2\cos\phi)^2 d\phi$

$$= \frac{1}{2} \int\limits_0^{\pi/2} (1+4\cos\phi+4\cos^2\phi) = \frac{1}{2} \int\limits_0^{\pi/2} (1+4\cos\phi+2(1+\cos2\phi))d\phi \quad \square$$

4.22 VARIATIONS ON EXERCISE 4.15

(1) $A = \int\limits_{-1}^2 (x^3+1)dx = 3 + \int\limits_{-1}^2 x^3 \, dx$

$$= 3 + 17/4 = 29/4.$$

(3) $A = \int\limits_a^b (-x^4+16-1.5^x)dx$

$$= 16(b-a) - \frac{b^5-a^5}{5} - (1.5^a-1.5^b)/\ln 1.5.$$

Note $1.5^a = 16-a^4$, $1.5^b = 16 - b^4$.

$$\therefore A = 16(b-a) - \frac{b^5-a^5}{5} + \frac{b^4-a^4}{\ln 1.5}$$

where $a \approx -1.986$, $b \approx 1.928$.

$$\therefore A \approx 52.820.$$

(4) $x = t - \sin t$, $y = 1 - \cos t$.

$\therefore x'(t) = 1 - \cos t \geq 0$, $\therefore x(t)$ is an increasing function.

$$\therefore A = \int\limits_0^{2\pi} (1-\cos t)d(t-\sin t) = \int\limits_0^{2\pi} (1-\cos t)^2 \, dt$$

$$= \int\limits_0^{2\pi} \left(1-2\cos t + \frac{1+\cos t}{2}\right)dt = \frac{3}{2} \cdot 2\pi = 3\pi.$$

(5) $A = \int\limits_{0}^{\pi/2} \frac{1}{2} \cos^2(2\phi)d\phi = \int\limits_{0}^{\pi/2} \frac{1}{4} (1+\cos4\phi)d\phi.$

4.23 VARIATIONS ON EXERCISE 4.15

(1) $A = \int\limits_{-1}^{0} \left(\frac{x^4}{4} - \frac{2x^3}{3} \right)dx + \int\limits_{0}^{\frac{8}{3}} \left(\frac{2x^3}{3} - \frac{x^4}{4} \right)dx.$

(3) $y = -\frac{x^4}{4} - 2x^3 + 4x^2 - 2.$

Compute $x_1 \approx -9.649,\ x_2 = -0.623$

$\therefore A = \int\limits_{x_1}^{x_2} \left(-\frac{x^4}{4} - 2x^3 + 4x^2 - 2 \right)dx$

$= \left[-\frac{x^5}{20} - \frac{x^4}{2} + \frac{4}{3} x^3 - 2x \right]_{x_1}^{x_2} \approx +1483.627.$

(4) $x = 1 + \sin t,\ y = \cos t\ (1+\sin t).\quad -\pi/2 < t < \pi/2$

$A = \int\limits_{-\frac{\pi}{2}}^{\frac{\pi}{2}} \cos t(1+\sin t)x'\ (t)dt$

$= \int\limits_{-\frac{\pi}{2}}^{\frac{\pi}{2}} \cos t(1+\sin t)\cos t\ dt = \int\limits_{-\frac{\pi}{2}}^{\frac{\pi}{2}} \left(\frac{1+\cos t}{2} \right)dt + \int\limits_{-\frac{\pi}{2}}^{\frac{\pi}{2}} \cos^2 t \sin t\ dt$

$= \frac{\pi}{2} + \frac{1}{4} \sin 2t \left.\Big|_{-\frac{\pi}{2}}^{\frac{\pi}{2}} \right. + \int\limits_{-\frac{\pi}{2}}^{\frac{\pi}{2}} -\cos^2 t\ d\cos t$

$= \frac{\pi}{2} + \frac{1}{2} + 0 = \frac{\pi+1}{2}.$

(5) $A = \int\limits_0^{\frac{\pi}{2}} \frac{1}{2} e^{2\phi/\pi} \, d\phi + \int\limits_\pi^{\frac{\pi}{2}} \frac{1}{2} e^{2-2\phi/\pi} \, d\phi$

$= \frac{\pi}{4} e^{2\phi/\pi} \Big|_0^{\frac{\pi}{2}} - \frac{\pi}{4} e^{2-2\phi/\pi} \Big|_{\frac{\pi}{2}}^{\pi}$

$= \frac{\pi}{4} (e-1) - \frac{\pi}{4} (1-e) = \frac{\pi}{2} (e-1).$

4.26 PARTIAL SOLUTIONS

(1) $\int \arctan x \, dx = x \arctan x - \int x(\arctan x)' \, dx$

$= x \arctan x - \int \frac{x}{1+x^2} \, dx = x \arctan x - \frac{1}{2} \ln(1+x^2) + C$

(3) $\int \sin(\ln x)dx = x \sin(\ln x) - \int x(\sin(\ln x)' \, dx$

$= x \sin(\ln x) - \int \cos(\ln x)dx$

$= x \sin(\ln x) - [x \cos(\ln x) - \int x(\cos(\ln x))' \, dx]$

$= x[\sin(\ln x) - \cos(\ln x)] - \int \sin(\ln x)dx$

therefore $\int \sin(\ln x)dx = \frac{x}{2} [\sin(\ln x) - \cos(\ln x)] + C.$

(5) $\int \sec^3(x)dx = \sec x \tan x - \int \tan^2 x \sec x \, dx$

$= \sec x \tan x - \int \sec^3 x \, dx + \int \sec x \, dx$

$\therefore \sec^3 x \, dx = \frac{1}{2} \sec x{\cdot}\tan x + \frac{1}{2} \int \sec x \, dx$

$\therefore \int \sec^3 x \, dx = \frac{1}{2} \sec x \tan x + \frac{1}{4}[\ln(1+\sin x) - \ln(1-\sin x)) + C.$

For $\int \sec x \, dx$. See next exercise set or Appendix 1 (Table of Integrals).

4.27 PARTIAL SOLUTIONS

(2) $\int \cos^3x \sin^2x \, dx = \int \cos^2x \sin^2x \, d\sin x = \int (1-\sin^2x)\sin^2x \, d\sin x$

$\qquad = \dfrac{1}{3} \sin^3x - \dfrac{1}{5} \sin^5x + C.$

(4) $\int \cos 5x \cos x \, dx = \int \dfrac{1}{2} [\cos 6x + \cos 4x]dx = \dfrac{1}{12} \sin(6x) + \dfrac{1}{8} \sin(4x) + C.$

4.28 PARTIAL SOLUTIONS

(3) Let $\theta = \arcsin\left(\dfrac{x}{3}\right)$ then $\sin\theta = \dfrac{x}{3}$ and $dx = 3\cos\theta \, d\theta.$

$\therefore \int \dfrac{dx}{(9-x^2)^{3/2}} = \int \dfrac{dx}{3^3\left(1-\left(\dfrac{x}{3}\right)^2\right)^{3/2}} = \int \dfrac{\cos\theta \, d\theta}{9(1-\sin^2\theta)^{3/2}} = \int \dfrac{d\theta}{9\cos^2\theta}$

$\qquad = \dfrac{1}{9} \int \sec^2\theta \, d\theta = \dfrac{1}{9} \tan\theta + C = \dfrac{1}{9} \cdot \dfrac{x}{\sqrt{9-x^2}}.$

(4) $\int (x^2+25)^{1/2}dx = \int 5\left(\left(1 + \left(\dfrac{x}{5}\right)^2\right)\right)^{1/2}dx.$

Let $\theta = \arctan\dfrac{x}{5}$ then $x = 5\tan\theta$, and $dx = 5\sec^2\theta \, d\theta, \ 1 + \left(\dfrac{x}{5}\right)^2 = \sec^2\theta.$

$\therefore \int (x^2+25)^{1/2}dx = \int 5\sec\theta \cdot 5\sec^2\theta \, d\theta = 25 \int \sec^3\theta \, d\theta$

$\qquad = 25 \int \dfrac{d\theta}{\cos^3\theta} = 25 \int \dfrac{d\sin\theta}{(1-\sin^2\theta)^2} = \dfrac{25}{4} \int \left(\dfrac{1}{1-\sin\theta} + \dfrac{1}{1+\sin\theta}\right)^2 d\sin\theta$

$\qquad = \dfrac{25}{4} \left[\int \dfrac{d\sin\theta}{(1-\sin\theta)^2} + \int \dfrac{d\sin\theta}{(1+\sin\theta)^2} + 2 \int \dfrac{d\sin\theta}{(1-\sin\theta)(1+\sin\theta)}\right]$

4.29 PARTIAL SOLUTIONS

(1) $\int \arcsin x \, dx = x \arcsin x - \int x(\arcsin x)' \, dx$

$\qquad = x \arcsin x - \int \dfrac{x}{\sqrt{1-x^2}} \, dx = x\arcsin x + \sqrt{1-x^2} + C.$

(3) $\int \cos(\ln x)dx = x \cos(\ln x) - \int x(\cos(\ln x))' \, dx$

$= x \cos(\ln x) + \int \sin(\ln x)dx$

$= x \cos(\ln x) + x \sin(\ln x) - \int \cos(\ln x)dx$

$\therefore \int \cos(\ln x)dx = \dfrac{x}{2} [\cos(\ln x)+\sin(\ln x)] + C.$

(5) $\int \csc^3 x \, dx = \int - \csc x \, d \cot x = - \csc x \cot x + \int \cot x \, (\csc x)' \, dx$

$= - \csc x \cot x - \int \csc x \cot^2 x \, dx$

$= - \csc x \cot x - \int \csc^3 x \, dx + \int \csc x \, dx$

$\therefore \int \csc^3 x \, dx = - \dfrac{1}{2} \csc x \cot x + \dfrac{1}{2} \int \csc x \, dx.$

Therefore: $\int \csc^3 x \, dx = - \dfrac{1}{2} \csc x \cot x + \dfrac{1}{4} \ln \left(\dfrac{\cos x-1}{\cos x+1} \right) + C.$

4.30 PARTIAL SOLUTIONS

(2) $\int \cos^3 x \sin^3 x \, dx = \int 2^{-3} \sin^3 2x \, dx$

$= 2^{-3} \int -2^{-1} \sin^2 2x \, d \cos 2x = - 2^{-4} \int (1-\cos^2 2x)d \cos 2x$

$= - \dfrac{1}{16}(\cos 2x - \dfrac{1}{3} \cos^3 2x) + C = \dfrac{1}{48} \cos^3 2x - \dfrac{1}{16} \cos 2x + C.$

(4) $\int \sin 5x \sin x \, dx = \int \dfrac{1}{2} (\cos 4x - \cos 6x)dx = \dfrac{1}{8} \sin 4x - \dfrac{1}{12} \sin 6x + C.$

4.31 PARTIAL SOLUTIONS

(1) Let $\theta = \arctan \dfrac{x}{5}$, then $x = 5\tan\theta, \, dx = 5\sec^2\theta \, d\theta$

$25 + x^2 = 25 \left(1 + \left(\dfrac{x}{5}\right)^2\right) = 25 \sec^2\theta$

$$\therefore \int \frac{dx}{(25+x^2)^2} = \int \frac{5\sec^2\theta \; d\theta}{25^2 \sec^4\theta} = \frac{1}{5^3} \int \cos^2\theta \; d\theta$$

$$= \frac{1}{5^3} \int \frac{1+\cos2\theta}{2} \; d\theta = \frac{1}{125}\left[\frac{1}{2}\theta + \frac{1}{4}\sin2\theta\right] + C$$

$$= \frac{1}{250}\left[\theta + \frac{\tan\theta}{1+\tan^2\theta}\right] + C = \frac{1}{250}\left[\arctan\frac{x}{5} + \frac{x/5}{1+(x/5)^2}\right] + C.$$

(3) Let $\theta = \arcsin\frac{x}{4}$, then $x = 4\sin\theta$, $dx = 4\cos\theta \; d\theta$ and

$$16\left(1 - \left(\frac{x}{4}\right)^2\right) = 16\cos^2\theta$$

$$\therefore \int \frac{dx}{(16-x^2)^{5/2}} = \int \frac{4\cos\theta d\theta}{4^5\cos^5\theta} = \frac{1}{4^4} \int \frac{d\theta}{\cos^4\theta} = 4^{-4} \int \sec^4\theta \; d\theta$$

Note that: $\int \sec^4\theta \; d\theta = \int \sec^2\theta \; d\tan\theta.$

(5) Let $\theta = \text{arcsec}\left(\frac{x}{3}\right)$. Then $x = 3\sec\theta$, $dx = 3\sec\theta\tan\theta \; d\theta$ and

$$x^2 - 9 = 9\left(\left(\frac{x}{3}\right)^2 - 1\right) = 9\tan^2\theta$$

$$\therefore \int \frac{dx}{(x^2-9)^2} = \int \frac{3\sec\theta\tan\theta d\theta}{9^2\tan^4\theta} = \frac{1}{27} \int \frac{\cos^2\theta}{\sin^3\theta} \; d\theta$$

$$= \frac{1}{27} \int \cot^2\theta \csc\theta \; d\theta.$$

Use $1 + \cot^2\theta = \csc^2\theta.$

4.32 PARTIAL SOLUTIONS

(1) $\int \arccos x \; dx = x\arccos x - \int x(\arccos x)^1 dx$

$$= x\arccos x + \int \frac{x}{\sqrt{1-x^2}} \; dx = x\arccos x - \sqrt{1-x^2} + C.$$

(3) $\frac{1}{2}$ arcsin x $-\frac{1}{2}$ x $\sqrt{1-x^2}$ + C.

(5) $\int \cos^2 x \, dx = \int \cos x \, d\sin x = \cos x \sin x - \int \sin x \, d\cos x$

$\qquad = \cos x \sin x + \int \sin^2 x \, dx = \cos x \sin x + x - \int \cos^2 x \, dx$

$\therefore \int \cos^2 x \, dx = \frac{1}{2} [\cos x \sin x + x] + C = \frac{1}{2} x + \frac{1}{4} \sin 2x + C.$ $\qquad\square$

4.33 PARTIAL SOLUTIONS

(1) $\int \tan x \, dx = \int \frac{\sin x}{\cos x} \, dx = \int \frac{d\cos x}{\cos x} = \ln |\cos x| + C.$

(3) $\int \sin^2 x \cos^4 x \, dx = \int \frac{1}{4} \sin^2 2x \cos^2 x \, dx = \frac{1}{8} \int \sin^2 2x (1+\cos 2x) dx$

$\qquad = \frac{1}{16} \int \sin^2 2x \, d(\sin 2x) + \frac{1}{8} \int \frac{1-\cos 4x}{2} \, dx$

$\qquad = \frac{1}{48} \sin^3 (2x) + \frac{1}{16} \left(x - \frac{1}{4} \sin 4x \right) + C.$

(5) $\int \sin x \cos 2x \, dx = \int (\sin 3x - \sin x) dx = \cos x - \frac{1}{3} \cos 3x + C.$

4.34 PARTIAL SOLUTIONS

(1) $\int \frac{dx}{36+x^2} = \frac{1}{36} \int \frac{dx}{1 + \left(\frac{x}{6}\right)^2} = \frac{1}{6} \arctan \frac{x}{6} + C$

(3) $\int \frac{dx}{(16-x^2)^{3/2}} = \int \frac{dx}{4^3 \left[1 - \left(\frac{x}{4}\right)^2 \right]^{3/2}} \quad x = 4\sin\theta \int \frac{4\cos\theta d\theta}{4^3 \cos^3\theta} = \frac{1}{16} \int \sec^2\theta d\theta$

$\qquad = \frac{1}{16} \tan\theta + C.$

(5) $\displaystyle \int \frac{dx}{(x^2-16)^3} = \int \frac{dx}{16^3 \left[\left(\frac{x}{4}\right)^2 -1\right]^3} = \int \frac{4\tan\theta\sec\theta d\theta}{16^3 \tan^6\theta}$, (let $x = 4\sec\theta$)

$\displaystyle = \frac{1}{4 \times 16^2} \int \cot^4\theta \csc\theta d\theta = \frac{1}{4 \times 16^2} \int \cot^3\theta \, d(-\csc\theta)$

$\displaystyle = \frac{1}{4 \times 16^2} [-\cot^3\theta\csc\theta - \int 3\cot^2\theta\csc^3\theta d\theta]$

$\displaystyle = \frac{1}{4 \times 16^2} [-\cot^3\theta\csc\theta - 3 \int \csc^5\theta d\theta + 3 \int \csc^3\theta d\theta]$

$\displaystyle = -\frac{1}{2} \csc\theta\cot\theta + \frac{1}{4} \ln\left(\frac{\cos\theta-1}{\cos\theta+1}\right) + C.$

Note $\displaystyle \int \csc^5\theta d\theta = \int \csc^3\theta d(-\cot\theta) = -\csc^3\theta\cot\theta - \int 3\csc^3\theta\cot^2\theta d\theta$

$\displaystyle = -\csc^3\theta\cot\theta - 3 \int \csc^5\theta d\theta + 3 \int \csc^3\theta d\theta$

$\displaystyle \therefore \int \csc^3\theta\cot \theta - 3 \int - \frac{1}{4} \csc^3\theta\cot\theta + \frac{3}{4} \int \csc^3\theta d\theta.$

4.35 PARTIAL SOLUTIONS

(1) $\displaystyle \int \text{arcsec } x \, dx = x \text{ arcsec } x - \int x(\text{arcsec } x)' \, dx$

$\displaystyle = x \text{ arcsec } x - \int \frac{1}{\sqrt{x^2-1}} \, dx$

Use Table of integrals or trigonometric substitution on $\displaystyle \int \frac{1}{\sqrt{x^2-1}} \, dx.$

(3) $\displaystyle \int e^x\sin x \, dx = \int \sin x \, de^x = e^x\sin x - \int e^x\cos x \, dx$

$\displaystyle = e^x\sin x - (e^x\cos x + \int \sin x \, e^x dx) = -\int e^x\sin x \, dx + e^x(\sin x - \cos x)$

$\displaystyle \therefore \int e^x\sin x \, dx = \frac{1}{2} e^x(\sin x - \cos x) + C.$

(5) $\displaystyle \int \sin^2 x \, dx = \int -\sin x \, d\cos x = -\sin x \cos x + \int \cos^2 x \, dx$

$$= -\sin x \cos x + \int (1-\sin^2 x)dx = x - \sin x \cos x - \int \sin^2 x \, dx$$

$$\therefore \int \sin^2 x \, dx = \frac{1}{2} (x - \sin x \cos x) + C.$$

$\square$

4.36 PARTIAL SOLUTIONS

(1) $\displaystyle \int \cot x \, dx = \int \frac{\cos x}{\sin x} \, dx = \int \frac{d\sin x}{\sin x} = \ln(\sin x) + C$

(3) $\displaystyle \int \sin^4 x \cos^4 x \, dx = \int \frac{1}{16} \sin^4 2x \, dx = \frac{1}{16} \int \left(\frac{1-\cos 2x}{2} \right)^2 dx$

$$= \frac{1}{64} \int (1-2\cos 2x + \cos^2 2x)dx = \frac{1}{64} [x - \sin 2x + \int \frac{1+\cos 4x}{2} \, dx]$$

$$= \frac{1}{64} [x - \sin 2x + \frac{1}{2} x + \frac{1}{8} \sin 4x] + C.$$

(5) $\displaystyle \int x\sin x^2 \cos 2x^2 dx = \int \frac{1}{2} \sin x^2 \cos 2x^2 dx^2$

$$= \frac{1}{4} \int (\sin 3x^2 - \sin x^2)dx^2 = \frac{1}{4} \cos x^2 - \frac{1}{12} \cos 3x^2 + C.$$

4.37 PARTIAL SOLUTIONS

(1) Let $x = 6\tan\theta$, i.e., $\theta = \arctan \frac{x}{6}$, then $dx = 6\sec^2\theta \, d\theta$

$$\therefore \int \frac{dx}{(36+x^2)^{5/2}} = \int \frac{6\sec^2\theta d\theta}{6^5 \sec^5\theta} = \frac{1}{6^4} \int \cos^3\theta \, d\theta$$

$$= \frac{1}{6^4} \int \cos^2\theta d\sin\theta = \frac{1}{6^4} \int (1-\sin^2\theta)d\sin\theta = \frac{1}{6^4} \left(\sin\theta - \frac{1}{3} \sin^3\theta \right) + C$$

$$= \frac{1}{6^4} \left(\frac{x}{\sqrt{x^2+36}} - \frac{1}{3} \frac{x^3}{(x^2+36)^{3/2}} \right) + C.$$

(3) Let $x = 6\sin\theta$, then $dx = 6\cos\theta d\theta$,

$$\int \frac{dx}{(36-x^2)^{5/2}} = \int \frac{6\cos\theta d\theta}{6^5\cos^5\theta} = \frac{1}{6^4} \int \sec^4\theta d\theta$$

Note $d\tan^3\theta = 3\tan^2\theta\sec^2\theta d\theta$

$$= 3(\sec^4\theta - \sec^2\theta)d\theta$$

$$\therefore \int \sec^4\theta d\theta = \frac{1}{3} \int d\tan^3\theta + \int \sec^2\theta d\theta = \frac{1}{3}\tan^3\theta + \tan\theta + C$$

$$\therefore \int \frac{dx}{(36-x^2)^{5/2}} = \frac{1}{6^4} \left[\frac{1}{3} \frac{x^3}{(36-x^2)^{3/2}} + \frac{x}{(36-x^2)^{1/2}} \right] + C.$$

(5) $\int \dfrac{dx}{(x^2-5)^2} = \int \left[\dfrac{1}{(x+\sqrt{5})(x-\sqrt{5})} \right]^2 dx = \int \left[\dfrac{1}{2\sqrt{5}}\left(\dfrac{1}{x-\sqrt{5}} - \dfrac{1}{x+\sqrt{5}} \right) \right]^2 dx$

$$= \frac{1}{20} \int \left[\frac{1}{(x-\sqrt{5})^2} + \frac{1}{(x+\sqrt{5})^2} - 2 \frac{1}{(x-\sqrt{5})(x+\sqrt{5})} \right] dx$$

$$= \frac{1}{20} \left[-\frac{1}{x-\sqrt{5}} - \frac{1}{x+\sqrt{5}} - \frac{1}{\sqrt{5}} \ln \frac{x-\sqrt{5}}{x+\sqrt{5}} \right] + C$$

$$= \frac{-1}{20} \left[\frac{2x}{x^2-5} + \frac{1}{\sqrt{5}} \ln \frac{x-\sqrt{5}}{x+\sqrt{5}} \right] + C. \qquad \qquad \square$$

4.38 HINTS

(1) Set $x = 4\tan\theta$.

(2) Use $\cos^6 x = (\cos^2 x)^3 = \dfrac{1}{8}(1+\cos 2x)^3$.

(3) Integrate by parts.

(4) Use $\cos^5 x\, dx = \cos^4 x\, d\sin x = (1-\sin^2 x)^2 d\sin x$.

(5) Use $x = 2\sec\theta$.

(6) Use $\sin x \cos 2x = \dfrac{1}{2}[\sin 3x - \sin x]$ or $\sin x \cos 2x\, dx = (1-\cos^2 x)d\cos x$.

(7) Integrate by parts twice.

(8) integrate by parts.

(9) Use $x = \sqrt{a} \sin \theta$.

(10) Integrate by.parts.

(11) Use $x = 6\sin\theta$.

(12) Use $\sin^2 x \cos^4 x = \dfrac{1}{4} \sin^2 2x \cos^2 x = \dfrac{1}{4} \left(\dfrac{1-\cos 4x}{2}\right)\left(\dfrac{1+\cos 2x}{2}\right)$ and   use

$\cos 4x \cdot \cos 2x = \dfrac{1}{2} (\cos 6x + \cos 2x)$.

(13) Use $\sin^5 x \, dx = -\sin^4 x \, d\cos x = -(1-\cos^2 x)^2 d\cos x$.

(14) Integrate by parts.

(15) Use $x = 3\sin\theta$.

(16) Use $x = a^2\sin\theta$.

(17) Use $\sin^3 x \, dx = (\cos^2 x - 1)d\cos x$.

(18) Let $y = 1-x$ and expand $x^3 = (1-y)^3 = 1 - 3y + 3y^2 - y^3$.

(19) Integrate by parts three times.

4.39 HINTS

(1) Use $\cos^4 x = \left(\dfrac{1+\cos 2x}{2}\right)^2$, and $\cos^2 2x = \dfrac{1+\cos 4x}{2}$

(2) Use $x = 4\sin\theta$.

(3) Integrate by parts.

(4) Use $y = x+3$, and expand $x^2 = (y-3)^2 = y^3 - 6y + 9$.

(5) Use $\sec x \, dx = \dfrac{1}{\cos x} \, dx = \dfrac{d\sin x}{1-\sin^2 x}$.

(6) Use $\cos 2x \cos 3x = \dfrac{1}{2} (\cos 5x + \cos x)$.

(7) Use $x = 4\sin\theta$.

(8) Integrate by parts.

(9) Use $\sin^5 x \, dx = -(1-\cos^2 x)^2 d\cos x$.

(10) Use $x = 4\sec\theta$.

(11) Integrate by parts.

(12) First use $\cos(3e^w)\sin(5e^w) = \dfrac{1}{2} [\sin(8e^x) + \sin(2e^w)]$ then note $e^w dw = de^w$.

(13) Use $x = 2\sin\theta$.

(14) Integrate by parts.

(15) Use $x = 5\tan\theta$.   ☐

4.42 SOLUTIONS

(1) $\dfrac{2x^2+3}{(2x+1)^5(x^2+x+1)^2} = \dfrac{A_1}{2x+1} + \dfrac{A_2}{(2x+1)^2} + \dfrac{A_3}{(2x+1)^3} + \dfrac{A_4}{(2x+1)^4} + \dfrac{A_5}{(2x+1)^5}$

$+ \dfrac{B_1x+C_1}{x^2+x+1} + \dfrac{B_2x+C_2}{(x^2+x+1)^2}.$

(2) $\dfrac{3x^3-8x^2+2x+3}{(3x-2)(x^2-2x-1)} = 1 + \dfrac{A_1}{3x-2} + \dfrac{B_1x+C_1}{x^2-2x-1}.$

In order to get A_1, B_1, C_1, multiply by $(3x-1))(x^2-2x-1))$ to get
$3x^3-8x^2+2x+3 = (3x-2))(x^2-2x-1)) + A_1(x^2-2x-1) + (3x-2)(B_1x+C_1)) =$
$3x^3 + (-8+A_1+3B_1)x^2+(1-2A_1+3C_1-2B_1)x+(2-A_1-2C_1).$

Comparing the coefficients on both sides, we get $A_1 + 3B_1 = 0$,
$1-2A_1+3C_1-2B_1 = 2$, $2-A_1-2C_1 = 3$.

Solving this linear system, $A_1 = \dfrac{-15}{17}$, $B_1 = \dfrac{5}{17}$, $C_1 = \dfrac{-1}{17}$.

so $\dfrac{3x^3-8x^2+2x+3}{(3x-2)(x^2-2x-1)} = 1 - \dfrac{15/17}{3x-2} + \dfrac{\dfrac{5}{17}x - \dfrac{1}{17}}{x^2-2x-1}.$

(3) Set $y = x^3-1$. Then $\dfrac{x^3+1}{x^3-1} = \dfrac{y+2}{y^2} = \dfrac{1}{y} + \dfrac{2}{y^2} = \dfrac{1}{x^3-1} + \dfrac{2}{(x^3-1)^2}.$ Write
$x^3-1 = (x-1)(x^2+x+1)$. We get

$\dfrac{x^3+1}{(x^3-1)^2} = \dfrac{-1/9}{x-1} + \dfrac{2/9}{(x-1)^2} + \dfrac{1/9}{(x^2+x+1)} + \dfrac{\dfrac{6}{9}(x+1)}{(x^2+x+1)^2}.$ $\square$

(4)

(a) Using the integral table, we find

$$\int \dfrac{dx}{(1+x+x^2)^3} = \dfrac{2x+1}{3}\left(\dfrac{1}{2(1+x^2)} + \dfrac{1}{1+x+x^2}\right) + \dfrac{2}{3}\int \dfrac{dx}{1+x+x^2}$$

$$= \dfrac{2x+1}{3}\left[\dfrac{1}{2(1+x+x^2)^2} + \dfrac{1}{1+x+x^2}\right] + \dfrac{2}{3}\left[\dfrac{2}{\sqrt{3}}\tan^{-1}\left(\dfrac{2x+1}{\sqrt{3}}\right)\right] + C.$$

(b) First use partial fractions (let $x^2+1 = y$, $x^4 = (y-1)^2$) to get

$$\frac{x^4}{(x^2+1)^2} = 1 + \frac{-2}{x^2+1} + \frac{1}{(x^2+1)^2}.$$

$$\Rightarrow \int \frac{x^4}{(x^2+1)^2}\, dx = x - 2\int \frac{dx}{x^2+1} + \int \frac{dx}{(x^2+1)^2}.$$

Expand $\dfrac{1}{(x^2+1)^2}$ by partial fractions.

(c) $\dfrac{x}{(x-1)^2(x+1)^2} = \dfrac{A_1}{x-1} + \dfrac{A_2}{(x-1)^2} + \dfrac{A_3}{x+1} + \dfrac{A_4}{(x+1)^2}.$ Solving gives

$$\frac{x}{(x-1)^2(x+1)^2} = \frac{1}{2}\left[\frac{-1}{x-1} + \frac{1}{(x-1)^2} - \frac{1}{x+1} - \frac{1}{(x+1)^2}\right]$$

$$\Rightarrow \int \frac{x}{(x-1)^2(x+1)^2} = \frac{1}{2}\left[-\ln(x-1) - \frac{1}{x-1} - \ln(x+1) + \frac{1}{x+1}\right] + C.$$

(d) $\dfrac{1}{(x-1)^2(x+1)^2} = \left[\dfrac{1}{2}\left(\dfrac{1}{x-1} - \dfrac{1}{x+1}\right)\right]^2$

$$= \frac{1}{4}\left[\frac{1}{(x-1)^2} + \frac{1}{(x+1)^2} - \left(\frac{1}{x-1} - \frac{1}{x+1}\right)\right]$$

$$\therefore \int \frac{dx}{(x-1)^2(x+1)^2} = \frac{1}{4}\left[-\frac{1}{x-1} - \frac{1}{x+1} - \ln\frac{x-1}{x+1}\right] + C. \qquad \square$$

4.43 HINTS

(1) $\dfrac{2x^2+3}{(2x+1)^5(x^2+x-1)^2} = \dfrac{A_1}{2x+1} + \dfrac{A_2}{(2x+1)^2} + \dfrac{A_3}{(2x+1)^3} + \dfrac{A_4}{(2x+1)^4} + \dfrac{A_5}{(2x+1)^5}$

$$+ \frac{B_1x+C_1}{(x^2+x-1)} + \frac{B_2x+C_2}{(x^2+x-1)^2}.$$

(2) $\dfrac{3x^3-8x^2+2x+3}{(3x-2)^2(x^2-2x-1)} = \dfrac{A_1}{(3x-2)} + \dfrac{A_2}{(3x-2)^2} + \dfrac{B_1x+C_1}{x^2-2x-1}.$

(3) $\dfrac{x^2+3}{(x^4+2x^2+1)^2} = \dfrac{x^2+3}{(x^2+1)^4} = \dfrac{1}{(x^2+1)^3} + \dfrac{2}{(x^2+1)^4}.$

(4)

(a) Note $1+2x+x^2 = (1+x)^2$.

(b) First compute $\dfrac{x^3}{(x^2+1)^2} = \dfrac{B_1x+C_1}{x^2+1} + \dfrac{B_2x+C_2}{(x^2+1)^2}$ then use the integral table.

4.44 HINTS

(1) $\dfrac{2x^9+3x^5+2x^2+1}{(2x+1)^5(x^2+x-1)^2} = \dfrac{1}{2^4} + \dfrac{A_1}{2x+1} + \dfrac{A_2}{(2x+1)^2} + \dfrac{A_3}{(2x+1)^3} + \dfrac{A_4}{(2x+1)^4} + \dfrac{A_5}{(2x+1)^5}$

$$+ \dfrac{B_1x+C_1}{x^2+x-1} + \dfrac{B_2x+C_2}{(x^2+x-1)^2} \quad .$$

(2) $\dfrac{3x^3-8x^2+2x+3}{(x^2+1)(x^2-2x+2)} = \dfrac{B_1x+C_1}{x^2+1} + \dfrac{B_2x+C_2}{x^2-2x+2} \quad .$

(3) $\dfrac{(x+1)^2}{(x^2+2x+2)^2} .= \dfrac{B_1x+C_1}{x^2+2x+2} + \dfrac{B_2x+C_2}{(x^2+2x+2)^2} \quad .$

(4)

(a) $\dfrac{x^4}{(1+x+x^2)^2} = 1 + \dfrac{B_1x+C_1}{1+x+x^2} + \dfrac{B_2x+C_2}{(1+x+x^2)^2} \quad ,$

then use the integral table involving $X = (1+x+x^2)$.

(b) Note $\dfrac{x^4}{(x^2+1)^2} = \dfrac{(x^2+1)^2-2x^2-1}{(x^2+1)^2} = 1 - \dfrac{2(x^2+1)-1}{(x^2+1)^2}$

$$= 1 - \dfrac{2}{(x^2+1)} + \dfrac{1}{(x^2+1)^2} \quad .$$

(c) First compute

$$\dfrac{x^4}{(x-1)^3(x+1)} = 1 + \dfrac{A_1}{x-1} + \dfrac{A_2}{(x-1)^2} + \dfrac{A_3}{(x-1)^3} + \dfrac{A_4}{x+1} \quad .$$

(d) Note

$$\frac{x^4}{(x-1)^2(x+1)^2} = 1 + \frac{A_1}{x-1} + \frac{A_2}{(x-1)^2} + \frac{A_3}{x+1} + \frac{A_4}{(x+1)^2} \quad .$$

4.58 SOLUTIONS

(1) The area of the isoceles triangle at x is

$$A(x) = \frac{1}{2} \cdot 3^x \cdot 2^x = \frac{1}{2} \cdot 6^x \quad .$$

So the volume of the solid between $x = 0$ and $x = 2$ is

$$V = \int_0^2 \frac{1}{2} 6^x \, dx = \frac{1}{2} \frac{1}{\ln 6} \cdot 6^x \Big|_0^2$$

$$= \frac{1}{2\ln 6} [6^2 - 6^0] = \frac{36}{2\ln 6} \quad ,$$

the total volume of the solid in the region $x < 0$ is

$$V = \int_{-\infty}^0 \frac{1}{2} \cdot 6^x \, dx = \frac{1}{2\ln 6} 6^x \Big|_{-\infty}^0 = \frac{1}{2\ln 6} \quad .$$

(2) $A(x) = \pi y^2(x) = \pi x^3$.

$\therefore V(x) = \int_0^x \pi x^3 \, dx = \frac{\pi}{4} x^4$.

(3) Using the method of cylinders, we have

$$V_x(y) = \int_0^y 2\pi y(x - y^{2/3}) dx = \pi x y^2 - \frac{3}{4} \pi y^{8/3}$$

$\therefore V(x) = V_x(x^{3/2}) = \pi x^4 - \frac{3}{4} \pi x^4 = \frac{1}{4} \pi x^4$.

(4) $L(x) = \int_0^x \sqrt{1 + y'(x)^2} dx = \int_0^x \sqrt{1 + \frac{9}{4} x} \, dx = \frac{4}{9} \cdot \frac{2}{3} \left(1 + \frac{9}{4} x\right)^{3/2} \Big|_0^x$

$= \frac{8}{27} \left[\left(1 + \frac{9}{4} x\right)^{3/2} - 1 \right] \quad .$

(5) $\quad S(x) = \int\limits_0^x 2\pi\, y(x)\, \sqrt{1+y'(x)^2}\ dx = \int\limits_0^x 2\pi \cdot x^{3/2}\, \sqrt{1 + \dfrac{9}{4}\, x}\ dx$

$\quad\quad = 2\pi \int\limits_0^x x\left(x + \dfrac{9}{4}\, x^2\right)^{1/2} dx$

$\quad\quad = 3\pi \int\limits_0^x x\left(x^2 + \dfrac{4}{9}\, x\right)^{1/2} dx \quad .$

Use Euler-transformation $\ z = \left(x^2 + \dfrac{4}{9}\, x\right)^{1/2} - x. \ \Rightarrow x = \dfrac{z^2}{\dfrac{4}{9} - 2z}$

$\Rightarrow S(x) = 3\pi \int\limits_0^x x(x+z)dx = \pi x^3 + 3\pi \int\limits_0^x \dfrac{1}{2}\, z\ dx^2$

$\quad\quad = \pi x^3 + \dfrac{3\pi}{2}\left[zx^2 - \int\limits_0^x x^2\, dz\right] = \pi x^3 + \dfrac{3\pi}{2}\, x^2\left[\left(x^2 + \dfrac{4}{9}\, x\right)^{1/2} - x\right]$

$\quad\quad - \dfrac{3\pi}{2} \int\limits_0^x \dfrac{z^4}{\left(\dfrac{4}{9} - 2z\right)^2}\, dz \quad .$

Now use partial fractions.

(6) $\quad L = \int\limits_0^\pi (r^2 + r'(\phi)^2)^{1/2}d\phi = \int\limits_0^\pi \left(2^\phi + \dfrac{ln^2 2}{4}\, 2^\phi\right)d\phi$

$\quad\quad = \left(1 + \dfrac{ln^2 2}{4}\right)^{1/2} \dfrac{2}{ln\, x}\left[2^{\frac{\pi}{2}} - 1\right] \quad .$

(7) $\quad S = \int\limits_0^\pi 2\pi \cdot 2^{\phi/2}\sin\phi \left(1 + \dfrac{ln^2 2}{4}\right)^{1/2} \cdot 2^{\phi/2}\, d\phi$

$\quad\quad = 2\left(1 + \dfrac{ln^2 2}{4}\right)^{1/2} \pi \int\limits_0^\pi 2^\phi \sin\phi\ d\phi$

$\quad\quad = 2\left(1 + \dfrac{ln^2 2}{4}\right)^{1/2} \cdot \pi \cdot \dfrac{1}{1 + ln^2 2}\, [2^\pi - 1] \quad .$

(8) $\quad V = \int\limits_0^\pi \int\limits_0^{r(\phi)} 2\pi\, r^2 \sin\phi\ drd\phi = \int\limits_0^\pi \dfrac{2\pi}{3}\, r^3 \sin\phi\ d\phi$

$\quad\quad = \int\limits_0^\pi \dfrac{2\pi}{3}\, 2^{3\phi/2} \sin\phi\ d\phi = \dfrac{2\pi}{3} \int\limits_0^\pi 2^{3\phi/2} \sin\phi\ d\phi \quad .$

Use Appendix 1 or using integration by parts, we have

$$\int_0^\pi 2^{3\phi/2} \sin\phi \, d\phi = \left[-2^{3\phi/2} \cos\phi + \frac{3}{2} \ln 2 \cdot \sin\phi \right]_0^\pi - \left(\frac{3}{2} \ln 2 \right)^2 \int_0^\pi 2^{3\phi/2} \sin\phi \, d\phi$$

$$\therefore \int_0^\pi 2^{3\phi/2} \sin\phi \, d\phi = \frac{2^{3\pi/2}+1}{1 + \left(\dfrac{3}{2} \ln 2 \right)^2}$$

$$\therefore V = \frac{2\pi}{3} \frac{2^{3\pi/2}+1}{1 + \left(\dfrac{3}{2} \ln 2 \right)^2} \cdot$$

$$\left[-\phi^3 \cos\phi + 3\phi^2 \sin\phi + 6\phi \cos\phi - 6 \sin\phi \right]_0^\pi = \frac{2\pi^2}{3} (\pi^2 - 6)$$

4.59 SOLUTIONS

(1) $\quad V = \int_1^2 2^x \log_2 x \, dx = 2^2 \log_2 x \bigg|_1^2 - \frac{1}{\ln 2} \int_1^2 \frac{2^x}{x} \, dx$

$$= 4 - \frac{1}{\ln 2} \left[\sum_{i=1}^{10} \frac{2^{1 + \frac{1}{10}}}{1 + \frac{1}{10}} \right] \cdot \frac{1}{10} \approx 1.51387$$

(2) $\quad V = \int_0^x \pi \sin^2 x \, dx = \int_0^x \frac{\pi}{2} (1 - \cos 2x) dx$

$$= \frac{\pi}{2} \left(x - \frac{1}{2} \sin 2x \right) \quad .$$

(3) $\quad V_x(y) = \int_0^y 2\pi \, y(x - \arcsin y) dy$

$$= \pi xy^2 - 2\pi \int_0^y y \arcsin y \, dy = \pi xy^2 - 2\pi \int_0^x \sin x \cdot x \, d\sin x$$

$$= \pi x \sin^2 x - 2\pi \int_0^x \frac{1}{2} x \sin 2x \, dx$$

$$= \pi x \sin^2 x - \pi \left(\frac{1}{4} \sin 2x - \frac{1}{2} x \cos 2x \right) = \frac{\pi}{2} \left(x - \frac{1}{2} \sin 2x \right)$$

(4) $L = \int\limits_0^\pi \sqrt{1+\cos^2 x}\ dx \approx \dfrac{\pi}{10}\left[\sqrt{2} + \sum\limits_{i=1}^9 \sqrt{1+\cos^2\left(\dfrac{i\pi}{10}\right)}\right] \approx 3.8202$.

(5) $A(x) = \int\limits_0^x 2\pi \sin x \sqrt{1+\cos^2 x}\ dx = -2\pi \int\limits_0^x (1+\cos^2 x)^{1/2}\ d\cos x$

$= -2\pi \cdot \dfrac{1}{2}\left[\cos x \sqrt{1+\cos^2 x} + \ln(\cos x + \sqrt{1+\cos^2 x})\right]_0^\infty$

$= \pi[\sqrt{2} + \ln(1 + \sqrt{2}) - \cos x \sqrt{1+\cos^2 x} - \ln(\cos x + \sqrt{1+\cos^2 x})]$.

(6) $L = \int\limits_0^\pi \sqrt{r^2 + r'\ (\phi)^2}\,d\phi = \int\limits_0^\pi \sqrt{\phi^2 + 1}\ d\phi$

$= \dfrac{1}{2}\left[\phi \sqrt{\phi^2+1} + \ln(\phi + \sqrt{\phi^2+1})\right]_0^\pi$

$= \dfrac{1}{2}[\pi \sqrt{\pi^2+1} + \ln(\pi + \sqrt{\pi^2+1})]$.

(7) $A = \int\limits_0^\pi 2\pi \phi \sin \phi \sqrt{\phi^2+1}\ d\phi \approx \dfrac{2\pi^2}{10}\left[\sum\limits_{i=1}^9 \dfrac{i\pi}{10} \sin\left(\dfrac{i\pi}{10}\right) \sqrt{\left(\dfrac{i\pi}{10}\right)^2+1}\right] \approx 41.78705$.

(8) $V = \int\limits_0^\pi \int\limits_0^r 2\pi\, r^2 \sin \phi\ dr d\phi = \int\limits_0^\pi \dfrac{2\pi}{3} \phi^3 \sin \phi\ d\phi$

$= \dfrac{2\pi}{3}\left[-\phi^3 \cos \phi + 3\phi^2 \sin \phi + 6\phi \cos \phi - 6 \sin \phi\right]_0^\pi = \dfrac{2\pi^2}{3}\ (\pi^2 - 6)$ $\square$

4.60 SOLUTIONS

(1) $V = \int\limits_0^2 x^2 \cdot 2^x\ dx$

$= \left[\dfrac{1}{\ln 2} x^2 2^x - \left(\dfrac{1}{\ln 2}\right)^2 \cdot 2x \cdot 2^x + \left(\dfrac{1}{\ln 2}\right)^3 2 \cdot 2^x\right]_0^2$

$= \left(\dfrac{2}{\ln 2}\right)^3 - \dfrac{2}{\ln^3 2} = \dfrac{6}{\ln^3 2}$.

(2)
$$V = \int_0^x \pi\, y^2(x)dx = \int_0^x \pi\cdot 4^x dx = \frac{\pi}{\ln 4}\,[4^x-1]$$

(3) $$V_x(y) = \int_1^y 2\pi\, y(x - \log_2 y)dy + \pi\cdot 1^2\cdot x$$

$$= \left[\pi\, xy^2 - \frac{2\pi}{\ln 2}\cdot\frac{1}{2}\, y^2\Big(\ln y - \frac{1}{2}\Big)\right]_1^y + \pi x$$

$$\therefore\ V_x(2^x) = \frac{\pi}{\ln 4}\,(4^x-1).$$

(4) $$L(x) = \int_0^x \sqrt{1+y'(x)^2}dx = \int_0^x \sqrt{1+\ln^2 2\cdot 4^x}\ dx$$

$$= \int_0^x \frac{1+\ln^2 2\cdot 4^x}{\sqrt{1+\ln^2 2\cdot 4^x}}\ dx$$

$$= \int_0^x \frac{dx}{\sqrt{1+\ln^2 2\cdot e^{x\ln 4}}} + \int_0^x \frac{1}{\ln 4}\,(1+\ln^2 2\cdot 4^x)^{-1/2}d(1+\ln^2 2\cdot 4^x)$$

$$= \frac{1}{\ln 4}\left[\ln\left(\sqrt{1+\ln^2 2\cdot 4^x} - 1\right) - \ln\left(\sqrt{1+\ln^2 2\cdot 4^x} + 1\right)\right]_0^x + \frac{2}{\ln 4}\left.\sqrt{1+\ln^2 2\cdot 4^x}\ \right|_0^x$$

$$= \frac{1}{\ln 4}\left[\ln\frac{\sqrt{1+\ln^2 2\cdot 4^x-1}}{\sqrt{1+\ln^2 2\cdot 4^x+1}} - \ln\frac{\sqrt{1+\ln^2 2-1}}{\sqrt{1+\ln^2 2+1}}\right] + \frac{2}{\ln 4}\left[\sqrt{1+\ln^2 2\cdot 4^x} - \sqrt{1+\ln^2 2}\right]$$

(5) $$A(x) = \int_0^x 2\pi 2^x \sqrt{1+\ln^2 2\cdot 4^x}\ dx = \int_0^x \frac{2\pi}{\ln^2 2}\sqrt{1+(2^x\cdot\ln 2)^2}\ d(2^x\ln 2)$$

$$= \frac{2\pi}{\ln^2 2}\cdot\frac{1}{2}\left[2^x\ln 2\sqrt{1+(2^x\ln 2)^2} + \ln(2^x\ln 2 + \sqrt{1+(2^x\ln 2)^2}\ \right]_0^x$$

$$= \frac{\pi}{\ln^2 2}\,[2^x\ln 2\cdot\sqrt{1+(2^x\ln 2)^2} + \ln(2^x\ln 2 + \sqrt{1+(2^x\ln 2)^2})]$$

$$- \frac{\pi}{\ln^2 2}\,[\ln 2\sqrt{1+\ln^2 2} + \ln(\ln 2 + \sqrt{1+\ln^2 2})]\quad .$$

(6) $$L = \int_0^\pi (r^2+r'(\phi)^2)^{1/2}d\phi = \int_0^\pi (\sin^2\phi+\cos^2\phi)^{1/2}d\phi = \pi\quad .$$

(7) $$A = \int_0^\pi 2\pi \sin^2\phi\cdot 1\ d\phi = 2\pi \int_0^\pi \frac{1-\cos 2\phi}{2}\ d\phi = \pi^2\quad .$$

(8) $\quad V = \int_0^\pi \int_0^r 2\pi r^2 \sin\phi \, dr \, d\phi = \int_0^\pi \frac{2\pi}{3} r^3 \sin\phi \, d\phi = \int_0^\pi \frac{2\pi}{3} \sin^4\phi \, d\phi$

$$= \frac{2\pi}{3} \int_0^\pi \frac{(1-\cos 2\phi)^2}{4} \, d\phi + \frac{\pi}{6} \int_0^\pi \left[1 - 2\cos 2\phi + \frac{1+\cos 4\phi}{2} \right] d\phi = \frac{\pi^2}{4} \quad .$$

5.17 SOLUTIONS, PARTIAL SOLUTIONS, AND HINTS

(1)

    (a)   Divide numerator and denominator by n^5. Sequence converges to $4/7$.

    (b)   Divide numerator and denominator by n^3. Sequence converges to 0.

    (c)   Divide numerator and denominator by n^n. Sequence is bounded divergent.

    (d)   The sequence $(1+n^{-1})^n$ converges to e. Compare this problem with $\left(\frac{n}{e}\right)^n$.

    (e)   Divide numerator and denominator by n^3. Use fact that $\frac{(\ln(n))^5}{n^3}$ converges to 0.

    (f)   Compute $\frac{n^2+1}{n+2} - n = \frac{1-2n}{n+2}$. Thus $\pi \frac{n^2+1}{n+2} = \pi \left(n + \frac{1}{n+2} - \frac{2n}{n-2}\right)$. Compare the sequence with $\cos(\pi(n-2))$.

    (g)   Note that $\log_2(n^5) = 5\log_2(n) = 5\log_2(3)\log_3(n)$.

    (h)   Using L'Hopital's rule, show that $\lim\limits_{x\to\infty} \frac{\log_2(x)}{x^{.001}} = 0$.

    (i)   Note that $\lim\limits_{x\to\infty} \frac{\log_2(x)}{x} = 0$ and $\log_2(n)$ goes to ∞ as n goes to ∞.

    (j)   Write $\frac{n^2+1}{n+2} = n - \frac{1}{n+2} - \frac{2n}{n+2}$.

    (k)   converges to e^2.

5.18 SOLUTIONS, PARTIAL SOLUTIONS, AND HINTS

(1)

    (a) Would this limit be any different if $\ln(n)$ were replaced by n?

    (b) Multiply numerator and denominator by n^4.

    (c) Divide numerator and denominator by n^n.

    (d) Consider $\ln(1+n)^{1/n} = \dfrac{\ln(1+n)}{n}$.

    (e) Divide numerator and denominator by n^3.

    (f) Write $\dfrac{\pi^2 n^2 + 1}{\pi n + 2} = \pi n + \dfrac{1}{\pi n + 2} - \dfrac{2\pi n}{\pi n + 2}$.

    (g) Write $n^5 + \log_2(n) = n^5\left(1 + \dfrac{\log_2(n)}{n^5}\right)$. Thus $\log_2(n^5 + \log_2(n)) = \log_2(n^5) + \log_2\left(1 + \dfrac{\log_2(n)}{n^5}\right)$.

    (h) Replace n in denominator by $e^{\ln(n)}$.

    (i) Use $\log_2(x) = \log_2(3)\log_3(x)$ twice, first with $x = n$ and then with $x = \log_3(n)$.

    (j) Write $\sin(2n) = 2\sin(n)\cos(n)$. Every intnerval of the form $[k\pi + \pi/4 - 1/2, k\pi + \pi/4 + 1/2]$ must contain an integer n.

    (k) $\ln\left(1 + \dfrac{\ln(n)}{n}\right)^n = n\ln\left(1 + \dfrac{\ln(n)}{n}\right)$. Apply L'Hopital's rule to $\dfrac{\ln\left(1 + \dfrac{\ln(x)}{x}\right)}{1/x}$.

5.44 SOLUTIONS, PARTIAL SOLUTIONS, AND HINTS

(1)

(a) Compare with k^{-2}, $k > 5$.

For root test use Stirling's formula to get $\lim_{k \to \infty} a_k^{1/k} = 0$.

For ratio test, $\dfrac{a_{k+1}}{a_k} = \dfrac{k}{(k+1)(k-3)}$ tends to 0. The series converges by all three tests.

(b) The terms of this series tend to infinity (make sure you can prove this). Thus, this series diverges. This may be regarded as comparison with the series with all terms equal to 1.

For root test, $a_k^{1/k}$ converges to $2^{1/2} > 1$.

For ratio test, $\dfrac{a_{k+1}}{a_k}$ converges to $2^{1/2} > 1$.

(c) The terms of this series tend to $1/2$ so the series diverges. The root test and the ratio test are inconclusive.

(d) Compare with series with terms $(1.01)^{-k/2}$ to get convergence.

For $a_k = \dfrac{k^{200}}{(1.01)^k}$, $a_k^{1/k}$ converges to $\dfrac{1}{1.01} < 1$, so root test implies convergence. The ratio $\dfrac{a_{k+1}}{a_k}$ also converges to $\dfrac{1}{1.01}$.

(2)

(a) Comparison with $\dfrac{1}{k^{1.0001}}$ shows convergence.

(b) Comparison with $\dfrac{1}{2^k}$ shows convergence.

(c) Comparison with $\dfrac{1}{k^{3/4}}$ shows convergence.

(d) Multiply numerator and denominator by $(k+1)^{1/4} + (k-1)^{1/4}$. Compare with $\dfrac{1}{k^{3/4}}$.

(3)

    (a) Let $b_k = \dfrac{(-1)^k}{k}$ and $a_k = \left(\dfrac{1}{2(\ln(2))^2} + \cdots + \dfrac{1}{k(\ln(k))^2}\right)$. Show that $\displaystyle\int_2^\infty \dfrac{dx}{x(\ln(x))^2} < \infty$ and hence that a_k, which is monotonically increasing, converges. Apply Abel's test.

    (b) Let $b_k = (-1)^k$ and $a_k = \dfrac{1}{k}\left(\dfrac{(\ln(1))^2}{1} + \cdots + \dfrac{(\ln(k))^2}{k}\right)$. By computing $a_k - a_{k-1}$, show a_k is monotonoically decreasing. Using $\displaystyle\int \dfrac{(\ln(x))^2}{x}\,dx = \dfrac{(\ln(x))^3}{3}$ and $\displaystyle\lim_{k\to\infty}\dfrac{(\ln(k))^3}{k} = 0$, show $\displaystyle\lim_{k\to\infty} a_k = 0$. Apply Dirichlet's test.

(4) Compare with $r^{(\ln(k))^2}$.

(5)

    (a) Converges. Apply Dirichlet's test.

    (b) Converges. Apply Dirichlet's test.

    (c) Converges. Show series with terms $(-1)^k \cos(k)$ has bounded partial sums, using the fact that the series with terms $\cos(\alpha k + \beta)$ has bounded partial sums. Apply Abel's test.

    (d) Converges. The series with terms $\sin(k)\cos(k) = (1/2)\sin(2k)$ has bounded partial sums. Apply Abel's test.

5.45 SOLUTIONS, PARTIAL SOLUTIONS, AND HINTS

(1)

    (a) Diverges. Compare with $1/k$.

    (b) Diverges. Write $k^2 = 2^{2\ln(k)}$.

    (c) Diverges.

    (d) Converges.

(2)

    (a) Diverges

    (b) Diverges

    (c) Converges

    (d) Converges

(3)

    (a) Let $a_k = e^{-1/k} - 1 + 1/k$. Using $e^{-x} - 1 + x = \dfrac{x^2}{2!} - \dfrac{x^3}{3!} + \dfrac{x^4}{4!} \cdots$ we see that $0 < a_k \leqslant \dfrac{1}{2k^2}$. Show that $e^{-1/k} - 1 + 1/k$ tends <u>monotonically</u> to 0. Apply Dirichlet's test.

    (b) Diverges.

(4) Write $k^{\alpha(\ln(k))^\beta} = e^{\alpha(\ln(k))^{\beta+1}} = r^{(\ln(k))^{\beta+1}}$ with $r = e^\alpha$.

(5)

    (a) Diverges if $\beta \leqslant 0$. Converges if $0 < \beta < 1$. Does not converge absolutely for any $\beta \leqslant 1$ because $\lim\limits_{x \to \infty} \dfrac{1/x}{\text{arccot}(x)} = -1$. Show this and explain why this means no absolute convergence.

    (b) Choose constants β and α such that $\beta \sin(k+\alpha) = \sin(k) - \cos(k)$. Apply Dirichlet's test.

    (c) Converges conditionally. Use Dirichlet's test.

    (d) Converges conditionally. Use Dirichlet's test.

5.70 SOLUTIONS, PARTIAL SOLUTIONS, AND HINTS

(5) Most of the work needed here has already been done in connection with EXERCISE 5.42. You need to determine, if you have not already done so in connection with EXERCISE 5.42, $\operatorname*{limit}_{,\to\infty} |a_k|^{1/k}$. The main difficulties of this exercise set involve issues of conditional convergence:

(i) $R = \operatorname*{limit}_{k\to\infty} \left(\frac{1}{k}(1+2^{-2} + \cdots + k^{-2})\right)^{1/k}$

$$= \left[\operatorname*{limit}_{k\to\infty}\left(\frac{1}{k}\right)^{1/k}\right]\left[\operatorname*{limit}_{k\to\infty}(1+2^{-2} + \cdots + k^{-2})^{1/k}\right] = \operatorname*{limit}_{k\to\infty}(1+2^{-2} + \cdots + k^{-2})^{1/k}.$$

But the series $1 + 2^{-2} + \cdots + k^{-2} + \cdots$ converges and thus its partial sums $1 + 2^{-2} + \cdots + k^{-2}$ are bounded. In fact, they are bounded by $1 + \int\limits_{1}^{\infty} x^{-2}dx = 2.$ Thus,   taking   logarithms, $\operatorname*{limit}_{k\to\infty}(1/k)\ln(1+2^{-2} + \cdots + k^{-2}) = 0$. So $\operatorname*{limit}_{k\to\infty}(1+2^{-2} + \cdots + k^{-2}))^{1/k} = 1$.   Hence, $R = 1$.   At $x = 1$ the   series   with   terms $\frac{(-x)^k}{k}(1+2^{-2} + \cdots + k^{-2})$ converges conditionally as was shown in EXERCISE 5.42. At $x = -1$ it diverges (compare with harmonic series).

(j) Define $a_k = \frac{1}{k}(1+\frac{1}{2} + \cdots + \frac{1}{k})$. $\operatorname*{limit}_{k\to\infty}(a_k^{1/k}) = $ $[\operatorname*{limit}_{k\to\infty}\left(\frac{1}{k}\right)^{1/k}][\operatorname*{limit}_{k\to\infty}(1+\frac{1}{2} + \cdots + \frac{1}{k})^{1/k}]$ $= \operatorname*{limit}_{k\to\infty}(1+\frac{1}{2} + \cdots + \frac{1}{k})^{1/k}.$ Taking logarithms gives $\operatorname*{limit}_{k\to\infty}\frac{1}{k}\ln(1+\frac{1}{2} + \cdots + \frac{1}{k})$. Show, this limit is zero. The idea is that $1 + \frac{1}{2} + \cdots + \frac{1}{k}$ is approximately $\ln(k)$. We know that $\frac{\ln(\ln(k))}{k}$ tends to zero. Thus $R = 1$ in this problem. The series converges conditionally at $x = 1$ and diverges at $x = -1$.

INDEX